METHODS IN MOLECULAR BIOLOGY

Series Editor
John M. Walker
School of Life and Medical Sciences
University of Hertfordshire
Hatfield, Hertfordshire, AL10 9AB, UK

For further volumes:
http://www.springer.com/series/7651

Bacteriophages

Methods and Protocols, Volume 3

Edited by

Martha R.J. Clokie

Department Infection, Immunity & Inflammation, University of Leicester School of Medicine, Leicester, UK

Andrew M. Kropinski

University of Guelph, Guelph, ON, Canada

Rob Lavigne

KU Leuven, Laboratory of Gene Technology, Leuven, Belgium

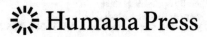 Humana Press

Editors
Martha R.J. Clokie
Department Infection, Immunity &
Inflammation
University of Leicester School
of Medicine
Leicester, UK

Andrew M. Kropinski
University of Guelph
Guelph, ON, Canada

Rob Lavigne
KU Leuven
Laboratory of Gene Technology
Leuven, Belgium

ISSN 1064-3745   ISSN 1940-6029   (electronic)
Methods in Molecular Biology
ISBN 978-1-4939-8458-9   ISBN 978-1-4939-7343-9   (eBook)
https://doi.org/10.1007/978-1-4939-7343-9

This Humana Press imprint is published by Springer Nature
The registered company is Springer Science+Business Media, LLC
The registered company address is: 233 Spring Street, New York, NY 10013, U.S.A.

Preface

Since the publication of the First Edition of *Bacteriophages: Methods and Protocols* in 2009, the field of bacterial virus research has evolved extensively. This can be readily observed from the fact that this latest volume contains all new chapters addressing newly emerging themes and methodologies.

One of the first key trends is the successful and broad-scale introduction of phage-based teaching innovation tools within the field of phage biology and beyond. Most notable in this regard is the "phage hunting program" from the University of Pittsburgh, headed by Dr. Graham Hatfull and the Science Education alliance (SEA). This program has exposed university and high school students across the United States to the scientific method and the joy of bacteriophage discovery.

Phage research is undergoing a clear shift from the microbiological and genomic to the postgenomic era. New phage genome sequences and metavirome analyses are flooding public databases and are revealing new insights into the field of ecology, all supported by new bioinformatics approaches and tools. This type of research has now also reshaped bacterial virus taxonomy from the morphology-driven classification (originally introduced by Professor Hans-Wolfgang Ackermann) to an integrated genome-driven taxonomy, which has gradually been implemented in the last decade. Affordable high-throughput sequencing is now also opening the door to systematic transcriptome analysis using RNAseq, introducing new standard towards experimental validation of gene predictions, genome organization, and the importance of ncRNAs.

This postgenomic era is driven by curiosity of the vast numbers of unknown gene products encoded by phage, also termed the "viral dark matter." The functional elucidation of the function of these proteins using new state-of-the-art approaches is rekindling research questions which have driven the "Golden Age" of phage research and have led to key advances in biotechnology between the 1950s and 1970s. One may argue that a new generation of researchers is currently emerging which may hopefully lead us into a "Second Golden Age" of phage research. Indeed, the discovery and impact of the CRISPR/cas system and its derived biotechnological techniques is yet again a driving force impacting entire research fields. The CRISRP/cas genome editing tools are just a single example of the impact of phage research on synthetic biology. The advances in our ability to engineer phage in various bacterial hosts provide a scaffold for new and innovative antibacterial design strategies.

Indeed, the last decade has also resulted in the re-evaluation of phage and phage-derived strategies to combat multidrug resistant human pathogens and approaches for the food and agriculture industry. Companies driven/supported by academic research have emerged and develop phage-based antimicrobials (phage therapy, endolysins, Artilysins™), the first of which have now entered the market in agriculture and food industry and for human applications (diagnostics, ongoing clinical trials).

All of these developments in phage research have been supported by initiatives from within the research society to organize and broaden its scope. The establishment of the "International Society for Viruses of Microbes" has expanded the community, made it more tightknit, and is coming together through social media initiatives (e.g., PhageBook, "A smaller flea" blog). We hope this edition of *Bacteriophages: Methods and Protocols* will like the previous volumes assist both the established and novice phage scientist.

Leicester, UK   *Martha R.J. Clokie*
Guelph, ON, Canada   *Andrew M. Kropinski*
Leuven, Belgium   *Rob Lavigne*

Contents

Preface .... *v*

Contributors .... *ix*

PART I KEY BASICS OF PHAGE BIOLOGY

1 Basic Phage Mathematics 1
  Stephen T. Abedon and Tena I. Katsaounis

2 Analysis of Host-Takeover During SPO1 Infection of *Bacillus subtilis* 31
  Charles R. Stewart

3 Practical Advice on the One-Step Growth Curve 41
  Andrew M. Kropinski

4 Iron Chloride Flocculation of Bacteriophages from Seawater 49
  Bonnie T. Poulos, Seth G. John, and Matthew B. Sullivan

5 Purification of Bacteriophages Using Anion-Exchange Chromatography 59
  Dieter Vandenheuvel, Sofie Rombouts, and Evelien M. Adriaenssens

6 Encapsulation Strategies of Bacteriophage (Felix O1) for Oral Therapeutic
  Application 71
  Golam S. Islam, Qi Wang, and Parviz M. Sabour

7 Encapsulation of *Listeria* Phage A511 by Alginate to Improve
  Its Thermal Stability 89
  Hanie Ahmadi, Qi Wang, Loong-Tak Lim, and S. Balamurugan

8 Application of a Virucidal Agent to Avoid Overestimation of Phage
  Kill During Phage Decontamination Assays on Ready-to-Eat Meats 97
  Andrew Chibeu and S. Balamurugan

PART II SEQUENCING ANALYSIS OF BACTERIOPHAGES

9 Sequencing, Assembling, and Finishing Complete Bacteriophage Genomes 109
  Daniel A. Russell

10 Identification of DNA Base Modifications by Means of Pacific
   Biosciences RS Sequencing Technology 127
   Philip Kelleher, James Murphy, Jennifer Mahony, and Douwe van Sinderen

11 Analyzing Genome Termini of Bacteriophage Through High-Throughput
   Sequencing 139
   Xianglilan Zhang, Yahui Wang, and Yigang Tong

12 Amplification for Whole Genome Sequencing of Bacteriophages
   from Single Isolated Plaques Using SISPA 165
   Derick E. Fouts

13 Genome Sequencing of dsDNA-Containing Bacteriophages Directly
   from a Single Plaque 179
   Witold Kot

14 Preparing cDNA Libraries from Lytic Phage-Infected Cells
    for Whole Transcriptome Analysis by RNA-Seq............................. 185
    Bob Blasdel, Pieter-Jan Ceyssens, and Rob Lavigne

PART III   PHAGE-RELATED BIOINFORMATICS TOOLS

15 Essential Steps in Characterizing Bacteriophages: Biology, Taxonomy,
    and Genome Analysis... 197
    *Ramy Karam Aziz, Hans-Wolfgang Ackermann, Nicola K. Petty,
    and Andrew M. Kropinski*

16 Annotation of Bacteriophage Genome Sequences Using DNA Master:
    An Overview 217
    Welkin H. Pope and Deborah Jacobs-Sera

17 Phage Genome Annotation Using the RAST Pipeline 231
    *Katelyn McNair, Ramy Karam Aziz, Gordon D. Pusch, Ross Overbeek,
    Bas E. Dutilh, and Robert Edwards*

18 Visualization of Phage Genomic Data: Comparative Genomics
    and Publication-Quality Diagrams 239
    *Dann Turner, J. Mark Sutton, Darren M. Reynolds, Eby M. Sim,
    and Nicola K. Petty*

PART IV   BACTERIOPHAGE GENETICS

19 Transposable Bacteriophages as Genetic Tools............................. 263
    Ariane Toussaint

20 Applications of the Bacteriophage Mu In Vitro Transposition Reaction
    and Genome Manipulation via Electroporation of DNA
    Transposition Complexes 279
    Saija Haapa-Paananen and Harri Savilahti

21 Use of RP4::Mini-Mu for Gene Transfer.................................... 287
    Frédérique Van Gijsegem

22 Muprints and Whole Genome Insertion Scans: Methods
    for Investigating Chromosome Accessibility and DNA Dynamics
    using Bacteriophage Mu 303
    N. Patrick Higgins

Index 315

Contributors

STEPHEN T. ABEDON • *Department of Microbiology, The Ohio State University, Columbus, OH, USA*

HANS-WOLFGANG ACKERMANN • *Department of Microbiology, Immunology, and Infectiology, Faculty of Medicine, Université Laval, Quebec, QC, Canada*

EVELIEN M. ADRIAENSSENS • *Laboratory of Gene Technology, KU Leuven, Leuven, Belgium; Centre for Microbial Ecology and Genomics, University of Pretoria, Hatfield, South Africa; Institute of Integrative Biology, University of Liverpool, Liverpool, UK*

HANIE AHMADI • *Guelph Research and Development Centre, Agriculture and Agri-Food Canada, Guelph, ON, Canada; Department of Food Science, University of Guelph, Guelph, ON, Canada*

RAMY KARAM AZIZ • *Department of Microbiology and Immunology, Faculty of Pharmacy, Cairo University, Cairo, Egypt; Argonne National Laboratory, Argonne, IL, USA*

S. BALAMURUGAN • *Guelph Research and Development Centre, Agriculture and Agri-Food Canada, Guelph, ON, Canada; Agropur Dairy Cooperative, Saint-Hubert, QC, Canada*

BOB BLASDEL • *Laboratory for Gene Technology, KU Leuven, Leuven, Belgium*

PIETER-JAN CEYSSENS • *Laboratory for Gene Technology, KU Leuven, Leuven, Belgium; Bacterial Diseases, Unit Antibiotic Resistance, Scientific Institute of Public Health, Brussels, Belgium*

ANDREW CHIBEU • *Guelph Food Research Centre, Agriculture and Agri-Food Canada, Guelph, ON, Canada; Agropur Dairy Cooperative, Saint-Hubert, QC, Canada*

BAS E. DUTILH • *Theoretical Biology and Bioinformatics, Utrecht University, Utrecht, The Netherlands; Centre for Molecular and Biomolecular Informatics, Radboud Institute for Molecular Life Sciences, Radboud University Medical Centre, Nijmegen, The Netherlands*

ROBERT EDWARDS • *Computational Sciences Research Center, San Diego State University, San Diego, CA, USA; Department of Biology, San Diego State University, San Diego, CA, USA; Department of Computer Science, San Diego State University, San Diego, CA, USA*

DERICK E. FOUTS • *J. Craig Venter Institute, Rockville, MD, USA*

FRÉDÉRIQUE VAN GIJSEGEM • *Institute of Ecology and Environmental Sciences of Paris, INRA UMR1392, UPMC barre 44-45 CC 237, Paris Cedex, France*

SAIJA HAAPA-PAANANEN • *Division of Genetics and Physiology, Department of Biology, University of Turku, Turku, Finland*

N. PATRICK HIGGINS • *Department of Biochemistry and Molecular Genetics, The University of Alabama at Birmingham, Birmingham, AL, USA*

GOLAM S. ISLAM • *Guelph Research and Development Center, Agriculture and Agri-Food Canada, Guelph, ON, Canada*

DEBORAH JACOBS-SERA • *Department of Biological Sciences, University of Pittsburgh, Pittsburgh, PA, USA*

SETH G. JOHN • *Department of Earth Sciences, University of Southern California, Los Angeles, CA, USA*

TENA I. KATSAOUNIS • *Department of Mathematics, The Ohio State University, Columbus, OH, USA*

PHILIP KELLEHER • *School of Microbiology, University College Cork, Cork, Ireland*

WITOLD KOT • *Department of Environmental Science, Aarhus University, Roskilde, Denmark*

ANDREW M. KROPINSKI • *Department of Molecular and Cellular Biology, University of Guelph, Guelph, ON, Canada; Department of Pathobiology, University of Guelph, Guelph, ON, Canada; Department of Food Science, University of Guelph, Guelph, ON, Canada*

ROB LAVIGNE • *KU Leuven, Laboratory for Gene Technology, Leuven, Belgium*

LOONG-TAK LIM • *Department of Food Science, University of Guelph, Guelph, ON, Canada*

JENNIFER MAHONY • *School of Microbiology, University College Cork, Cork, Ireland; Alimentary Pharmabiotic Centre, University College Cork, Cork, Ireland*

KATELYN MCNAIR • *Computational Sciences Research Center, San Diego State University, San Diego, CA, USA*

JAMES MURPHY • *School of Microbiology, University College Cork, Cork, Ireland*

ROSS OVERBEEK • *Argonne National Laboratory, Argonne, IL, USA*

NICOLA K. PETTY • *The ithree Institute, University of Technology Sydney, Sydney, NSW, Australia*

WELKIN H. POPE • *Department of Biological Sciences, University of Pittsburgh, Pittsburgh, PA, USA*

BONNIE T. POULOS • *Department of Ecology and Evolutionary Biology, University of Arizona, Tucson, AZ, USA*

GORDON D. PUSCH • *Argonne National Laboratory, Argonne, IL, USA*

DARREN M. REYNOLDS • *Centre for Research in Biosciences, Faculty of Health and Applied Sciences, University of the West of England, Bristol, UK*

SOFIE ROMBOUTS • *Laboratory of Gene Technology, KU Leuven, Leuven, Belgium*

DANIEL A. RUSSELL • *Department of Biological Sciences, University of Pittsburgh, Pittsburgh, PA, USA*

PARVIZ M. SABOUR • *Guelph Research and Development Center, Agriculture and Agri-Food Canada, Guelph, ON, Canada*

HARRI SAVILAHTI • *Division of Genetics and Physiology, Department of Biology, University of Turku, Turku, Finland*

EBY M. SIM • *The ithree institute, University of Technology Sydney, Sydney, NSW, Australia*

DOUWE VAN SINDEREN • *School of Microbiology, University College Cork, Cork, Ireland; Alimentary Pharmabiotic Centre, University College Cork, Cork, Ireland*

CHARLES R. STEWART • *Department of BioSciences, Rice University, Houston, TX, USA*

MATTHEW B. SULLIVAN • *Department of Ecology and Evolutionary Biology, University of Arizona, Tucson, AZ, USA; Department Microbiology, The Ohio State University, Columbus, OH, USA*

J. MARK SUTTON • *Public Health England, Salisbury, Wiltshire, UK*

YIGANG TONG • *State Key Laboratory of Pathogen and Biosecurity, Beijing Institute of Microbiology and Epidemiology, Beijing, People's Republic of China*

ARIANE TOUSSAINT • *Laboratoire de Génétique et Physiologie bactérienne (LGPB), Université Libre de Bruxelles, Charleroi, Belgium*

DANN TURNER • *Centre for Research in Biosciences, Faculty of Health and Applied Sciences, University of the West of England, Bristol, UK*

DIETER VANDENHEUVEL • *Laboratory of Gene Technology, KU Leuven, Leuven, Belgium*

QI WANG • *Guelph Research and Development Centre, Agriculture and Agri-Food Canada, Guelph, ON, Canada*

YAHUI WANG • *State Key Laboratory of Pathogen and Biosecurity, Beijing Institute of Microbiology and Epidemiology, Beijing, People's Republic of China*

XIANGLILAN ZHANG • *State Key Laboratory of Pathogen and Biosecurity, Beijing Institute of Microbiology and Epidemiology, Beijing, People's Republic of China*

Part I

Key Basics of Phage Biology

Chapter 1

Basic Phage Mathematics

Stephen T. Abedon and Tena I. Katsaounis

Abstract

Basic mathematical descriptions are useful in phage ecology, applied phage ecology such as in the course of phage therapy, and also toward keeping track of expected phage–bacterial interactions as seen during laboratory manipulation of phages. The most basic mathematical descriptor of phages is their titer, that is, their concentration within stocks, experimental vessels, or other environments. Various phenomena can serve to modify phage titers, and indeed phage titers can vary as a function of how they are measured. An important aspect of how changes in titers can occur results from phage interactions with bacteria. These changes tend to vary in degree as a function of bacterial densities within environments, and particularly densities of those bacteria that are susceptible to or at least adsorbable by a given phage type. Using simple mathematical models one can describe phage–bacterial interactions that give rise particularly to phage adsorption events. With elaboration one can consider changes in both phage and bacterial densities as a function of both time and these interactions. In addition, phages along with their impact on bacteria can be considered as spatially constrained processes. In this chapter we consider the simpler of these concepts, providing in particular detailed verbal explanations toward facile mathematical insight. The primary goal is to stimulate a more informed use and manipulation of phages and phage populations within the laboratory as well as toward more effective phage application outside of the laboratory, such as during phage therapy. More generally, numerous issues and approaches to the quantification of phages are considered along with the quantification of individual, ecological, and applied properties of phages.

Key words Adsorption rate, Efficiency of plating, Killing titers, Multiplicity of infection, Phage ecology, Phage population growth, Phage therapy modeling, Poisson distribution, Titer determination

1 Introduction

The relative simplicity of phage biology, especially as observed on a whole organismal level—i.e., as free phages, phage-infected bacteria, or not-yet infected bacteria—lends itself to simple mathematical description. The simplest of these descriptors involves the counting of phage numbers. Once a means of counting is achieved, then it is only natural to consider how these counts can change over time. For most phages, it requires only straightforward manipulation to assure that counts mostly correspond to *individual* phages, whether as free phages or instead as phages that are infecting

Martha R.J. Clokie et al. (eds.), *Bacteriophages: Methods and Protocols, Volume 3*, Methods in Molecular Biology, vol. 1681, https://doi.org/10.1007/978-1-4939-7343-9_1, © Springer Science+Business Media LLC 2018

bacteria at low multiplicities, that is, low ratios of phages that have adsorbed or are infecting to total numbers of adsorbable bacteria, thereby assuring that most phage-infected bacteria are singly phage infected. Changes in free phage counts are seen when phages adsorb bacteria, which one observes simultaneously as declines in free phage numbers, as increases in adsorbed-phage counts, and as increases in numbers of phage-infected bacteria. The latter additionally can give rise to reductions in bacterial counts and, after a time lag, increases in phage counts as well. Those increases in phage counts are an additional and fundamental mathematical descriptor, one that corresponds to what is known as the phage burst size.

We can then consider rates of change, whether this is the rate that free phages are lost to adsorption—as well as, somewhat equivalently, uninfected bacteria are lost to phage adsorption—or instead the *rate* corresponding to the lag between phage adsorption and subsequent burst (i.e., the phage latent period). These effectively are physiological or ecological phenomena. Also potentially of interest are evolutionary rates such as the occurrence within individual phages or phage populations of spontaneous mutations or, instead, rates of deterministic changes in phage allele frequencies as described, for example, by the relative Darwinian fitness associated with different phage genotypes.

In this chapter we consider mathematical descriptions of phages that are applicable to the manipulation of phages as whole organisms, especially in the laboratory, and consider in particular descriptions that are more ecological (numbers and interactions) rather than evolutionary (changes in allele frequencies as a function of time). We thus consider, mostly in verbal form, the mathematics of phage titers, phage efficiency of plating, ratios of phages to bacteria, Poisson distributions, virion adsorption rates, and phage population growth as well as mention of the modeling of phage-bacterial interactions as they can occur such as within chemostats, some consideration of the impact of spatial structure on phage presence (e.g., as during phage plaque growth), and also mathematics in considering the phage impact on bacteria during phage-mediated biocontrol (aka phage therapy). We then turn to consideration of basic statistical underpinnings of phage enumeration such as via plaque count determination. Our overall goal is to provide an introduction to mathematical concepts that may otherwise have become less appreciated in the decades since their introduction. We also hope to motivate an at least intuitive appreciation of basic phage mathematics such as can be useful while manipulating phages in the laboratory or while studying their interactions with bacteria under less controlled circumstances.

2 Phage Enumeration

The near-term impact of phages on bacteria tends to increase as a function phage concentrations within a given environment, particularly concentrations of virions. Phage concentrations typically are considered in terms of some measure of their numbers per unit volume, often as phages or the equivalent per mL. When the phages involved are free virions, then we can describe these concentrations as titers. A related concept is that of plaque-forming units or PFUs, where PFUs typically are a measure of virion numbers, and thus a titer can presented as PFUs/mL. Alternatively, when plaquing it can be difficult to distinguish between free phages and phage-infected bacteria, in which case the term infective center can be employed instead. An infective center, that is, is essentially a plaque-forming unit that consists either of a free phage or phage-infected bacterium, since both will tend to give rise to only a single plaque.

A complication especially on these concepts is that of virion clumping, where multiple otherwise free phages can give rise to only a single plaque. Phage secondary adsorption, secondary infection (superinfection), and/or coinfection [1] all represent, as well, essentially bacteria-mediated phage "clumping," where multiple virions adsorbed to or infecting a single bacterium also give rise to only a single plaque. Multiple phage adsorptions to multiple, clumped bacteria such as found within a bacterial arrangement also will, upon plating, give rise to only a single plaque. Indeed, even a productive phage infection can be viewed as a form of phage clumping since such infective centers can give rise to only a single plaque regardless of the number functional virions that currently may be present within a prelysis phage-infected bacterial cell.

Additional means of phage enumeration exist besides plaquing. These include via a transmission electron or epifluorescence microscopy [2]. Phage numbers can also be counted via additional detection methods that do not involve microscopic visualization [3, 4]. It is possible as well to determine phage titers using liquid growth media rather than plaquing, resulting in a phage equivalent to the most-probable number (MPN) method employed to determine bacterial counts. One such alternative approach to phage enumeration is based on the phage ability to kill target bacteria, such as one sees with the use of killing titers, as discussed following consideration of the Poisson distribution. What most of these latter methods have in common is that they are not direct measures of phage viability, that is, ability to productively infect target bacteria. As a result, we can determine ratios of viable phages present within a sample to total numbers of phages that are present. Since viability is typically determined using plaque assays [5–7], that is, a form of "plating," discrepancies between total phage counts and viable counts—something that ecologically is sometimes described as the "great plaque count anomaly," e.g., [8]—one describes more prosaically in terms of what is known instead as efficiency of plating or EOP [9].

2.1 Efficiency of Plating

Efficiency of plating (EOP) can be distinguished into absolute efficiency of plating, relative efficiency of plating, and also what can be described as efficiency of center of infection, or ECOI (aka efficiency of infective center formation). Absolute efficiency of plating is the ratio of viable phage counts to total counts, the latter as determined using methods that do not distinguish between viable (i.e., active) versus nonviable (inactive) phages. This value is simply the ratio of especially plaque counts (i.e., PFUs) to measures of total phage presence. An absolute EOP of 1.0 means that every phage that is known to be present by means other than plaque formation, and particularly in terms of total rather than viable counts, not only is able to productively infect a target bacterium but is able to do so with sufficient vigor so as to give rise to a visible plaque [10].

Relative efficiency of plating, by contrast, is a comparison between plaque counts using one host as the indicator bacterium versus another (or otherwise plaquing under different conditions). In this case the EOP in principle could be represented as a number that is either higher than or lower than 1.0, though typically instead it is shown as lower than one. Thus, if the number of plaques from a given phage stock that form upon plating on a lawn of bacterial strain A is 100 and the number that form on a lawn of bacterial strain B is 50, then we can say that the relative efficiency of plating on strain B is 0.5 or 50% of that on strain A. Typically when the EOPs are stated without qualification it is this latter, that is, *relative* connotation that is implied. One often sees EOP measurements in phage host range determinations such that a higher EOP on a given host is taken as an indication of greater potential by a given phage type to propagate on that host, e.g., [11].

Plaques are more complicated phenomena than many appreciate and so as a consequence it may *not* be reasonable to attempt to infer phage inviability based solely upon a lack of plaque formation [10]. That is, an ability to form plaques can require greater phage "performance" (or "infection vigor") than simply an ability to productively infect a given bacterial strain, where to productively infect is to produce virion progeny following virion adsorption of a host cell. To distinguish between phages that are able to form plaques from phages that can productively infect but nevertheless are unable to form plaques, e.g., due to a display of relatively small burst sizes, we can employ ECOI determinations instead of EOP. ECOI determinations [12] involve virion preadsorption to a given host bacterium, removal of remaining free phages, and then plating using as indicator bacteria an alternative host type upon which the phage is known to plaque with high efficiency. Any infections of the preadsorbed host that are productive, over reasonable time frames, should result in plaque formation on the sensitive host. With careful accounting of numbers of phages plated it can be possible to compare efficiencies of productive infection of a test host strain versus productive infection of the known sensitive host.

2.2 Additional Aspects of Enumeration Efficiency

An ability to productively infect requires greater phage performance than simply phage adsorption. Phage adsorption that results in bacterial killing, but not necessarily productive infection, is measurable instead using the above-noted killing-titer procedure. In general we can predict that phage total counts as determined via microscopy (e.g., transmission electron microscopy) will be greater than or equal to phage killing titers, which will be greater than or equal to phage titers that are determined via ECOI-like procedures, which in turn should be greater than or equal to phage titers that are determined via the standard plaquing approaches used for EOP determinations. That is, absolute numbers of phage particles (microscopically determined) may be greater than or equal to numbers of phage particles that are capable of adsorbing and then killing bacteria (killing titer, KT, determined), which in turn should be greater than (or equal to) the number of those killing phage particles that are capable to giving rise to a productive phage infection (ECOI determined). This latter number should be greater than (or equal to) the number that also can form plaques (EOP determined).

At the other extreme, virus-like particles (VLPs) as may be observed within a given environment, as may be determined via epifluorescence microscopy, can be greater than or equal in number to counts of confirmed phage particles (PPs) or even confirmed virus particles (VPs), such as determined electron microscopically [13]. Thus, at least in terms of numbers of bactericidal viruses:

VLP ≥ VP ≥ PP ≥ KT ≥ viable phages ≥ infective centers ≥ plaques,

where the latter two are what one quantifies in ECOI and EOP determinations, respectively. As noted above, infective-center or plaque counts can be lower than numbers of viable phages as a consequence of various forms of phage clumping.

3   Ratios of Phages to Bacteria

The rate at which bacteria become phage adsorbed is proportional to free phage densities (i.e., phage titers) whereas the rate at which phages are lost to adsorption, within a given environment, is proportional to concentrations of phage susceptible bacteria. Thus, more bacteria means faster phage loss to adsorption whereas more free phages means faster rates at which bacteria become phage adsorbed. If we hold phage densities constant, which can be approximated over finite time periods when ratios of phages to bacteria are high, then phage adsorption of bacteria should occur at a constant rate, at least until successive phage adsorption becomes physically difficult on a per-bacterium basis, that is, exceeds a bacterium's physical capacity [14] to support phage adsorption. Alternatively, if ratios of phages to bacteria are not terribly high and, in addition, either bacterial densities are relatively high or sufficient time is allowed, then phage numbers may decline appreciably. At the point at which 100% of accounted for free phages have succeeded in adsorbing then we have achieved what is known as a multiplicity of infection or MOI, that is, the ratio of infecting phage particles to total numbers of bacteria present. This is true even if not all free phages present or added to a given environment have succeeded in adsorbing, e.g., they instead could be removed or inactivated prior to adsorbing, just so long as we take into account only those phages that are thought or known to have succeeded in adsorbing when making MOI calculations. Also relevant to issues of phage multiplicity of infection are those of Poisson distributions and phage killing titers.

3.1 Multiplicity of Infection

The concept of multiplicity of infection likely has never been without conflict between those for whom precision is preferable to expediency and those with the opposing proclivity. Unfortunately for the former, not only has the precision of MOI determinations tended to decline through the decades but the very concept of multiplicity of infection has become less precise as well [15]. With precision, MOI is determined from the number of phages added to a bacterial culture over a given time period minus the number of phages that have failed to adsorb, with effort necessarily made to prevent phage bursts from occurring over the time period of interest. That is, in MOI determinations one generally takes into account only those phages that were explicitly added to cultures and which have succeeded in adsorbing. At a minimum, this measure requires some means of post-adsorption-interval enumeration of remaining free phages—abbreviated here as R, as measured in units of phages/mL, and as distinct from phage-infected bacteria—as well as an account of the total number of free phages (P) that have been added to or at least have been present within an adsorption vessel over the internal of interest (P also is in units of phages/mL or, more precisely, the phage density present at the beginning of an adsorption interval may be described as P_0). In addition, one needs a measure of the density of bacteria, i.e., number of bacteria/mL (B), to which P_0 phages have been added. Thus, $\text{MOI} = (P_0-R)/B$.

Because phage adsorption often occurs fairly quickly when bacterial densities are relatively high (e.g., $>10^7$/mL) and phages adsorb with mostly consistent ability across a phage population (though this is not always the case), then a standard "cheat" is to assume that 100% phage adsorption has occurred soon following phage addition to bacteria, that is, without actually confirming that 100% efficiency in adsorption has actually occurred. Thus, in this less precise approach, $\text{MOI} = P_0/B$. Note regardless that P_0/B should exist as the limit of $(P_0-R)/B$ as time (t) goes to infinity and thereby R goes to zero. Nevertheless, if MOI determinations are important—e.g., killing titers, as can be relevant to phage-mediated biocontrol of bacteria, or phage therapy—then it can be relevant to MOI calculations for R in fact to succeed in declining to some approximation of zero over appropriate time frames, which typically are multiple minutes or, outside of the laboratory, at most a few hours. It nevertheless is possible to predict multiplicities of adsorption/infection as functions of phage density and time and to do so even independently of knowledge of bacterial density [16–19].

Alternatively, imprecision in use of MOI has cropped into the phage literature in what in fact is alarming frequency [15, 20]. This imprecision has involved what appears to be either an intentional ignoring or instead ignorance of the issue of R. Semantically, this has resulted effectively, if implicitly, in the concept of multiplicity of *infection* being replaced with that of multiplicity of *addition*, which especially at lower bacterial densities (e.g., $<10^6/$ mL) can wildly exceed actual MOIs over meaningful time frames [15, 20–24]. An additional issue, relevant to issues of phage performance, stems from the distinction between phage adsorption and phage infection. The concept of multiplicity of infection predates adequate understanding of the distinction between phage adsorption and infection, where the former need not always give rise to the latter [1]. Consequently, it can be more precise to consider the concept of multiplicity of *adsorption*. Such usage, even if not explicitly employed, nonetheless might allow for a better appreciation of what MOI is not, that is, it is *not* a multiplicity solely of phage *addition*. An online MOI calculator can be found here [25].

3.2 The Poisson Distribution

Multiplicity of infection is relevant in a number of contexts as it is a direct measure of the number of phages that, on average, have adsorbed each bacterium within a culture. If MOI is less than 1.0, for example, then a fair number of bacteria will remain unadsorbed within a culture. If MOI is greater than 1.0, then not only will more bacteria have been phage adsorbed, but more bacteria will have been adsorbed by more than one phage. Adsorption by more than one phage can have a number of consequences. It can give rise to coinfection, genetic recombination, multiplicity reactivation, and/ or interference between phages [1] as well as phage "clumping" more generally, i.e., as considered above. Coinfection can result in higher likelihoods of temperate phage infections resulting in lysogeny [26] while multiple adsorptions can switch the T-even phage genetic program from one of rapid lysis to that of lysis inhibition [16, 26], or even to what is known as lysis from without if the total number of adsorbing phages per bacterium is sufficiently high [26, 27]. In addition, in terms of phage-mediated biocontrol, not only do lower MOIs imply that fewer bacteria are being subject to bactericidal phage infections but, at higher MOIs, phage adsorptions are literally being wasted since usually only a single phage adsorption is required to kill an adsorbed bacterium. Lastly, if we can calculate phage MOIs from the total number of bacteria that have been killed upon exposure to a phage population then, assuming 100% phage adsorption has occurred, we can calculate what is known as a phage stock's killing titer [14, 28–30].

In these examples it often is not sufficient to consider solely the average ratios of adsorbed phages to bacteria, i.e., MOI, but instead to be aware of the distribution of adsorbed phages to phage-susceptible bacteria. That distribution is usually assumed to be Poissonal [31]. For general discussion of the Poisson distribution, *see* for example [32]. Because phages adsorb with a Poisson *distribution*, this means especially that the number of multiply adsorbed bacteria or the number of not-adsorbed bacteria, within a population, may be greater than one would expect based solely on average ratios. In practical terms, this means that there will be more multiple phage adsorptions to individual bacteria than one might expect given low MOIs as well as greater bacterial survival than one might expect given higher MOIs.

The Poisson distribution as applied to phages is a function of MOI, that is, the average number of phages that have adsorbed per bacterium. The equivalent value for generation of Poisson distributions is typically abbreviated as λ, though we will use the abbreviation, m, to avoid confusion with the well-known phage λ. For a given MOI (m) we can then describe the probability of bacterial adsorption by a specific number of phages (n) as $f(n,m)$. Rather than starting with the generalized equation, instead we start with the probability that a given bacterium will have been adsorbed by $n = 0, 1, 2$, and 3 phages (we use n here rather than, for example, k, to avoid confusion with the phage adsorption rate constant, which often is abbreviated also as k). Note that, formally speaking, n as used here must be a nonnegative integer. To further simplify things, let us define MOI (m) as equal to 1. The associated probability (P) that $n = 0$ can be written as $P(n = 0) = f(0,1) = e^{-1} = 0.37$.

In addition, $P(n = 1) = f(1,1) = e^{-1} = 0.37$, $P(n = 2) = f(2, 1) = e^{-1}/2 = 0.18$, and $P(n = 3) = f(3, 1) = e^{-1}/6 = 0.061$. You will note in these examples that the probability that a bacterium has *not* been phage adsorbed ($n = 0$) and the probability that a bacterium has been adsorbed by one phage ($n = 1$) are expected to be equal in magnitude and this is so even though on *average* 1 phage is adsorbed per bacterium (i.e., MOI $= 1 = m$). The probability that a bacterium has been adsorbed by 2 phages, in turn, is half that

(0.37/2), and the probability that a bacterium has been adsorbed by 3 phages is one-third of the probability of adsorption by two (0.37/6).

The above ratios we present as examples and are what one observes given an MOI of 1, i.e., where $m = 1$, but will differ for different MOIs. Specifically, the generalized representation of a Poisson distribution is $f(n,m) = m^n e^{-m}/n!$, that is multiplicity, m, raised to the category, n, that we are interested in ($n = 0$ phages adsorbed, $n = 1$ phage adsorbed, etc.) multiplied by the base of the natural logarithm, e, raised to the multiplicity, m, and then divided by the category of interest, factorial ($n!$). That is, $0! = 0$, $1! = 1$, $2! = 2$, $3! = 6$, etc. For the example given in the previous paragraph, $m^n = 1$ no matter what the value of n since $m = 1$, and also e^{-m} is a constant. Therefore, the difference in values for $f(n,m)$ is determined solely by $n!$, as described. When MOI is greater than or less than 1, however, then this shortcut of holding m^n constant and equal to 1 is not applicable and hence, as noted, ratios between categories will differ from those calculated for $m = 1$.

3.2.1 Solving the Poisson Distribution (Using Excel)

Using Microsoft Excel one can employ the POISSON.DIST function (or simply POISSON). The arguments are X, Mean, and Cumulative. X is equivalent to n as defined above and Mean is equivalent to m, that is, to the multiplicity of infection. Cumulative, by contrast, is a logical operator. It should be set equal to False to generate the probabilities described above. Thus, POISSON.DIST(2,1,False) = 0.18. Alternatively, POISSON.DIST(2,1,True) = 0.92 = POISSON.DIST(0,1, False) + POISSON.DIST(1,1,False) + POISSON.DIST(2,1,False), which is the cumulative probability, $P(n \le 2) = P(n = 0) + P(n = 1) + P(n = 2)$. Do not forget to place an equal sign in front of expressions when evaluating them using Excel. Thus, the proper syntax to place in the formula bar, in the first example, would be "= POISSON.DIST(2,1,False)" (without the quotation marks), which would return 0.18.

Though the probability that a bacterium will not be phage adsorbed given a known multiplicity of infection is straightforward to predict on its own, as it is equal simply to e^{-m}, nonetheless using Excel we can go through the exercise of calculating this probability—that is, that bacteria will remain unadsorbed—for $m = 1, 2, 3, 5$, or 10. These are 0.37, 0.14, 0.05, 0.01 (the latter as equivalent to 1%), and 4.5×10^{-5} (0.0045%), respectively. Thus, even with a confirmed multiplicity of 5, and assuming that all bacteria are equally susceptible to phage adsorption, still approximately 1% of bacteria are predicted to remain unadsorbed despite 100% adsorption by accounted for phages.

One can also calculate the probability that bacteria are adsorbed by more than one phage, which is equal to 1—POISSON.DIST(0, m,False)—POISSON.DIST(1,m,False) = 1—POISSON.DIST(1, m,True). Thus, for $m = 0.01, 0.1, 1$, and 2, these numbers are

0.00005, 0.005, 0.26, and 0.59, respectively (0.0005%, 0.5%, 26%, and 59%). For the latter, only a little more than one half of the bacteria are predicted to be adsorbed by more than one phage despite a multiplicity of 2. Alternatively, a facile conclusion that one can make from, for example, the $m = 0.01$ calculation, is that very few bacteria are multiply adsorbed at this low phage multiplicity. This latter inference, however, simultaneously is both true and misleading.

3.2.2   Per Infection Versus per Bacterium

The misleading aspect results from our often being more concerned with what fraction of *adsorbed* bacteria are predicted to have been adsorbed by more than one phage rather than what fraction of *total* bacteria have been multiply adsorbed. For determination of total number of multiple phage adsorptions as a function of total number phage-adsorbed bacteria, the calculation instead is (1–POISSON.DIST(1,m,True))/(1–POISSON.DIST(0,m,True)). Again for $m = 0.01$, 0.1, 1, and 2, these numbers are 0.5%, 4.9%, 42%, and 69%, respectively.

In other words, even with on average only 1 phage adsorbed per 100 bacteria, approximately 0.5% of those bacteria that are phage infected will have been adsorbed by more than one phage, which is 1000-fold larger in magnitude than as calculated as a function of all bacteria present (0.5% versus 0.0005%). On the other hand, even with an MOI of 2, nearly one-third of bacteria that have been phage adsorbed will have been adsorbed by only one phage, though that fraction drops to below 5% with an MOI of 5. Thus, to fully appreciate what fraction of bacteria are or are not expected to become phage adsorbed, or what fraction of phage infections have been multiply phage adsorbed, it is crucial to understand how Poisson probabilities can be used as well as the true meaning of MOI as the ratio of *adsorbed* phages to total bacteria. For consideration of how differences in the relative adsorbability of different bacteria can affect assumptions of Poissonal distribution, *see* ref. [31].

3.3   Killing Titers

As the Poisson distribution is calculated based upon MOIs, so too can MOIs be calculated based upon Poisson distributions. The latter is most easily accomplished as a function of the number of unadsorbed bacteria that remain within a culture following phage exposure since this number can be determined simply by plating for bacterial viable counts. As the formula predicting this number as based on MOI is e^{-m}, we can predict MOI based on the fraction of a known quantity of bacteria that remain viable following phage exposure. That is, m, or MOI, must be equal to the opposite of the natural logarithm of the fraction of bacteria that have *not* been killed. In Excel this would be $= -$LN([CFUs post phage]/[CFUs pre phage]) where CFU stands for Colony Forming Unit. Thus, if 1% of bacteria remain able to form colonies, then the multiplicity of infection must have been -ln(0.01) or 4.5. As noted above, that is, a multiplicity of approximately 5 is necessary to eliminate 99% of bacteria through phage adsorption alone, i.e., through what is known as passive treatment [19, 33, 34]. Implicit to this concept, however, is that adsorption of one phage to one CFU will result in loss of that CFU, even if a CFU should consist of multiple cells. An online killing titer calculator can be found here: [35] For mathematical consideration of the ecological implications of this latter point, *see* ref. [36, 37].

The phage killing titer, taking these various considerations into account, is equal to the thus calculated MOI multiplied by the number of bacteria that were originally present prior to phage addition, that is, pre-phage-addition CFUs. For example, this would be equal to 4.5×10^7 phages/mL added and adsorbed given a bacterial culture that is reduced from 10^7/mL to 10^5/mL, with that reduction occurring through phage adsorption alone (since a 100-fold reduction in bacterial densities results from a phage multiplicity of 4.5 and 4.5×10^7 phages/mL is literally that ratio of phages to bacteria, in units of phages/bacteria, multiplied by the density of bacteria in units of bacteria/mL). As noted above, however, this value could very well be in excess of the number of phages that are able to successfully form plaques upon plating or even in excess of the number that can productively infect bacteria, as phages can be inactivated such as due to DNA damage in such a way that they can retain both their adsorption ability and subsequent bactericidal action while losing their ability to produce new phages.

By combining the concept of phage killing titers with that of rates of phage adsorption to bacteria we can predict the *rapidity* with which a given titer of phages may be able to reduce, through adsorption alone, a bacterial population by a given amount. Alternatively, this would be the titer of phages that can *initiate* that reduction should CFUs consist of multiple bacterial cells, thereby potentially requiring in situ generation of new phages via lytic cycles to subsequently remove those bacteria that are in addition to the number which were initially adsorbed [36, 37]. That is, the rate that phages can eliminate bacteria from environments can be determined from a knowledge of phage titers, as giving rise to a phage multiplicity of *addition*, in combination with what is known as the phage adsorption rate constant.

4   Phage Adsorption

The rate of attachment of free virions to bacteria [38] is typically described as a rate of phage adsorption [37, 39]. The more phage-susceptible bacteria that are present within a given environment then the faster a phage population will become adsorbed. At the same time, the more phages that present within a given environment then the faster a bacterial population will become phage adsorbed. The rate at which adsorbed and/or infected bacteria are created in absolute terms via phage–bacterial adsorption, however, is a function of both phage and bacterial densities as well as phage-virion diffusion rates, virion attachment affinity for bacteria, and bacterial target size. The latter together define what is known as the phage adsorption rate constant [14]. Thus, we can predict how fast free phages should be lost, how fast uninfected bacteria should be lost, and also how fast adsorbed and infected bacteria should be created, all as functions in part of phage adsorption rates. In addition is the complication of the potential for multiple phage adsorptions per bacterium, which has the effect of reducing the rate that bacteria are killed per phage adsorption, and also that phage infections are created as a function of the absolute rate of phage adsorptions to uninfected bacteria. For simplicity, in all cases we assume that adsorptions are occurring within a well-mixed fluid environment, though if that assumption is not valid then some bacteria may be more likely to become phage adsorbed than others and so too some phages may be more likely to adsorb a bacterium than others.

4.1 Rate of Phage Adsorption

The rate at which phages are lost to adsorption is directly proportional to the concentration of phage-susceptible bacteria that are found within an environment. That is, if one should double the number of bacteria present then that should result also in a doubling of the rate at which phages encounter those bacteria as well as a doubling of the likelihood that a given phage will adsorb per some unit time. The actual rate that a phage will adsorb a single bacterium found in this environment is equal to the phage adsorption rate constant, or k. This constant is the likelihood of a single phage adsorbing a single bacterium per unit volume per unit time. The phage adsorption rate constant often is described using units of mL/min (or $mL^{-1} min^{-1}$) though alternatively units of mL/h. can be seen, which is a 60-fold larger value that otherwise conveys the same information.

As the rate of adsorption of one phage to one bacterium is defined by k, the rate at which phages are lost to adsorption of a given target population consisting of more than one bacterium is equal simply to Bk, where B is the concentration of bacteria. As noted, the phage adsorption rate constant is a function phage virion diffusion rates, bacterial target size, and the likelihood of phage adsorption given phage collision with a bacterium. Stent [14], for example, describes an adsorption rate constant for phage T4 of 2.5×10^{-9} mL/min. In an environment consisting of 10^5 bacteria/mL, we as a consequence would expect a loss of 2.5×10^{-4} of those free phages present per min ($=10^5 \times 2.5 \times 10^{-9}$). If there are 10^9 phages present per mL then this would be 2.5×10^5 phages lost per minute ($= 10^9 \times 10^5 \times 2.5 \times 10^{-9}$), or more phages lost per minute than there are bacteria that are present (for an MOI of 2.5 achieved every minute $= 10^9 \times 10^5 \times 2.5 \times 10^{-9}/10^5$). If there are only 10^7 phages present per mL, then only 2.5×10^3 phages will be lost to adsorption per minute, which corresponds instead to a per-minute multiplicity of 0.025 ($= 10^7 \times 10^5 \times 2.5 \times 10^{-9}/10^5$).

4.2 Rate of Bacterial Adsorption

The latter calculations are indicative of the dependency on phage titers of the rate that bacteria are adsorbed by phages. Consistently, the calculation for the rate at which bacteria are lost to phage adsorption, as a function of phage density, is equivalent to that provided in the previous paragraph, i.e., it is Pk, where P is the phage titer. This calculation implies that for every tenfold increase in phage titer one should observe a tenfold increase in the rate at which bacteria are adsorbed by phages. The converse, however, is also true, and that is that the rate at which bacteria become phage adsorbed should decrease tenfold for every tenfold decrease in phage titer. The difference between the rates that bacteria are adsorbed given titers of 10^8 phages/mL thus should be 100-fold greater than rates at which bacteria are adsorbed given 10^6 phages/mL. While it should take 4 min to achieve an MOI of 1 given 10^8 phages/mL, assuming Stent's adsorption rate constant, the same multiplicity starting with 10^6 phages/mL should require instead 400 min to achieve, which is over 6 h.

4.3 Rate of Infection Creation

The rate at which phage infections are created via phage adsorptions is both a continuation of the above idea and, as a consequence of phages adsorbing over a Poisson distribution, more complicated. The complication, however, is often ignored, and to a degree legitimately so [40]. Thus, we will focus on the simple case where essentially multiple phage adsorption of bacteria is ignored, meaning in effect that every phage adsorption is considered to result in one bacterium being converted from not yet phage adsorbed to, instead, no longer *not* phage adsorbed. We will then speak to the occurrence of multiple adsorptions. Ignoring multiple adsorption, the rate at which new infections are created, I in units of, e.g., infections/mL, is equal to PBk, that is, the phage titer (in free phages) times the bacterial density times the phage adsorption rate constant. The rate at which infections are created, in other words, increases as either phage or bacterial densities increase, or both. A complication, however, is that P will decline as a consequence of this adsorption. A further complication is that so too will B decline. Declines in P means that rates of infection creation will also decline. The same is true for declines in B, though declines in B will *not* immediately result in declines in rates of phage adsorption if multiple adsorptions in fact are allowed, as usually phage-infected bacteria will continue to be able to adsorb phages. Rates of phage loss, in other words and allowing for multiple phage adsorptions per bacterium, should be equal more or less to $k(B + I)$.

4.4 Rate of Secondary Adsorption

Consistent with the previous calculation, the rate of phage adsorption to already phage-infected bacteria is simply kI. This calculation does not quite describe how many infected bacteria end up being adsorbed by more than one phage, however, because infected bacteria are not themselves distinguished into those that have been adsorbed by a second phage and those that have not. That is, phage adsorption to an already multiply phage adsorbed bacterium does not have the effect of increasing the number of multiply phage adsorbed bacteria. In addition, infected bacteria should eventually lyse (assuming lytic infections), so for any given infected bacterium there is only a limited span over which secondary adsorption might occur.

To solve the question of to what degree phage infections experience subsequent adsorptions, we have to turn again to the Poisson distribution. The expression IkL, where L is the phage latent period, describes the fraction of phages that we expect would adsorb to a given quantity of already phage-infected bacteria, I, over one latent period. Multiplying this fraction by the number of phages present, holding that density of phages constant as a function of time for the sake of simplifying the calculation, and then dividing this number of phage adsorptions by the number of infections—i.e., I—provides a multiplicity of secondary adsorption or infection, and this is equal to simply PkL. The fraction of infected bacteria that we would expect would be adsorbed by additional phages prior to lysing therefore is simply $1 - e^{-PkL}$, where e^{-PkL} is the zero-adsorption fraction (i.e., $1 - \text{POISSON.DIST}(0, PkL, \text{False})$). The case where phage density is not held constant is considered elsewhere [16] and a number of additional comparisons, both with and without holding phage densities constant over time and especially as applicable to phage therapy, have also been explored [19].

Note that the longer the interval between phage adsorption and phage-induced bacterial lysis then the more phage secondary adsorptions that will occur per already infected bacterium. The more phage secondary adsorptions then the more phages that will be lost to adsorption per bacterium killed, again since only a single phage adsorption typically is necessary to result in bacterial death; that is, phages display single-hit killing kinetics [41]. Also, and obviously, the more secondary adsorptions then the more phage secondary adsorption-associated phenomena that will occur within a culture, such as increased rates of phage induction of lysogenic cycles or of lysis inhibition [26].

5   Phage Population Growth

Phage population growth can described as a reaction-diffusion process. The reaction component is the phage infection, and it has the effect of generating more phages. The diffusion component describes the means by which new infections are acquired. The processes together define the phage generation, e.g., as starting with phage adsorption and ending with phage adsorption, the latter by phage progeny. The infection also gives rise to a phage burst size, that is, the release from phage-infected bacteria of multiple phage progeny. The shorter the generation time and/or the larger the burst size, then, all else held constant, the greater the rate of phage population growth. An important simplification in considering that growth is to assume that bacterial densities remain constant and, further, that the volumes within which phage population growth occurs are infinite. The result is that we can describe a rate of phage population growth that exists at a given, constant bacterial density. In mathematical terms, we assume simply that phages adsorb bacteria with bacterial replacement. In other words, for every unit of I created an equivalent unit of B is created.

These growth rates can be complicated by multiple phage adsorptions to individual bacteria, but we can ignore that by assuming that not only are environments infinite but that they are infinitely rapidly mixed as well, such that the potential for a free phage to adsorb an already phage-infected bacterium is equal essentially to zero. An additional complication, not as readily ignored, is that phage adsorption occurs in a manner that is equivalent to a free phage-population exponential decline. Thus, the interval over which phages adsorb is not constant from phage virion to virion and for an infinitely large phage population in fact overall, that is, for the entire phage population, is infinitely long in duration. Interestingly, though, the interval over which the phage infection occurs, the latent period, *can* be assumed to be constant but actually supplies an even greater complication toward the modeling of rates of phage population growth. This latter issue sometimes is ignored in models of phage population growth. The result of ignoring latent period in terms of rates of phage population growth is biologically unrealistic, however, so will not be considered here. The interval over which phage infections occur, i.e., the phage latent period, itself is not a constant in practice but instead exists over a distribution [42, 43], but this too we will ignore, though not before noting that while there can be a temptation to assume that shorter phage latent periods over this distribution might give rise to smaller phage burst sizes (as shorter individual infections would have less time to generate phage progeny intracellularly over shorter infection intervals), in fact there is little evidence for a universality to this assertion and indeed there exist logical arguments for why it might not be the case. For example, it is known that phages that as a population display metabolically more robust infections can display both shorter latent periods *and* larger burst sizes [44].

5.1 The Phage Generation Time

Rates of phage population growth at a minimum are a function of the length of the phage infection period (the latent period or L), the phage burst size (β in phages released/infection), and the interval over which adsorption occurs (which we call A). If for the moment we hold the latter constant, then the phage generation time is equal to $L + A$. If this interval is doubled, then the rate of phage population growth should decline by half (i.e., it should take twice as long to reach a given phage density). Latent periods can be reduced via so-called phage clock mutations [45], but a hard limit exists as to how short latent periods can be since lysis that occurs prior to the production of progeny phages—which are not produced (by definition) until the end of what is known as the phage eclipse [46]—will not contribute to phage population growth, including in terms of plaque formation [47]. A further complication is that shorter phage latent periods, all else held constant and thus not contradicting the Hadas et al. [44] observation, really should (and do) result in smaller phage burst sizes. We will ignore these complications as well, however, and focus instead on the A term, noting that, no matter how short the rate that phages find bacteria to infect, it is not possible for the phage generation time to be reduced to below L. Indeed, L represents a limit that phage generation times in principle can approach but nevertheless cannot reach since phages are unable to find and then adsorb bacteria instantaneously.

Since A becomes shorter as bacterial densities increase—recall that the rate of phage adsorption is equal to Bk—this means that greater bacterial densities will result in shorter phage generation times. Phage generation times, however, will nevertheless display saturation kinetics as a function of bacterial densities, with high bacterial densities resulting in generation times approaching L just as high substrate densities will result in enzyme-mediated reaction rates that approach an enzyme's turnover rate, i.e., Michaelis–Menten enzyme kinetics [48]. Consistently, while a doubling of bacterial densities at lower bacterial densities will result in an approximate doubling of rates of phage population growth, at already high bacterial densities instead there may be little impact, even ignoring the potential negative consequences of higher bacterial densities on the physiology of phage infections. Lastly, at the end of one generation one phage should, with this simple model, give rise to β new phages for a net gain of $\beta - 1$, where the 1 represents the phage lost to the initial adsorption.

5.2 Modeling Population Growth

As noted, the exponential-decline nature of free phage loss to adsorption in combination with the need to account for the phage latent period greatly complicates modeling of phage population growth. We, as a consequence, present but otherwise do not greatly consider this model here. Nevertheless, we use this opportunity to distinguish between continuous and iterated models. Both approaches offer advantages. Continuous models, if one succeeds in incorporating biologically relevant details, are more realistic than iterated models. Alternatively, continuous models can be more difficult to solve, often for phage population growth are solved as iterated models, and are more difficult to introduce additional potentially biologically relevant details into. Nonetheless, models of phage population growth often are presented as continuous models so it can be important to recognize these, as well as appreciate what they mean. Caution, though, as these models involve use of calculus. See Stopar and Abedon [49] for further discussion.

The basic continuous model of phage population growth can be written as follows:

$$\frac{dP}{dt} = \beta k \widehat{B} \widehat{P} - kBP$$

Here β is the phage burst size, B is bacterial density, P is phage density, k is the phage adsorption rate constant, and $\widehat{B}$ and $\widehat{P}$ are the bacterial and phage densities, respectively, but from one phage latent period (L) earlier. Use of the latter two terms results in phage infections being an implicit part of the model rather than explicitly shown, which is possible since phage adsorptions of already infected bacteria are being ignored. Thus, in words, the instantaneous change in free phage density as a function of time is equal to the rate of lysis of bacteria that had become phage infected L minutes previously as multiplied by the phage burst size, with the rate of phage loss due to bacterial adsorption subtracted from this amount (the latter, also, has the effect of implicitly increasing the density of phage-infected bacteria). Note the simplifying absence of additional terms such as any for phage loss or gain other than in association with bacterial adsorption and infection, particularly such as phage inactivation or, instead, outflow or inflow to or from another environment.

The iterated model follows this same pattern:

$$P_{t+1} = P_t + \beta k B_{t-L} P_{t-L} - kB_t P_t$$

This can be simplified, by assuming that bacterial density is held constant, to the following (for continuous and iterated models, respectively):

$$\frac{dP}{dt} = Bk\left(\beta\widehat{P} - P\right)$$

$$P_{t+1} = P_t + Bk(\beta P_{t-L} - P_t)$$

As presented it would appear that rates of phage population growth should increase as a direct function of bacterial density, which to a degree is true. Even ignoring issues of declining bacterial physiology, however, at higher bacterial densities phage population growth will display saturation kinetics as bacterial densities increase, with phage generation times limited at their lower end by the duration of the phage latent period. This is, as noted above, just as enzymes are limited in their activity at very high substrate densities by their turnover rate [48].

The impact of increasing phage adsorption rate constants is equivalent to that of increasing bacterial density. With larger adsorption rate constant values (as analogous to enzyme affinity for substrate) or with greater bacterial densities (as analogous to densities of enzyme substrate), then there will be diminishing returns in terms of phage population growth rates (as analogous to increases in enzyme activity) that are associated with absolute increases in rates of phage adsorption (as analogous to rates of formation of enzyme-substrate complexes). Strictly speaking, however, note that we are assuming that any increases in the adsorption rate constant is in terms of the rate of phage encounter and then reversible attachment to target bacteria and not in terms of the rate of transition from reversible to irreversible phage adsorption. In continuing the analogy to enzyme kinetics, the reversible-irreversible transition would be equivalent to formation of an activated complex, which serves as a component of enzyme turnover rate, that is, a component of the phage latent period, where latent period length is held constant in the above scenario.

With iterated models, loss and gain functions occur over short but nevertheless finite intervals. These processes are quite tractable and in fact may be solved using simple computer code or even spreadsheet programs such as Microsoft Excel. The continuous models, by contrast, typically will be solved employing a differential equation solving program, particularly what is known as Berkeley Madonna, which ultimately also generates an iterative solution. In either case, calculations over finite intervals are inherently imprecise since, in this case, phage losses will not occur among those phages which are not gained until the end of intervals, that is, rather than both gains and losses occurring continuously throughout intervals. The impact of these imprecisions can be small, however. In particular, one can test whether they make much of a difference by varying the interval length that one employs to solve an iterated model to see whether there is a substantial impact on modeling outcomes. Generally shorter intervals will provide greater precision but at the cost of greater amounts of computer processing per modeled unit time. For models of phage population growth, an interval length of 1 min can be a good place to start in establishing associated modeling costs and benefits.

5.3 Modeling Phages Within Chemostats

Modeling the population dynamics of interactions between phages and bacteria, as they can occur within chemostats, has received a fair amount of attention and this is in part because it is possible to perform relatively complex corresponding experiments. It is outside of the scope of this chapter to consider the modeling of bacterial growth and resource utilization components of these interactions, but *see* Stopar and Abedon [49] as well as Abedon [40] for discussion of how this is accomplished. Instead, we focus on the impact of washout on phage population densities within chemostats. A chemostat is a laboratory apparatus involving a constant rate of addition of fresh media into a well-stirred growth chamber from which partially spent media washes out at a constant and equivalent rate. What is kept track of are densities within the growth chamber, which for free phages increase as a consequence of bacterial-infection burst and are lost due to bacterial adsorption, with both infections and free phages fractionally lost from the growth chamber at the same rate that media is lost.

The continuous model is as follows:

$$\frac{dP}{dt} = e^{-L\omega}\beta k\widehat{B}\widehat{P} - kBP - \omega P$$

where ω is the washout rate and L is the phage latent period. The two expressions involving ω describe phage losses. These losses occur in terms of phage-infected bacteria and free phages, respectively (going from left to right), with the more elaborate nature of the first expression ($e^{-L\omega}$) a consequence of its describing what is happing to a cohort of phage infections over an entire latent period rather than instantaneously. Thus, phage infections are initiated and for L min are lost from chemostats at some constant rate (ω) while free phages are simply lost at a constant rate (also ω). In both cases these losses represent an exponential decline in the densities of a given cohort.

Typically in such models at least two additional equations will be employed. One that keeps track of instantaneous changes in bacterial densities and the other instantaneous changes in the density of limiting nutrients that control rates of bacterial replication. As noted, however, these additional details go beyond the scope of this chapter.

5.4 Modeling Phages Within Plaques

Phage population growth is not limited to environments containing randomly dispersed, well-mixed populations of bacteria but also can occur in environments in which the movement of bacteria as well as that of free phages is constrained. Most notably, in nature these spatially constrained environments include those of bacterial biofilms. Serving as a potential model for phage population growth within biofilms are phage population dynamics as they can occur during the formation of phage plaques [15, 33, 50]. The biology of phage plaques has been reviewed by Abedon and Yin [10, 51]. A review of mathematical models of phage plaque formation is presented by Krone and Abedon [52].

Phage population growth within broth cultures can be viewed as consisting of alternating phage infection of stationary bacteria (that is, not moving rather than necessarily stationary *phase*) and then free phage dispersal. During phage plaque formation, however, a free phage is limited, spatially, in terms of that dispersal. Particularly, as distances between free phages and bacteria increase so too does the instantaneous likelihood of subsequent phage adsorption decline. The models employed to approximate the resulting interactions are quite a bit more complicated than those used to model broth–growth interactions, including as occurs within chemostats, and this increase in complexity is the case even though bacterial densities traditionally have been held constant within these models. See the above-cited reviews for further consideration.

6   Phage-Therapy Relevant Models

Many of the ideas considered above can be employed to gain a better appreciation of phage–bacterial interactions as they can occur in the course of phage-mediated biocontrol of bacteria, i.e., such as is seen with phage therapy. Though attempts have been made to model these interactions as based on assumptions of equivalency to phage–bacterial interactions as they can occur within chemostats [53], in fact it is questionable to what degree even qualitative predictions can be reached using that approach. In particular, an assumption of bacterial population growth occurring over the course of phage application may be overly complicating while an assumption of broth-like environments may be overly simplifying. More recent consideration has been of how in vitro phage properties which are potentially mathematically modellable may or may not be predictive of experimental phage therapy success [11, 54, 55].

With simpler models, better predictability may be possible, though even here these models should be viewed as descriptions of what *should* be happening during treatment efforts rather than predictions of what actually will occur or is occurring. Thus, simple models of phage and bacterial population dynamics during phage therapy should be viewed more as a means of applying theory toward planning and/or "debugging" protocols rather than as accurate representations. That is, in principle, knowledge of what should be happening during phage treatments can be helpful, even if what actually occurs is not identical to what is predicted. Particularly, the concepts of adsorption rates and killing titers can be helpful toward gaining a more intuitive appreciation of phage–bacterial population interactions, e.g., such as in terms of the degree to which a given protocol may be relying on phage population growth and therefore so-called active treatment to achieve bacterial eradication [56].

A great deal of the original work in this area of applying such simpler mathematical models toward gaining a better understanding of the dynamics of phage therapy can be traced to that of Payne and Jansen [57–60], with elaboration especially in Abedon and Abedon-Thomas [33]. *See* also Levin and Bull [54, 61–63] as well as Cairns et al. [64] and Gill [53]. *See* particularly Abedon [15] for review, summary, critique, and extension of various calculations. Included there are discussions of phage minimum inhibitory concentration (MIC) and minimum bactericidal concentration (MBC), with consideration of why MIC can be difficult to measure or even define for phages. MBC, by contrast, can be approximated in terms of phage killing titers, though even here calculations are not necessarily straightforward. Additional concepts that can be addressed to at least a first approximation using mathematical models, and which also can be of relevance to phage therapy, are phage-mediated decimal reductions times, time until bacterial eradication, and consideration of the frequency of phage dosing.

7 Accuracy and Precision in Phage Enumeration

Modeling aside, the most basic application of mathematics to phage biology is seen in determinations of phage titers. While at the beginning of this chapter we considered the basic mathematics of titer determination, here we address instead the minimization of errors in those determinations. Such errors can come about due to four basic causes: (1) apparatus/technician/operator error, (2) inherent phage biological properties, (3) sampling error (including random sampling error but also bias), and (4) statistical dependence. Errors in laboratory technique are beyond the scope of this chapter though nonetheless we address their consequences in the case of failure to achieve statistical independence of data. Problems stemming from phage biological properties to a degree have been addressed (e.g., efficiency of plating) but also help to define reasonable plate-count upper limits (i.e., Too Numerous To Count, TNTC). Sampling error is seen particularly in terms of Too Few To Count (TFTC) and has implication in statistical inference, as too does the issue of the noted statistical independence. Here we briefly consider TNTC, TFTC, and statistical independence toward phage titer determination with emphasis on plaque count determinations, all as considered from the perspectives of accuracy, i.e., avoiding biases in measurement, as well as achieving reasonable levels of precision in terms of margin of error (which measures random sampling error).

The accuracy or bias of a measurement is defined in terms of its closeness to an actual, expected value. Precision, by contrast, addresses the degree to which the spread of a distribution about its center, the mean value, has been minimized. Note that a measurement can be accurate (unbiased) without being precise (broad data point spread). Similarly, it is possible to attain a precise measurement that nonetheless is not highly accurate (narrow data point spread around something other than the expected value). As an example, if an actual value is 5, a measurement of 4 would be considered to be more accurate than one of 9. If three measurements were taken, values of 3, 5, and 7 would be more precise (smaller random sampling error) than ones of 1, 4, and 10 (larger random sampling error), even though the sample means of both data sets are the same. Similarly, values such as 7, 9, and 11 display similar random sampling error as 3, 5, and 7 but nevertheless would be less accurate (more biased) given an actual value of 5. We now consider issues of TNTC, TFTC, and statistical independence in terms of their impacts on the accuracy or precision of measurements, or both.

7.1 Too Numerous to Count

The issue of TNTC is its impact on accuracy. Specifically, the intention of declaring a plate count as TNTC, besides avoiding operator fatigue, is to avoid under counts, that is, where the actual number of PFUs is greater than the amount that can possibly be determined. The issue here is that the higher the density of plaques found on a plate then the lower the potential to spatially resolve what appears to be individual plaques into what instead are closely spaced plaques. Mathematics could be applied to determine the likelihood of such overlap based on individual plaque area and overall plaque numbers, but a much simpler issue results simply from the elimination of a count as TNTC, that is, without eliminating as well the entire dilution within which that count was determined.

The problem here is that eliminating high values, and only high values, even if done for valid reasons, has the effect of reducing mean counts, potentially resulting in an *underestimation* of measurements. A solution to this issue is to employ a trimmed mean. Thus, if among five determinations values of 100, 200, 300, 400, 500, and TNTC were obtained, a trimmed mean based solely on 200, 300, and 400 could be employed instead (that is, "trimming" the lower 20% and upper 20% of data points). An even simpler and still valid approach would involve employing the median value, essentially serving as the ultimate trimmed mean (trimming the lower 50% *and* the upper 50% of the data). In both cases, with this example, the sample would be assumed to have given rise to a plate count centering on 300. An online trimmed mean-determining titering calculator which takes into account TNTC entries can be found here [65]

7.2 Too Few to Count

A similar argument based on accuracy can be made for TFTC, though here it is an overestimation that may be avoided through the use of trimmed means. Thus, given five plate counts of, for example, 20, 30, 40, 50, and 60, instead of declaring a count of 20 as TFTC and thereby discarding that value, a far better approach would be to simply calculate the mean based on all five values, which here would be 40 (rather than 45 if calculated without the 20 value). Alternatively, one could employ a trimmed mean as based on 30, 40, and 50, or instead simply the median value, of 40. Note, though, that an equivalence between a given trimmed mean and median value will not always be the case, e.g., values of 25, 30, 45, and 60 have a mean of 40, a median of $37.5 = (30 + 45)/2$, and a 25% trimmed mean also of 37.5 (i.e., as equal to the mean of 30 and 45).

The Microsoft Excel functions which may be employed to calculate these values are AVERAGE, MEDIAN, and TRIM-MEAN. These functions are not directly applicable to TNTC calculations, however, since they require an input of actual values, which is possible when plates counts are very low but either is not or is less possible (and/or desirable) when plate counts are too high. Thus with the TNTC example we trimmed 20% of the data points whereas with the Excel function the "trimming" is of a percentage of the distribution as estimated based upon knowledge of the actual value of all data points (though note that a mock high value could be substituted for TNTC when using MEDIAN or TRIMMEAN, e.g., 1000 plaques, since in this case the value is not directly involved in mean calculation but instead is present solely for the sake of being trimmed). In addition, note that the Excel TRIMMEAN function uses a different convention to calculate the trimmed mean, i.e., with a 25% trimmed mean as described above indicated as 0.5, that is, the sum of both trims, rather than as 0.25. Keep in mind also that statistical comparisons are preferably done using means rather than medians, though they can be done using less fully trimmed means.

A way to view the issue of precision, as random sampling error, is that more data is preferable to less. The problem of having too few data is that random sampling error can be higher and thereby precision lower. In the earlier example, if the true value is 5 and instead of three measurements of 7, 9, and 11 we collect six of 7, 7, 9, 9, 11, and 11, then we have not corrected the problem of bias since, even though the latter data set possesses a smaller margin of error, it is still just as far from the actual value in terms of accuracy. In terms of TFTC, throwing out too low values of a total of many determinations, rather than factoring those low values into means or trimmed means, has the effect not only of overestimating the resulting mean but of lowering its precision as well.

Relying on lower versus higher numbers of plaques per plate for titer determinations nevertheless does have the effect of decreasing the precision of the resulting mean. This loss of precision in fact serves as the reason for invoking TFTC, that is, higher expected error given lower counts, but still should not serve as a justification for completely ignoring TFTC results since, as noted, doing so results simultaneously in a reduction in accuracy (i.e., as resulting in overestimation) and reduction in precision of the resulting aggregate measure. Alternatively, the expected loss of precision does serve as reason for ignoring the results of a *dilution* that resulted in lower counts in favor of one that instead results in higher counts, even though the latter can require greater operator effort to obtain. Thus, if a tenfold lower dilution results in counts of 25, 30, and 35 while the tenfold higher dilution results in counts of 290, 305, and 315, the recorded titer, all else equivalent, should be based on latter data set.

7.3 Statistical Independence

Of these various issues, that of statistical independence is the least straightforward. The goal again is one of achieving accuracy, with reduction especially in the impact of operator error. In particular, given statistical independence between measurements then we can reasonably assume a greater precision of the mean for a set of measurements. The mean of independent observations, that is, is more accurate than any of the individual measurements. The simplest scenario is where two independent measurements are made, both of which possess some random sampling error. Three measurements can be preferable, however, based simply on the idea that with very small samples, e.g., such as two or three, use of the median as an estimation of the true value again, as above, can be preferable to use of the mean. With two measurements, the mean and the median are generated based on the same calculation whereas with three measurements the calculations can differ in a useful manner. Another way of viewing the same issue is that with three measurements it is possible that the smallest and/or the highest of three values may represent outliers, and use of the median rather than the mean to make this calculation allows one to drop from the calculation those values that are furthest from the mean, and this can be done without requirement for further justification. Use of four measurements, or more, represents further improvement, but is less of a qualitative leap than the jump from two measurements to three.

A much more complex case occurs when error is inherent to a given sampling technique or even a specific laboratory, in which case the assumption of independence might be questionable. In this case greater accuracy may be attained instead by comparing repeated measurements obtained "independently," e.g., via different techniques or by comparing "independent" measurements that have been arrived at using the same technique but in different locations. In between these extremes are measurements taken within a single experiment done on a given day, which most likely are not independent, and measurements obtained during separate experiments performed on different days (or using different stock solutions, etc.), which can more reasonably be expected to be independent. In terms of individual phage titer determinations, the implication is that multiple titer determinations obtained from a single phage dilution series inherently are less independent than multiple titer determinations made from multiple dilution series. Thus, for example, if a dilution series itself is off by 10%, then you should expect that a mean obtained solely from that dilution series will also be off by 10%. As a consequence, if for the sake of enhancing precision you take the time to make multiple titer determinations, it similarly can be useful to keep those individual measurements statistically independent.

With relatively simple as well as inexpensive experiments, single measurements per individual data point nevertheless may be justified, with experiments as a whole repeated numerous times in order to achieve a desired level of precision of means. Alternatively, if experiments are difficult and/or expensive, then it can pay to take multiple, redundant measurements to determine individual data points. This should be at least three statistically independent measures per mean determined, or more if individual data points are easily obtained or inexpensive relative to the experiment as a whole, and this would be done in conjunction with fewer repeats of the overall experiment.

8   Concluding Statement

The application of mathematics to the study of phage biology has a long and in fact illustrious history, for example *see* Luria & Delbrück [66] but also Schlesinger [67] (the latter as originally published in 1932). At first the tradition was one of employing math toward better understanding phage molecular or organismal biology. From that a tradition began in which mathematics instead was used toward gaining a better understanding of phage ecology, e.g., Campbell [68]. Subsequently these considerations have been harnessed toward gaining a greater appreciation of the applied ecology of phage-mediated antibacterial biocontrol, i.e., phage therapy. Because phages are whole organisms and often handled in the laboratory as populations, a mathematical appreciation of phages continues to be relevant, especially toward phage phenotypic characterization—i.e., as an extension of the concept of "to really know them, you have to grow them" [69]—even as molecular techniques have become ever more powerful. In this chapter we provide an overview of the most basic means by which mathematics may be employed in the laboratory as well as toward better understanding the natural world and even to some degree the impact of phages in the clinic. It is not our expectation that mathematics will suddenly become central to research programs as a consequence of these efforts, but instead that the utility of simple mathematical modeling toward better intuitive understanding of phage-related phenomena might at least become better appreciated.

Acknowledgment

T.I.K. provided consulting to the sections considering statistics, "The Poisson Distribution" and "Accuracy and Precision in Phage Enumeration."

References

1. Abedon ST (2015) Bacteriophage secondary infection. Virol Sin 30:3–10

2. Ortmann AC, Suttle CA (2009) Determination of virus abundance by epifluorescence microscopy. Methods Mol Biol 501:87–95

3. Brussaard CP (2009) Enumeration of bacteriophages using flow cytometry. Methods Mol Biol 501:97–111

4. Anderson B, Rashid MH, Carter C, Pasternack G, Rajanna C, Revazishvili T et al (2011) Enumeration of bacteriophage particles: comparative analysis of the traditional plaque assay and real-time QPCR- and nanosight-based assays. Bacteriophage 1:86–93

5. Kropinski AM, Mazzocco A, Waddell TE, Lingohr E, Johnson RP (2009) Enumeration of bacteriophages by double agar overlay plaque assay. Methods Mol Biol 501:69–76

6. Mazzocco A, Waddell TE, Lingohr E, Johnson RP (2009) Enumeration of bacteriophages by the direct plating plaque assay. Methods Mol Biol 501:77–80

7. Mazzocco A, Waddell TE, Lingohr E, Johnson RP (2009) Enumeration of bacteriophages using the small drop plaque assay system. Methods Mol Biol 501:81–85

8. Serwer P, Hayes SJ, Thomas JA, Hardies SC (2007) Propagating the missing bacteriophages: a large bacteriophage in a new class. Virol J 4:21

9. Kutter E (2009) Phage host range and efficiency of plating. Methods Mol Biol 501:141–149

10. Abedon ST, Yin J (2009) Bacteriophage plaques: theory and analysis. Methods Mol Biol 501:161–174

11. Henry M, Lavigne R, Debarbieux L (2013) Predicting *in vivo* efficacy of therapeutic bacteriophages used to treat pulmonary infections. Antimicrob Agents Chemother 57:5961–5968

12. Sing WD, Klaenhammer TR (1990) Characteristics of phage abortion conferred in lactococci by the conjugal plasmid pTR2030. J Gen Microbiol 136:1807–1815

13. Forterre P, Soler N, Krupovic M, Marguet E, Ackermann HW (2012) Fake virus particles generated by fluorescence microscopy. Trends Microbiol 21:1–5

14. Stent GS (1963) Molecular biology of bacterial viruses. WH Freeman and Co., San Francisco, CA

15. Abedon ST (2011) Bacteriophages and biofilms: ecology, phage therapy, plaques. Nova Science Publishers, Hauppauge, New York

16. Abedon ST (1990) Selection for lysis inhibition in bacteriophage. J Theor Biol 146:501–511

17. Abedon ST (1999) Bacteriophage T4 resistance to lysis-inhibition collapse. Genet Res 74:1–11

18. Abedon ST (2008) Phage population growth: constraints, games, adaptation. In: Abedon ST (ed) Bacteriophage ecology. Cambridge University Press, Cambridge, UK, pp 64–93

19. Abedon S (2011) Phage therapy pharmacology: calculating phage dosing. Adv Appl Microbiol 77:1–40

20. Abedon ST (2016) Phage therapy dosing: the problem(s) with multiplicity of infection (MOI). Bacteriophage 6:e1220348

21. Kasman LM, Kasman A, Westwater C, Dolan J, Schmidt MG, Norris JS (2002) Overcoming the phage replication threshold: a mathematical model with implications for phage therapy. J Virol 76:5557–5564

22. Abedon ST (2008) Phage, bacteria, and food. Appendix: rate of adsorption is function of phage density. In: Abedon ST (ed) Bacteriophage ecology. Cambridge University Press, Cambridge, UK, pp 321–324

23. Hagens S, Loessner MJ (2010) Bacteriophage for biocontrol of foodborne pathogens: calculations and considerations. Curr Pharm Biotechnol 11:58–68

24. Abedon ST (2011) Envisaging bacteria as phage targets. Bacteriophage 1:228–230

25. Abedon ST (2017) Multiplicity of infection calculator. http://moicalculator.phage.org . Accessed 26 Aug 2017

26. Abedon ST (2017) Commentary: communication between viruses guides lysis-lysogeny decisions. Front Microbiol 8:893

27. Abedon ST (2011) Lysis from without. Bacteriophage 1:46–49

28. Dulbecco R (1952) Mutual exclusion between related phages. J Bacteriol 63:209–217

29. Carlson K, Miller ES (1994) Enumerating phage: the plaque assay. In: Karam JD (ed) Molecular biology of bacteriophage T4. ASM Press, Washington, DC, pp 427–429

30. Carlson K (2005) Working with bacteriophages: common techniques and methodological approaches. In: Kutter E, Sulakvelidze A (eds) Bacteriophages: biology and application. CRC Press, Boca Raton, Florida, pp 437–494

31. Dulbecco R (1949) Appendix: on the reliability of the Poisson distribution as a distribution of the number of phage particles infecting individual bacteria in a population. Genetics 34:122–125

32. DeGroot MH (1984) Probability and statistics. Addison-Wesley, Boston, MA

33. Abedon ST, Thomas-Abedon C (2010) Phage therapy pharmacology. Curr Pharm Biotechnol 11:28–47

34. Abedon ST (2014) Phage therapy: eco-physiological pharmacology. Scientifica (Cairo) 2014:581639

35. Abedon ST (2017) .Killing titer calculator. http://killingtiter.phage-therapy.org/calculator.html. Accessed 26 Aug 2017

36. Abedon ST (2012) Spatial vulnerability: bacterial arrangements, microcolonies, and biofilms

37. Abedon ST (2017) Active bacteriophage biocontrol and therapy on sub-millimeter scales towards removal of unwanted bacteria from foods and microbiomes. AIMS Microbiol 3:649–688

38. Kropinski AM (2009) Measurement of the rate of attachment of bacteriophage to cells. Methods Mol Biol 501:151–155

39. Abedon ST (2017) Phage adsorption theory. http://adsorption.phage.org. Accessed 26 Aug 2017

40. Abedon ST (2009) Deconstructing chemostats towards greater phage-modeling precision. In: Adams HT (ed) Contemporary trends in bacteriophage research. Nova Science Publishers, Hauppauge, New York, pp 249–283

41. Bull JJ, Regoes RR (2006) Pharmacodynamics of non-replicating viruses, bacteriocins and lysins. Proc R Soc Lond B Biol Sci 273:2703–2712

42. Storms ZJ, Brown T, Cooper DG, Sauvageau D, Leask RL (2014) Impact of the cell life-cycle on bacteriophage T4 infection. FEMS Microbiol Lett 353:63–68

43. Baker CW, Miller CR, Thaweethai T, Yuan J, Baker MH, Joyce P, Weinreich DM (2016) Genetically determined variation in lysis time variance in the bacteriophage fX174. G3 (Bethesda) 6:939–955

44. Hadas H, Einav M, Fishov I, Zaritsky A (1997) Bacteriophage T4 development depends on the physiology of its host *Escherichia coli*. Microbiology 143:179–185

45. Bläsi U, Young R (1996) Two beginnings for a single purpose: the dual-start holins in the regulation of phage lysis. Mol Microbiol 21:675–682

46. Hyman P, Abedon ST (2009) Practical methods for determining phage growth parameters. Methods Mol Biol 501:175–202

47. Johnson-Boaz R, Chang C-Y, Young R (1994) A dominant mutation in the bacteriophage lambda *S* gene causes premature lysis and an absolute defective plating phenotype. Mol Microbiol 13:495–504

48. Abedon ST (2009) Kinetics of phage-mediated biocontrol of bacteria. Foodborne Pathog Dis 6:807–815

49. Stopar D, Abedon ST (2008) Modeling bacteriophage population growth. In: Abedon ST (ed) Bacteriophage ecology. Cambridge University Press, Cambridge, UK, pp 389–414

50. Gallet R, Shao Y, Wang I-N (2009) High adsorption rate is detrimental to bacteriophage

as responses to low rather than high phage densities. Virus 4:663–687

fitness in a biofilm-like environment. BMC Evol Biol 9:241

51. Abedon ST, Yin J (2008) Impact of spatial structure on phage population growth. In: Abedon ST (ed) Bacteriophage ecology. Cambridge University Press, Cambridge, UK, pp 94–113

52. Krone SM, Abedon ST (2008) Modeling phage plaque growth. In: Abedon ST (ed) Bacteriophage ecology. Cambridge University Press, Cambridge, UK, pp 415–438

53. Gill JJ (2008) Modeling of bacteriophage therapy. In: Abedon ST (ed) Bacteriophage ecology. Cambridge University Press, Cambridge, UK, pp 439–464

54. Bull JJ, Gill JJ (2014) The habits of highly effective phages: population dynamics as a framework for identifying therapeutic phages. Front Microbiol 5:618

55. Lindberg HM, McKean KA, Wang I-N (2014) Phage fitness may help predict phage therapy efficacy. Bacteriophage 4:e964081

56. Abedon S.T. (2017) Expected efficacy: applying killing titer estimations to phage therapy experiments. http://killingtiter.phage-therapy.org. Accessed 26 Aug 2017

57. Payne RJH, Phil D, Jansen VAA (2000) Phage therapy: the peculiar kinetics of self-replicating pharmaceuticals. Clin Pharmacol Ther 68:225–230

58. Payne RJH, Jansen VAA (2001) Understanding bacteriophage therapy as a density-dependent kinetic process. J Theor Biol 208:37–48

59. Payne RJH, Jansen VAA (2002) Evidence for a phage proliferation threshold? J Virol 76:13123

60. Payne RJH, Jansen VAA (2003) Pharmacokinetic principles of bacteriophage therapy. Clin Pharmacokinet 42:315–325

61. Levin BR, Bull JJ (1996) Phage therapy revisited: the population biology of a bacterial infection and its treatment with bacteriophage and antibiotics. Am Nat 147:881–898

62. Bull JJ, Levin BR, DeRouin T, Walker N, Bloch CA (2002) Dynamics of success and failure in phage and antibiotic therapy in experimental infections. BMC Microbiol 2:35

63. Levin BR, Bull JJ (2004) Population and evolutionary dynamics of phage therapy. Nat Rev Microbiol 2:166–173

64. Cairns BJ, Timms AR, Jansen VA, Connerton IF, Payne RJ (2009) Quantitative models of in vitro bacteriophage-host dynamics and their application to phage therapy. PLoS Pathog 5: e1000253

65. Abedon ST (2017) Titering calculator. http://www.phage.org/calculators/titering.html . Accessed 26 Aug 2017

66. Luria SE, Delbrück M (1943) Mutations of bacteria from virus sensitivity to virus resistance. Genetics 28:491–511

67. Schlesinger M (1960) Adsorption of bacteriophages to homologous bacteria. II. Quantitative investigations of adsorption velocity and saturation. Estimation of particle size of the bacteriophage [translation]. In: Stent GS (ed) Bacterial viruses. Little, Brown and Co, Boston, pp 26–36

68. Campbell A (1961) Conditions for the existence of bacteriophages. Evolution 15:153–165

69. Reguera G. (2017) "The great plate count anomaly" that is no more. http://schaechter.asmblog.org/schaechter/2014/12/the-great-plate-count-anomaly-that-is-no-more.html. Accessed 26 Aug 2017

Chapter 2

Analysis of Host-Takeover During SPO1 Infection of *Bacillus subtilis*

Charles R. Stewart

Abstract

When *Bacillus subtilis* is infected by bacteriophage SPO1, the phage directs the remodeling of the host cell, converting it into a factory for phage reproduction. Much synthesis of host DNA, RNA, and protein is shut off, and cell division is prevented. Here I describe the protocols by which we have demonstrated those processes, and identified the roles played by specific SPO1 gene products in causing those processes.

Key words Host-takeover, Host-shutoff, Macromolecular syntheses, Cell division, Resistant mutants, Bacteriophage SPO1, *Bacillus subtilis*

1 Introduction

When *Bacillus subtilis* is infected by bacteriophage SPO1, the phage directs the remodeling of the host cell, converting it into a factory for phage reproduction. Much synthesis of host DNA, RNA, and protein is shut off, presumably to prevent those syntheses from competing with the corresponding phage biosyntheses for materials, energy and access to biosynthetic machinery [1–3]. Host cell division is inhibited, possibly to prevent separation of phage components synthesized in different parts of the cell [4]. Here I describe the protocols that we have followed in elucidating these and other elements of the host-takeover process, including: (a) Assay of shutoff of host DNA synthesis. (b) Assays of shutoff of host RNA and protein synthesis. (c) Selection of bacterial mutants resistant to the cytotoxic effects of host-takeover gene products. This makes use of the fact that, since some SPO1 gene products inhibit essential functions of the host cells, expression of certain SPO1 genes in uninfected cells has a cytotoxic effect. (d) Effect of SPO1 infection on host cell division.

Measurement of rates of host DNA, RNA, and protein synthesis is done by pulse-labeling with tritiated thymidine, uridine, or

Martha R.J. Clokie et al. (eds.), *Bacteriophages: Methods and Protocols, Volume 3*, Methods in Molecular Biology, vol. 1681,
https://doi.org/10.1007/978-1-4939-7343-9_2, © Springer Science+Business Media LLC 2018

leucine, respectively. Comparing these syntheses in uninfected and infected cells shows that SPO1 infection strikingly reduces each of those syntheses. Comparing the effects of wild-type SPO1 with those of specific SPO1 mutants has identified specific genes that play specific roles in accomplishing these shutoffs [1–3]. Measurement of the effect on cell division is shown by observing microscopically the effect of SPO1 infection on average cell size, and observing the prevention of that effect by a specific gene mutation [4].

2 Materials

2.1 B. subtilis Strains

CB313 and CB10 are the SU$^+$ (suppressor plus) and Su$^-$ strains, respectively [5]. The suppressor in CB313 inserts lysine at nonsense codons [6].

2.2 Cloning Vector

The primary *B. subtilis/E. coli* shuttle vector that we have used is pPW19, described by Wei and Stewart [1]. It has a selectable chloramphenicol-resistance gene, and an IPTG-inducible promoter just upstream of its polylinker.

2.3 Growth Media

1. TSA plates: 40 gm Trypticase Soy Agar (BBL) in 1 L of water. This is used for plating for colonies and as bottom agar for plating phage plaques.
2. TC plates: TSA plates containing 10 μg/mL of chloramphenicol (Cm).
3. TCI plates: TC plates containing 4 μg/mL IPTG (*see* **Note 1**).
4. TSA top agar: 12 g Trypticase Soy Agar (BBL) in 1 L of water. This is used as top agar for plating phage for plaque formation.
5. VY broth: 25 g Veal Infusion Broth (Difco), 5 g Yeast Extract (Difco), 1000 mL water.
6. Penassay Broth: 17.5 g Antibiotic Medium 3 (Difco), 1000 mL water.
7. NY Broth: 8 g Nutrient Broth (Difco), 5 g Yeast Extract (Difco), 1000 mL water.
8. NY plus Cm: NY Broth containing 5 μg/mL chloramphenicol.

2.4 Solutions

1. Amino Acid Group I. 1 g arginine, 1 g methionine, 1 g cysteine, 1 g lysine, 100 mL water.
2. Amino Acid Group III. 1 g tryptophan, 1 g phenylalanine, 1 g tyrosine, 100 mL water.
3. Amino Acid Group IV. 1 g threonine, 1 g histidine, 1 g glutamic acid, 1 g aspartic acid, 100 mL water.

4. Amino Acid Group V. 1 g serine, 1 g alanine, 1 g glycine, 100 mL water.

5. Spizizen's minimal medium: 200 mg $MgSO_4 \cdot 7H_2O$, 1 g sodium citrate.$2H_2O$, 2 g ammonium sulfate, 14 g K_2HPO_4, 6 g KH_2PO_4, 1000 mL water [7].

6. 40% glucose: 40 g glucose, 60 mL water (*see* **Note 2**).

7. C4 medium: 96.6 mL Spizizen's minimal medium, 0.6 mL of each of Amino Acid groups I, III, IV, and V, 1.0 mL 40% Glucose (*see* **Note 3**).

8. Adsorption buffer: 0.05 M Tris–pH 7.5, 0.1 M NaCl, 0.01 M $MgSO_4$.

9. 4 μCi/mL tritiated thymidine: 4 μL of 1.0 mCi/mL Methyl-3H thymidine (MP Biomedicals), 996 μL water.

10. 4 μCi/mL tritiated uridine: 4 μL of 1.0 mCi/mL 5-3H uridine (MP Biomedicals), 996 μL water.

11. 4 μCi/mL tritiated leucine: 4 μL of 1.0 mCi/mL 4,5-3H leucine (MP Biomedicals), 996 μL water.

12. 10% TCA: 10 g trichloroacetic acid (TCA), 90 mL water.

13. CSC (concentrated saline citrate): 1.5 M NaCl, 0.15 M Sodium Citrate.

14. DSC (dilute saline citrate): 0.015 M NaCl, 0.0015 M Sodium Citrate).

15. 0.3% TCA in DSC: 30 mL 10% TCA, 10 mL CSC, 960 mL water.

16. SP (saline phosphate): 48.75 mL 0.2 M NaH_2PO_4, 76.25 mL 0.2 M Na_2HPO_4, 8.8 g NaCl, 875 mL water.

3   Methods

3.1 Assay of Shutoff of Host DNA Synthesis

Pulse labeling must be done in medium lacking unlabeled precursors, which would compete with the labeled precursors for incorporation into macromolecules. For this reason, all of the pulse-labeling experiments are done in C4 medium, and the phage to be used for these experiments are first resuspended in adsorption buffer.

1. The infecting lysate is prepared by doing a large volume infection under optimal conditions (*see* **Note 4**), and resuspending the resulting lysate in adsorption buffer. Grow 100 mL of CB313 in VY at 37° to a cell density of about 2×10^8/mL. (*see* **Note 5**). Infect with 10 mL of the desired SPO1 strain, having a titer of about 2×10^{10}/mL. (*see* **Note 6**) Continue shaking under the same conditions until lysis. Centrifuge for 5 min at RCF = 4080 × G (5000 rpm in Sorvall GSA rotor), to pellet unlysed cells. Centrifuge the supernatant for 150 min at

RCF $= 13{,}200 \times$ G (9000 rpm in Sorvall GSA rotor) to pellet the phage. After discarding supernatant, add 25 mL of adsorption buffer, and allow pellet to resuspend overnight at 4°. Swirl the bottle gently to complete the resuspension and produce a homogeneous culture. Plate a 10^{-7} dilution for pfu to determine titer of phage. (*see* **Notes 7** and **8**).

2. For each assay, a culture of CB10 in 5 mL of C4 medium is grown at 30° with vigorous shaking to a cell density about 1×10^{8}/mL. (*see* **Notes 5** and **9**) That culture is infected with enough of the phage lysate (as prepared in the preceding paragraph) to give $>5 \times 10^{8}$ phage/mL in the infected culture (i.e., MOI > 5). Continue shaking at 30°. At varying times, remove 0.5 mL aliquots for pulse labeling. In a typical experiment, we would pulse-label at 0, 5, 12, 25, and 45 min after infection, with the 0 time aliquot being taken before infection.

3. To check the MOI, plate the culture for colonies on TSA plates at 10^{-4} and 10^{-5} dilutions, just before infection, and again between 5 and 10 min after infection.

4. At the time of each pulse, transfer 0.5 mL of the culture from the flask to a 20×150 mm culture tube containing 0.1 mL of 4 µCi/mL tritiated thymidine, and continue shaking both flask and tube at 30°. (*see* **Note 10**) After 5 min, place the tube on ice and add 0.6 mL cold 10% TCA to precipitate macromolecular DNA. Mix by swirling the tube. Hold on ice at least 15 min (no more than 2 h). Pour the contents of the tube through a 0.45 µm nitrocellulose filter, which retains the precipitate and allows unincorporated precursors to pass through (*see* **Note 11**). Wash the filter with 30 mL cold 0.3% TCA in DSC. The counts per minute (cpm) retained on the filter is determined by liquid scintillation counting (*see* **Note 12**).

5. The cpm determined at each time point is proportional to the rate of host DNA synthesis at that time point. Since SPO1 DNA has hydroxymethyluracil in place of thymine in its DNA, the labeled thymidine is not incorporated into phage DNA, so its rate of incorporation measures only the rate of host DNA synthesis, which is shut off almost completely by 25 min after SPO1 infection.

3.2 Assays of Shutoff of Host RNA and Protein Synthesis

For assaying the rate of RNA synthesis, the same procedure is used, except that tritiated uridine is used instead of thymidine. For assaying the rate of protein synthesis, the same procedure is used except that tritiated leucine is used instead of thymidine, and that, after the tubes have been on ice for 15 min after adding 10% TCA, they are heated in a boiling water bath for 15 min and then held on ice for at least 10 more minutes before filtration (*see* **Note 13**).

Measurement of shutoff of host RNA and protein synthesis is not as precise as that of host DNA synthesis for two reasons: First, host RNA and protein syntheses are not entirely shut off during SPO1 infection. Ribosomal RNA and protein synthesis continue, and constitute a significant fraction of total RNA and protein synthesis [8, 9]. Secondly, 5-3H uridine and 4,5-3H leucine are also incorporated into phage RNA and protein, so each data point from infected cells represents the sum of host and phage RNA or protein synthesis. Nevertheless, reasonable estimates of the extent of host-shutoff can be made, because total RNA and protein synthesis in uninfected cells is substantially greater than that in infected cells. The measured decrease in total RNA or protein synthesis provides a minimum estimate of the extent to which host RNA or protein synthesis was shut off in each culture [3].

3.3 Selection of Bacterial Mutants Resistant to Cytotoxic Effects of Host-Takeover Gene Products

Since some host-takeover gene products inhibit essential functions of the host cells, expression of such genes in uninfected cells is inhibitory to cell growth. The cellular target of such a cytotoxic gene can tentatively be identified by selecting mutants resistant to its cytotoxic effect, and identifying the gene in which the resistant mutation is located. Since the cloning vector used for this purpose, pPW19, replicates in both *B. subtilis* and *E. coli*, and since many of the host-takeover genes have similar effects on both bacteria, resistant mutants can frequently be selected in either *B. subtilis* or *E. coli*. Here, I just describe the procedure for selecting and isolating resistant mutants of *B. subtilis*. Procedures for *E. coli* are analogous.

1. Clone the gene to be tested. The cloning vector, pPW19, is described above. We have usually made the fragment to be cloned by PCR amplification from an SPO1 genomic DNA template. The region amplified includes the ribosome-binding site(s) of the gene(s) to be cloned, but not their natural promoter, so their expression would be dependent upon the IPTG-inducible promoter on the vector. We include the restriction site to be used for cloning near the 5′ end of each PCR primer.

2. The cloned gene is tested for cytotoxic activity by plating a culture of cells carrying the cloned gene on TC and TCI plates. Failure to form colonies on TCI indicates a strongly inhibitory effect of the expressed gene product, which can be used to select for resistant mutants.

3. Grow an overnight culture of CB10 carrying the plasmid with the cytotoxic gene, in NY plus Cm. Plate at 10^{-5} dilution on TC plates and at 10^{-5}, 10^{-3}, 10^{-1}, and 10^{-0} dilutions on TCI plates. Incubate at least 24 h at 37 °.

4. The rare colonies that grow on TCI plates are due to either chromosomal mutations (which presumably alter the cellular

target to make it resistant) or plasmid mutations (which inactivate the cytotoxic gene product). Label up to 20 of these colonies A, B, C, etc.

5. Grow overnight cultures of each of these resistant colonies in 2 mL NY + Cm. Make plasmid minipreps from each (*see* **Note 14**).

6. Transform each of the plasmids into CB10 cells. Include, as controls, pPW19 and the above plasmid carrying the cytotoxic gene. Plate each transformation culture at 10^{-2} and 10^{-0} dilutions on TC and TCI plates. Those that form substantial numbers of colonies on TCI no longer have their cytotoxic effect and are therefore plasmid mutants, which are discarded.

7. For those transformants that did not form colonies on TCI, the plasmid retained its cytotoxic effects and the resistance must have been provided by a chromosomal mutation. Refer back to their original colonies. Grow overnight cultures of these colonies in NY medium without chloramphenicol. Many of the cells in these cultures will have been cured of the plasmid. Plate these cultures on TSA plates at 10^{-5} dilution.

8. Replica streak about 20 individual colonies from each plate onto TSA and TC plates. Strains which have been cured of the plasmid should not grow on TC. Of the colonies that did not grow on TC, grow overnight cultures from the corresponding TSA streak to make competent cells for transformation.

9. Transform each of the putatively resistant cultures from **step 8** with the wild-type plasmid carrying the cytotoxic gene. Plate each transformation culture at 10^{-2} and 10^{-0} dilutions on TC and TCI plates. If colonies grow on TCI, the strain being transformed is a resistant strain, resulting from a chromosomal mutation, which presumably alters the cellular target of the cytotoxic gene product.

3.4 Effect of SPO1 Infection on Host Cell Division

In analyzing the cytotoxic genes discussed in the preceding section, it is often desirable to distinguish between bacteriostatic and bactericidal effects. To do this, a culture carrying the cytotoxic gene is grown in liquid medium (NY plus Cm) at 37° with vigorous shaking. At cell density about 1×10^{7}/mL, the culture is divided into two halves, 1.0 mM IPTG is added to one half, and the cultures continue shaking at 37°. At regular time intervals, the cultures are tested for viable count by plating appropriate dilutions on TC plates, and for turbidity by measurement in the Klett-Summerson colorimeter. A substantial decrease in viable count after adding IPTG indicates a bactericidal effect.

When this assay was performed with SPO1 gene 56, the turbidity continued to increase despite the loss of viability [4], suggesting that gp56 may prevent cell division without preventing cell growth. This was confirmed by showing microscopically that expression of gene 56 caused formation of long filaments. The following is the procedure that we followed to show that gp56 also prevents cell division during SPO1 infection.

1. Cultures of CB10 are grown in VY at 30° to a cell density of about $7 \times 10^7/\text{mL}$ and are then infected with the appropriate SPO1 strain at MOI about 3.5. A parallel culture remains uninfected. The cultures continue shaking vigorously at 30°. At 30, 50, and 70 min after infection, 0.5 mL samples are taken from each flask for microscopy.

2. The samples are centrifuged for 15 s at 13,200 rpm in a microfuge, resuspended in 0.5 mL of 0.3% Triton X-100, 4.5 M paraformaldehyde in saline phosphate (SP), and incubated for 20 min at room temperature.

3. Coverslips are prepared ahead of time by spreading a 10 μL aliquot of 0.01% poly-L-lysine onto each coverslip and allowing it to dry. The coverslips are rinsed by dipping in sterile water, and allowed to dry.

4. Each 0.5 mL sample is spread onto one of the polylysine-coated coverslips, allowed to stand for 30 min, and the coverslip then rinsed by dipping in water and allowed to dry at room temperature.

5. For each sample, a drop of 2.5 μg/mL Hoechst 33342 in 50% glycerol is placed on a microscope slide (*see* **Note 15**). The dry coverslip containing the cells is inverted over that drop, and the edges sealed with nail polish.

6. The slides are observed under 63× and 100× oil immersion objective lenses on a Zeiss Axioplan 2 microscope equipped with a digital CoolSnap camera (Photometrics). Cell images are viewed using differential interference contrast (DIC) optics. Images are recorded, and cell length is measured using Meta-Morph software (Molecular Devices).

7. The average length of cells infected with wild-type SPO1 increased with time after infection, showing that infection permitted continued cell growth but prevented cell division. That increase was not observed during infection with a gene 56 mutant, showing that the gene 56 product is responsible for the inhibition of cell division [4].

4　Notes

1. Neither chloramphenicol nor IPTG can tolerate autoclaving. They must be dissolved in media that are already sterile. For supplementation of TSA plates, the TSA solution is autoclaved

and then placed in a 50° water bath until its temperature has equilibrated at 50°. A sterile solution of Cm or IPTG is added and mixed with gentle swirling to avoid bubble formation, before pouring into petri dishes.

2. To dissolve glucose at this high concentration, it is best to add the water to the glucose just before putting it in the autoclave. Autoclave for no more than 15 min, to minimize caramelization of the glucose.

3. Prepare C4 medium in a sterile vessel. Each of the components must be sterile when added. Do not autoclave the prepared C4.

4. Conditions for maximal burst size of SPO1 are growth in VY medium at 37° with vigorous shaking.

5. We estimate cell density by measurements of turbidity, using a side-arm flask (Klett flask), and a Klett-Summerson colorimeter. A cell density of 1×10^8/mL yields a reading of about 50 on our colorimeter, which corresponds to an optical density at 500 nm of about 0.9.

6. 2×10^{10}/mL is a typical titer for a fresh lysate of an SPO1 strain prepared under optimal conditions. It is usually not necessary actually to measure the titer, since large deviations from the targeted MOI of 10 will still yield successful infections.

7. For plating for colonies or plaques, we typically plate 0.1 mL of the indicated dilution and incubate the plates at 37°.

8. Usually, a significant number of phage is lost in the centrifugation process, which is the reason for resuspending in a smaller volume.

9. Infection at 30° approximately doubles the latent period, in comparison with 37°, without substantially affecting the burst size [1]. We do these experiments at 30°, so experimental operations can be done at a slower pace, and also because the differential effects of mutations are enhanced at 30°. To decrease the time of preliminary growth, the culture may be grown at 37°, and shifted to 30° 10 min before infection. For multiple assays, a single culture may be grown in the preliminary flask, and distributed in 5 mL aliquots to 125 mL Erlenmeyers at the time of shift to 30°.

10. The tubes of tritiated precursors should be prepared ahead of time and kept on ice. 5–10 min before use, the tube is placed into the 30° water bath shaker, to prewarm it to 30°. Flask or tube should be removed from the shaker for the minimum time necessary to make the transfer, as it is important to maintain temperature and aeration.

11. We filter through a circular filter, 25 mm in diameter, into a filter flask, with suction provided by a faucet aspirator.

12. The procedure for scintillation counting would vary with the facilities available. We place each filter into the top of a 10 mL filmware bag (Nalgene). When all filters are ready, they are dried for 30 min under a Fisher "Infra-Rediator" drying lamp. After cooling, the filters are pushed to the bottom of the filmware bags, and 1.5 mL of scintillation fluid (Insta-Fluor Plus, PerkinElmer) is added, completely covering the filter. The bags are sealed and placed in counting vials, and cpm is determined in a Packard Tri-Carb 2100TR Liquid Scintillation Analyzer.

13. This is for the purpose of removing 3-H leucine from leucyl tRNA molecules, which would also be precipitated by the TCA.

14. We use a Qiaprep Spin Miniprep kit from Qiagen for our plasmid preparations.

15. The Hoechst was included in all of our slides to stain the bacterial nucleoids. Its presence is irrelevant for the current protocol.

References

1. Wei P, Stewart CR A cytotoxic early gene of *Bacillus subtilis* bacteriophage SPO1. J Bacteriol 175:7887–7900

2. Sampath A, Stewart CR (2004) Roles of genes 44, 50 and 51 in regulating gene expression and host takeover during infection of *Bacillus subtilis* by bacteriophage SPO1. J Bacteriol 186:1785–1792

3. Stewart CR, Yip TKS, Myles B, Laughlin L (2009) Roles of genes 38, 39, and 40 in shutoff of host biosynthesis during infection of *Bacillus subtilis* by bacteriophage SPO1. Virology 392:271–274

4. Stewart CR, Deery WJ, Egan ESK, Myles B, Petti AA (2013) The product of SPO1 gene 56 inhibits host cell division during infection of *Bacillus subtilis* by bacteriophage SPO1. Virology 447:249–253

5. Glassberg JS, Franck M, Stewart CR (1977) Initiation and termination mutants of *Bacillus subtilis* phage SPO1. J Virol 21:147–152

6. Mulbry WW, Ambulos NP, Lovett PS (1989) *Bacillus subtilis* mutant allele *sup3* causes lysine insertion at ochre codons. J Bacteriol 171:5322–5324

7. Spizizen J (1958) Transformation of biochemically deficient strains of *Bacillus subtilis* by deoxyribonucleate. Proc Natl Acad Sci U S A 44:1072–1078

8. Shub, D.A. (1966) Functional stability of messenger RNA during bacteriophage development. Ph.D. Thesis. Massachusetts Institute of Technology

9. Gage LP, Geiduschek EP (1971) RNA synthesis during bacteriophage SPO1 development: six classes of SPO1 RNA. J Mol Biol 57:279–300

Chapter 3

Practical Advice on the One-Step Growth Curve

Andrew M. Kropinski

Abstract

The one-step growth experiment is fundamental to the description of a new bacteriophage. The following protocol is optimized for those working with rapidly growing bacterial cultures.

Key words Adsorption, One-step growth, Latent period, Burst size

1 Introduction

The one-step growth experiment [1] is one of the classic bacteriophage techniques used in the characterization of new isolates; and defining the latent period and the average burst size. Though the theory associated with this experiment was discussed in a previous chapter in this series [2] we have received numerous requests for specific instructions on how to carry out a one-step growth (OSG) experiment. The following protocol is based upon Symond's approach [3], and primary experience with *Escherichia coli* and *Pseudomonas* strains.

1.1 Preliminary Points

1. This experiment needs to be refined after the preliminary run which could be used just to determine the length of the latent period.

2. It is based upon the assumption that 90–95% of the phages adsorb to the host cells within 5 min. If they adsorb poorly alternative approaches have to be taken to remove unadsorbed phages (*see* **Note 1**).

3. While the equipment and medium should be sterile, strictly aseptic technique is not necessary if you are dealing with a rapidly growing bacterium.

4. Accuracy in pipetting is *essential*; as is following a rigid time frame.

Martha R.J. Clokie et al. (eds.), *Bacteriophages: Methods and Protocols, Volume 3*, Methods in Molecular Biology, vol. 1681, https://doi.org/10.1007/978-1-4939-7343-9_3, © Springer Science+Business Media LLC 2018

5. If your phage has, like coliphage T7, a very short latent period at 37 °C you might consider running the OSG experiment at 30 °C.

6. This protocol is set up to eliminate the need to carry out dilutions prior to plating.

7. This is not the type of experiment that you can do once and obtain publication quality results; and the more samples you take, the better.

2 Materials

1. A water bath set at optimum growth temperature of bacterium (or phage).

2. Log phase bacterial culture; and an ON culture.

3. Phage preparation diluted to 10^7 PFU/mL (for small plaques, 1 mm) or 5×10^6 PFU/mL for large plaque-forming phages (>2 mm; e.g., coliphage T7).

4. Four small sterile Erlenmeyer flasks—one empty; one with exactly 9.9 mL of broth; and, two with exactly 9.0 mL broth.

5. Numerous 1000 and 100 μL sterile pipette tips.

6. One empty 13 × 100 mm test tube.

7. One 13 × 100 mm tube containing 50 μL of chloroform in an ice bucket.

8. Numerous tubes of molten OVERLAY medium in 48 °C water bath (*see* **Note 2**).

9. Numerous fresh prewarmed agar plates (UNDERLAY PLATES; *see* **Note 2**).

10. Accurate timer.

11. Pasteur pipettes and bulbs.

3 Methods

1. Subculture host bacterium in medium of choice plus 2 mM $CaCl_2$ and grow to mid-log phase (ca. 0.5 OD_{650nm}).

2. Label the small flasks as indicated in the accompanying diagram (Fig. 1) and place in the water bath.

3. Pipette 9.9 mL of the log phage culture into the empty flask and place at the appropriate incubation temperature for 5 min.

4. Pipette a couple of mL of ON host culture into a 13 × 100 mm tube, place in rack with Pasteur pipette and bulb (PLATING HOST).

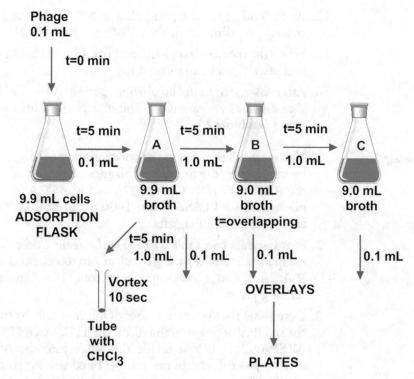

Fig. 1 OSG experiment—layout

5. Add 0.1 mL of phage preparation to the 9.9 mL culture (ADSORPTION FLASK), swirl gently and replace in incubator for 5 min. (N.B. this represents a 1/100 dilution of phage titer: 1×10^5 PFU/mL).

6. After 5 min remove 0.1 mL from this flask to 9.9 mL of fresh prewarmed medium (FLASK **A**. N.B. 1/100 dilution; titer: 1×10^3 PFU/mL); mix well.

7. Transfer 1.0 mL from FLASK **A** to the test tube containing chloroform, Vortex 10 s.; replace on ice (ADSORPTION CONTROL; *see* **Note 3**).

8. Transfer 1.0 mL from FLASK **A** to 9.0 mL of prewarmed medium (FLASK **B**. N.B. 1/10 dilution; titer: 1×10^2 PFU/mL); mix well.

9. Transfer 1.0 mL from FLASK **B** to 9.0 mL of prewarmed medium (FLASK **C**. N.B. 1/10 dilution; titer: 1×10^1 PFU/mL); mix well (*see* **Note 4**).

10. At various times remove 0.1 mL from the appropriate FLASK (**A**, **B** or **C**) add to the molten OVERLAY; add 1–3 drops of PLATING HOST; mix and pour on surface of UNDERLAY plates (*see* **Note 5**).

11. At the end of the sampling plate two 0.1 mL samples via the overlay procedure from the ADSORPTION CONTROL.

12. When the overlays have hardened (ca. 15 min) invert the plates and place them in an incubator.

13. After an appropriate incubation period (ON for *E. coli* or *Pseudomonas aeruginosa*) count the plaques on each of the plates (*see* **Note 6**).

3.1 Data Analysis

1. Normalize all the data to the concentration of phage in flask A, by multiplying the number of plagues obtained from sampling the ADSORPTION CONTROL and FLASK **A** by 10; multiply by 100 for FLASK **B** and 1000 for FLASK **C**. The results are expressed as PFU/mL.

2. Plot the data (*see* Table 1; PFU/mL) against time (*see* Fig. 2) using a commercial package such as Microsoft Excel or Graph-Pad Prism (http://www.graphpad.com/) or manually using semilog paper.

3. Determine the average number of infected cells by subtracting the number of phage in the ADSORPTION CONTROL from AVERAGE 1. If you divide this value into the AVERAGE 2 value you will obtain the average burst size for your phage.

4. Determine the intersect between the AVERAGE 1 line and the slope will give you the latent period of your phage.

4 Notes

1. Some phages, such as *Pseudomonas* phage Φ-S1 [4], while recognizing a common receptor such as LPS adsorb poorly to their hosts. In other cases such as pilus or flagella-specific phages the culture may not constitutively produce the receptor which also complicates the one-step growth experiment. Classically the presence of unadsorbed phages was circumvented by using anti-phage serum; which is probably now available to you. Alternative approaches include testing on different hosts to select for the strain to which your phage adsorbs the best. Or, after the 5 min adsorption period the contents of the ADSORPTION FLASK can be rapidly centrifuged and resuspended in fresh prewarmed medium; or, collected on a 0.45 μm low protein binding filter, which is then inverted and the cells washed off. Please note that both of these techniques need to be carried out quickly. We do not recommend chilling the phage-host mixture.

Table 1
Data from an OSG experiment

Time (min)	Plaque numbers (flask A)	PFU/mL	Plaque numbers (Flask B)	PFU/mL	Plaque numbers (Flask C)	PFU/mL
6	100	1000				
8	100	1000				
10	100	1000				
12	100	1000				
14	100	1000				
16	100	1000				
18	100	1000				
20	100	1000				
22	100	1000				
24	100	1000				
26	100	1000				
28	100	1000				
30	100	1000	10	1000		
32	300	3000	30	3000		
34	TMTC		80	8000		
36			110	10000		
38			200	20000	20	20000
40			300	30000	30	30000
42			TMTC		40	40000
44					50	50000
46					70	70000
48					90	90000
50					95	95000
52					100	100000
54					100	100000
56					100	100000
58					100	100000
60					100	100000

TMTC to many to count

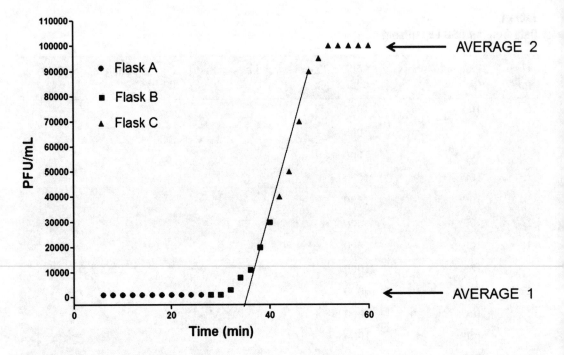

Fig. 2 OSG experiment—plotted results

2. The composition of the growth and phage overlay and under-lay media will depend on the host. We recommend the inclusion of 2 mM CaCl₂ since many phages require divalent ions for efficient adsorption [5]

3. The use of chloroform to kill phage-infected cells may not work since members of the families *Corticoviridae*, *Cystoviridae*, *Plasmaviridae*, and *Tectiviridae* contain lipids and are therefore chloroform sensitive. A significant number of members of the *Caudovirales* are also solvent sensitive (Ackermann, personal communication). In these cases a rapid spin in a microcentri-fuge or filtration through a 0.22–0.45 μm low protein binding filter will remove the phage-infected cells from the free phage particle.

4. The dilution series must be completed within 2 min.

5. If you are planning on just determining the latent period in the first experiment you do not need FLASKS **B** or **C**.

6. After running your experimental OSG experiment the first time you will be able to judge how often you should sample and from which flask. To make sampling easier, an overlapping timing of sampling from each flask is recommended.

References

1. Ellis EL, Delbrück M (1939) The growth of bacteriophage. J Gen Physiol 22:365–384

2. Hyman P, Abedon ST (2009) Practical methods for determining phage growth parameters. Methods Mol Biol 501:175–202

3. Symonds ND (1968) Experiment 14 – One-step growth curve and the Doermann experiment. In: Clowes RC, Hayes W (eds) Experiments in microbial genetics. Blackwell, Oxford, pp 75–78

4. Kelln RA, Warren RA (1971) Isolation and properties of a bacteriophage lytic for a wide range of pseudomonads. Can J Microbiol 17:677–682

5. Kropinski AM (2009) Measurement of the rate of attachment of bacteriophage to cells. Methods Mol Biol 501:151–155

Chapter 4

Iron Chloride Flocculation of Bacteriophages from Seawater

Bonnie T. Poulos, Seth G. John, and Matthew B. Sullivan

Abstract

Viruses influence ecosystem dynamics by modulating microbial host population dynamics, evolutionary trajectories and metabolic outputs. While they are ecologically important across diverse ecosystems, viruses are challenging to study due to minimal biomass often obtained when sampling natural communities. Here we describe a technique using chemical flocculation, filtration and resuspension to recover bacteriophages from seawater and other natural waters. The method uses iron to precipitate viruses which are recovered by filtration onto large-pore size membranes and then resuspended using a buffer containing magnesium and a reductant (ascorbic acid or oxalic acid) at slightly acid pH (6–6.5). The recovery of bacteriophages using iron flocculation is efficient (>90%), inexpensive and reliable, resulting in preparations that are amenable to downstream analysis by next generation DNA sequencing, proteomics and, in some cases, can be used to study virus–host interactions.

Key words Iron chloride flocculation, Ocean viruses, Bacteriophages, Viral ecology

1 Introduction

Viruses in the ocean are abundant [1, 2] and they play a significant role in ecosystem dynamics through lysing their microbial hosts [3–5], horizontally transferring genes [6], and encoding host genes that modulate the microbial metabolisms underlying ocean ecosystem function [7–12]. Because viral biomass in the oceans is relatively low, the ability to study ocean viruses is dependent on methods which efficiently and reproducibly concentrate them from large volumes of water. The method that has been used most frequently by aquatic virologists is tangential flow filtration (TFF) [13–15]. TFF is efficient for concentrating viruses from large sample volumes but the filters are expensive, the method is not well-suited to field use and, more importantly, the recovery of viruses may be highly variable due to numerous factors [16, 17]. The chemical flocculation method described here using iron chloride was developed as an alternative to TFF that can be easily applied to various field conditions, is inexpensive, and requires little technical expertise to perform. The technique was adapted from

Martha R.J. Clokie et al. (eds.), *Bacteriophages: Methods and Protocols, Volume 3*, Methods in Molecular Biology, vol. 1681,
https://doi.org/10.1007/978-1-4939-7343-9_4, © Springer Science+Business Media LLC 2018

wastewater treatment methods [18] and it has been used successfully to obtain ocean viruses for metagenomic sequencing [19]. When Fe is added to waters at circumneutral pH, the Fe flocculates as Fe oxyhydroxide particles and binds negatively charged viruses. These large flocculate particles can then be recovered on large pore size (1 μm) membranes Storing the filters moist at 4 °C until further processing protects the viruses from degradation and prevents the Fe from recrystallizing into different Fe oxyhydroxide minerals which are more difficult to dissolve. Viruses can later be recovered from the filters with a buffer that contains a reductant to dissolve the Fe oxyhydroxide minerals (ascorbic acid or oxalic acid), EDTA to chelate Fe in solution and prevent reprecipitation of Fe minerals, and magnesium (Mg) to maintain the integrity of virus particles. The resuspended virus particles can be used for downstream analyses including metagenomic sequencing [19, 20] and proteomics as well as infectivity assays if oxalic acid is used as the reductant in the resuspension buffer [17]. This method was used exclusively to isolate viruses for metagenomic sequencing underlying the temporally and spatially resolved Pacific Ocean Virome (POV) dataset [21], as well as the global surface waters from the TARA Oceans Expedition that are represented in the Tara Oceans Virome (TOV) dataset [22]. The resulting information derived from those virus samples has revealed new insights into the ecology of ocean viruses and has expanded the repertoire of bacteriophages that exist in the world's oceans ([22], reviewed in [23]).

2   Materials

All solutions should be prepared in water that has been purified from deionized water through a filtration system at >18 MΩ-cm or using "molecular biology" (MQ-H_2O) grade water purchased from a reliable vendor. Chemicals should be analytical grade or better. Where specified, solutions should be subsequently 0.2 μm filtered and then 0.02 μm (Whatman Anotop 25, Sigma-Aldrich) filtered if deep sequencing is planned to minimize contamination.

2.1 Precipitate Viruses from Seawater Using Iron Chloride

1. 10 g/L Fe stock solution: Weigh 4.83 g $FeCl_3.6H_2O$ and transfer to beaker with 100 mL MQ-H_2O. Stir until dissolved. Store at room temperature or 4 °C (*see* **Notes 1–4**).

2. Seawater: Collect seawater using acid-washed and rinsed collection vessels (*see* **Note 5**). The volume of water collected is dependent on the planned downstream analyses. Next-generation sequencing methods currently require a minimum of 10^9 viral-like particles for reliable construction of DNA sequencing libraries.

2.2 Filtration to Remove Bacteria	1. Two 142 mm filtering towers (*see* **Note 6** for alternative prefiltration method).
	2. Peristaltic pump with a pressure gauge (maximum pressure = 15 psi) and appropriate tubing (*see* **Note 7**).
	3. 150 mm Whatman GF/A prefilter (1.6 μm retention).
	4. 142 mm Millipore Express Plus filter (0.22 μm pore size; http://www.emdmillipore.com/).
	5. Acid-washed carboy for collecting filtered water (*see* **Note 5**).
2.3 Filtration to Collect Iron-Precipitated Viruses	1. One 142 mm filtering tower.
	2. Peristaltic pump with pressure gauge and appropriate tubing (*see* **Note 7**).
	3. 142 mm GE Waters (http://www.gelifesciences.com/) Polycarbonate membrane collection filter (1.0 μm pore size; available through Midland Scientific; http://www.midlandsci.com/) (*see* **Note 8**).
	4. 142 mm Supor-800 backing filter (0.8 μm pore size; Pall Corporation; http://www.pall.com/) (*see* **Note 9**).
	5. Sterile 50 mL centrifuge tubes, filter forceps, Parafilm.
2.4 Resuspension of Viruses	1. Resuspension buffer: 0.1 M EDTA-0.2 M $MgCl_2$-0.2 M ascorbic acid buffer. Make this buffer fresh on the day it is to be used. Dissolve 1.51 g Tris base (FW = 121.14) in 80 mL MQ-H_2O. Add 3.72 g disodium EDTA dihydrate (FW = 373.24) and stir until completely dissolved. Next, add 4.07 g $MgCl_2 \cdot 6H_2O$ (FW = 203.3) and stir until completely dissolved. Check pH and adjust to ~pH 6.5 with 5 M NaOH (prepare using 20 g NaOH per 100 mL MQ-H_2O), added dropwise in 0.5 mL volumes. Finally add 3.52 g ascorbic acid (FW = 176.12) and stir until completely dissolved. Adjust final pH to 6.0 with 5 M NaOH. Adjust final volume to 100 mL with MQ-H_2O (*see* **Note 10**).
	2. The viability of virus particles may be compromised when using ascorbic acid as the reductant. If viability is important, oxalic acid dihydrate (FW = 126.07) may be substituted for ascorbic acid, but the amount of $MgCl_3\text{-}6H_2O$ must be reduced by half (*see* **Note 11**). Store the resuspension buffer at room temperature, protected from light. It may be 0.2 μm filtered and may also be 0.02 μm filtered (Whatman Anotop syringe filters) if metagenomic sequencing is planned downstream.

3   Methods

3.1   Virus Precipitation

1. Assemble two 142 mm filtering towers and attach the filter apparatus to a peristaltic pump with a pressure gauge (*see* **Note 6** for alternative prefiltration method).

2. Wear gloves and use forceps for handling all filters.

3. Collect 20 L seawater and prefilter using a 150 mm Whatman GF/A filter followed by a 0.22 µm, 142 mm Millipore Express Plus filter into an acid-washed and rinsed carboy (*see* **Note 5**).

4. Treat the virus fraction (i.e., the 0.22 µm filtrate) with $FeCl_3$, to precipitate the viruses by adding 1 mL of 10 g/L Fe stock solution for each 20 L of filtered seawater. Immediately after adding Fe, shake gently for 1 min to mix.

5. Add an additional 1 mL of 10 g/L Fe stock solution for each 20 L of filtrate (for a total of 2 mL Fe stock solution per 20 L filtrate). Shake for 1 min. Repeat shaking several times.

6. Let the $FeCl_3$-treated filtrate sit for 1 h at room temperature.

7. Attach the filter apparatus to the peristaltic pump and filter the $FeCl_3$-treated filtrate through a 1.0 µm, 142 mm, polycarbonate (PC) membrane on top of a 0.8 µm, 142 mm, Supor support filter (Pall Corporation; *see* **Notes 8** and **9**).

8. Using forceps, place all of the polycarbonate filters into a 50 mL sterile centrifuge tube being careful not to scrape off any of the $FeCl_3$ on the edge of the tube (*see* **Note 13**). Discard the Supor support membrane.

9. Be sure the cap is on securely and seal the tube containing the filters with Parafilm. Store dark at 4 °C until ready to resuspend the precipitated viruses (*see* **Notes 14** and **15**).

3.2   Virus Resuspension

1. Prepare fresh 0.1 M EDTA-0.2 M $MgCl_2$–0.2 M ascorbic acid (or oxalic acid) buffer, pH 6.5 (*see* **Notes 10–12**).

2. Make sure the filter(s) is turned precipitate-side out in the 50 mL centrifuge tube to provide direct contact of the buffer with the precipitate (*see* **Note 16**).

3. Add 1 mL fresh resuspension buffer for each 1 mg Fe, which is 20 mL of buffer for 20 L seawater precipitate. Shake the tube vigorously.

4. Places the tubes on a tube rotator, or an orbital shaker, and resuspend overnight at 4 °C in the dark (*see* **Note 17**).

5. Recover the resuspended viral precipitate by pipetting off the liquid into a fresh tube. To recover the liquid remaining on the filters, cinch the filter 2–3 mm over the edge of the tube, and hold in place with the lid. Centrifuge gently ($500 \times g$) for 3–5 min so that the remaining liquid falls to the bottom of

the tube. If the filters still retain precipitate, additional resuspension buffer may be added to them and rotated for another 1–2 h (*see* **Note 18**). Repeat the centrifugation step to recover the liquid.

6. The viral suspension is now ready for downstream processing. Microscope counts may be performed to determine the quantity of viruses recovered. The viruses may be further purified by CsCl density centrifugation which will also act to remove traces of iron from the preparation. Alternatively, the viral nucleic acid may be extracted for sequencing (*see* **Notes 19** and **20**).

4   Notes

1. Iron chloride solution is acidic and should be handled with care.

2. The solution of ferric chloride hexahydrate is calculated based on the amount of iron, not on the amount of salt. The stock solution has 10 g Fe per liter. For precipitation, the final optimal concentration is 1 mg Fe per liter seawater, which is equal to 2.9 mg $FeCl_3$ per liter seawater or 4.83 mg $FeCl_3 \cdot 6H_2O$ per liter seawater.

3. Discard iron chloride solution if a cloudy precipitate forms. Do not dilute the solution as iron hydroxide precipitate will form quickly.

4. For use on board ship, it is best to preweigh iron chloride into 50 mL centrifuge tubes and add water as needed during the cruise: preweigh 0.966 g $FeCl_3 \cdot 6H_2O$ into several 50 mL tubes. Add 20 mL $MQ-H_2O$ when ready to precipitate viruses from seawater.

5. Use 1 M HCl to clean carboys and tubing: dilute concentrated HCl (12 M) 1:12 with $MQ-H_2O$. Place approximately 1 L in the carboy (or other vessel) to be cleaned, tightly close lid, shake acid around in carboy for 5 min. Decant acid back into acid container (it can be reused indefinitely). Rinse out acid from the carboy using $MQ-H_2O$ in a similar manner, discarding rinse water and repeating for a total of three rinses. For tubing, soak tubing 5–10 min in 1 M HCl and rinse with at least three changes of $MQ-H_2O$. Check tubing before use and replace if worn.

6. An alternative method for preparing seawater is to prefilter the seawater through a Whatman GF/D glass fiber filter to remove large particles (2.7 μm retention). The filtrate is filtered again using a Millipore Steripak GP10 (<10 L) or GP20 (<20 L) with 0.22 μm pore size. Larger capsule filters will easily filter 100–200 L (Pall Acropak 200, 500, 1000 and 1500). The

cellular fraction is retained on the filter while the virus fraction is in the filtrate.

7. Filtration should be done at a maximum pressure of 15 psi so as not to damage viral particles or cells. Peristaltic pumps (such as MasterFlex I/P with an Easy Load pump head) should be used with appropriate tubing (such as Platinum silicone tubing I/P 82 for the abovementioned pump) for optimal performance. Plastic hose clamps and connectors should fit tightly around the tubing. If something comes loose, turn off the pump first to prevent a flood.

8. The 1.0 μm polycarbonate membrane from GE Waters is the recommended membrane for collection of precipitated viruses. It is manufactured in Canada, available in the USA through Midland Scientific, and can take 4–5 weeks for delivery. Alternatively, if this membrane is not available, Millipore Isopore membrane 0.8 μm may be substituted. Although this membrane has not been evaluated for quantitative recovery of VLPs, successful sequencing libraries have been prepared using them. From surface ocean waters, it will take 1–3 filters to collect precipitate from 20 L starting volume.

9. The Supor-800 membrane is simply a backing filter to support the polycarbonate collection membranes in the filtration apparatus. The pore-size of this filter is not crucial, though it is advantageous to use a pore size larger than the prefilter to prevent clogging. Normally only one backing membrane is needed for filtering each batch of seawater (fresh polycarbonate membranes are placed on top) and it can be discarded when the filtration is complete.

10. The resuspension buffer should be prepared fresh since the reductant (ascorbic acid) is unstable and it is necessary for dissolving the iron precipitates. The buffer should be used within 24–48 h of preparation. The buffer may start to discolor after 24 h.

11. The original formulation for EDTA-Mg buffer used the chemical dimagnesium-EDTA from J.T. Baker ($C_{10}H_{12}Mg_2N_2O_8$, FW = 336.82), which is no longer available. The resuspension buffer was reformulated to use Na_2-EDTA·$2H_2O$ and $MgCl_2$·$6H_2O$. When preparing this buffer keep in mind that EDTA needs a pH above 8.0 to dissolve, and will come out of solution when the pH drops below about 5.0. The 0.125 M Tris base solution has a pH >10 which allows the EDTA to dissolve quickly. The $MgCl_2$ and the ascorbic acid will drop the pH. For this reason, it is best to adjust the pH close to the final desired pH (6.0–6.5) after the addition of each of these two chemicals.

12. The amount of reductant (ascorbic acid or oxalic acid) can vary between 0.125 M and 0.25 M; this formulation uses 0.2 M. With the reformulation of the buffer using Na_2- EDTA and $MgCl_2$, the addition of oxalic acid causes an irreversible precipitate to form. For this reason, when preparing the buffer with oxalic acid dihydrate (2.52 g per 100 mL for 0.2 M), the amount of $MgCl_2$-$6H_2O$ should be reduced by half (2.04 g per 100 mL) to prevent the precipitate from forming.

13. When handling the PC filters after filtration, it helps to have two pairs of forceps so that the membrane can be folded into quarters to put into the 50 mL centrifuge tube. Having the precipitate facing out aids in dissolving the precipitates, but be careful not to lose precipitate on the edge of the tube when doing this.

14. Do not shake off excess liquid; the filters must not be allowed to dry out before the precipitated viruses are resuspended and the excess liquid will create a sufficiently humid atmosphere in the sealed tube. If excess fluid is not present, adding 1–2 mL of sterile water or 0.22 μm filtered seawater to the tube will keep the filters moist.

15. Keep filters at 4 °C until processed. Processing is best done soon after collection, although we have successfully resuspended particles after several years if stored properly.

16. Turning the filters precipitate-side out for resuspension is critical for efficient resuspension. Wear gloves. Forceps for handling the filters should be disinfected in 10% bleach and rinsed in MQ-H_2O before touching the filters. Use of aluminum foil placed on the lab bench can provide a clean surface for opening up the folded filters; replace with a fresh piece of foil before each new sample.

17. The tubes containing filters and resuspension buffer can be packaged in aluminum foil and placed together on the rotator or shaker to keep them in the dark while resuspension is taking place. The goal is to remove all of the precipitate from the filters so constant contact of the filter with buffer is important.

18. The resuspension solution will change colors a number of times. This is due to the changing oxidation state of the Fe and is to be expected. However, if precipitates form in the liquid, there is a possibility that viruses are trapped or adhered to the particles. Remove as much liquid as you can and then gently centrifuge to pellet the precipitate. Add additional resuspension buffer to redissolve or at least wash the precipitate and recover the liquid phase which may contain residual viruses.

19. DNA extraction may be performed by various methods. This laboratory routinely treats the resuspended virus with DNaseI (to remove free DNA from the virus preparation), chemically

inactivates the enzyme with EDTA and EGTA (0.1 M final concentration) and then concentrates the volume using Amicon Ultra 100 kDa centrifugal concentrators (1000 × g for 10 min intervals). The DNA is extracted from this suspension using Wizard Prep resin and mini-columns from Promega (http://www.promega.com) at a ratio of 0.5 mL virus suspension to 1.0 mL Wizard Prep resin. DNA isolated in this manner has been successfully used for library preparation for next generation (i.e., 454 and Illumina) sequencing. The DNA is also now amenable to PCR testing due to the removal of EDTA and excess Mg that is contained in the resuspension buffer.

20. Updates to this protocol, as well as other protocols mentioned in this manuscript may be found at the protocols.io website https://www.protocols.io/view/Iron-Chloride-Precipitation-of-Viruses-from-Seawat-c2wyfd or at the Sullivan Lab webpage: http://u.osu.edu/viruslab/.

Acknowledgments

We thank Jennifer Brum and numerous viral ecologists for their feedback about this evolving protocol and its application across diverse sample types. This work was supported by funds from the Gordon and Betty Moore Foundation through a collaborative (GBMF3305) to M.B.S. and S.G.J., as well as methods development grants (GBMF2631 and 3790) to M.B.S.

References

1. Bergh O, Borsheim KY, Bratbak G, Heldal M (1989) High abundance of viruses found in aquatic environments. Nature 340:467–468

2. Proctor LM, Fuhrman JA (1990) Viral mortality of marine bacteria and cyanobacteria. Nature 343:60–62

3. Wommack KE, Colwell RR (2000) Viroplankton: viruses in aquatic ecosystems. Microbiol Mol Biol Rev 64:69–114

4. Suttle CA (2007) Marine viruses – major players in the global ecosystem. Nat Rev Microbiol 5:801–812

5. Breitbart M, Thompson LR, Suttle CA, Sullivan MB (2007) Exploring the vast diversity of marine viruses. Oceanography 20:135–139

6. Paul JH (1999) Microbial gene transfer: an ecological perspective. J Mol Microbiol Biotechnol 1:45–50

7. Mann NH, Cook A, Millard A, Bailey S, Clokie M (2003) Bacterial photosynthesis genes in a virus. Nature 424:741

8. Sullivan MB, Lindell D, Lee JA, Thompson LR, Bielawski JP, Chisholm SW (2006) Prevalence and evolution of core photosystem II genes in marine cyanobacterial viruses their hosts. PLoS Biol 4:e234

9. Anantharaman KA, Duhaime MB, Breier JA, Wendt K, Toner BM, Dick GJ (2014) Sulfur oxidation genes in diverse deep-sea viruses. Science 344:757–760

10. Sullivan MB, Huang KH, Ignacio-Espinoza JC, Berlin AM, Kelly L, Weigele PR, DeFrancesco AS, Kern SE, Thompson LR, Young S, Yandava C, Fu R, Krastins B, Chase M, Sarracino D, Osburne MS, Henn MR, Chisholm SW (2010) Genomic analysis of oceanic cyanobacterial myoviruses compared with T4-like myoviruses from diverse hosts and environments. Environ Microbiol 12:3035–3056

11. Hurwitz BL, Hallam SJ, Sullivan MB (2013) Metabolic reprogramming by viruses in the sunlit and dark ocean. Genome Biol 14:R123

12. Hurwitz BL, Brum JR, Sullivan MB (2015) Depth-stratified functional and taxonomic niche specialization in the core and flexible Pacific Ocean Virome. ISME J 9:472–484

13. Suttle CA, Chan AM, Cottrell MT (1991) Use of ultrafiltration to isolate viruses from seawater which are pathogens of marine phytoplankton. Appl Environ Microbiol 57:721–726

14. Wommack KE, Sime-Nogando T, Winget DM, Jamindar S, Helton RR (2010) Filtration-based methods for the collection of viral concentrates from large water samples. In: Wilheim SW, Weinbauer MG, Suttle CA (eds) Manual of aquatic viral ecology. American Society of Limnology and Oceanography, Waco, Texas, pp 110–117

15. Rohwer F (2013) How to set up a TFF. https://www.Youtube.Com/watch? V=d38ys3SxAZ0. Accessed 2 May 2015

16. Colombet J, Robin A, Lavie L, Bettarel Y, Cauchie HM, West S, Mann NH (2007) Viroplankton 'pegylation': use of PEG (polyethylene glycol) to concentrate and purify viruses in pelagic ecosystems. J Microbiol Methods 71:212–219

17. John SG, Mendez CB, Deng L, Poulos B, Kaufmann AKM, Kern S, Brum J, Polz MF, Boyle EA, Sullivan MB (2011) A simple and efficient method for concentration of ocean viruses by chemical flocculation. Environ Microbiol Reports 3:195–202

18. Zhu B, Clifford DA, Chellam S (2005) Virus removal by iron coagulation-microfiltration. Water Res 39:5153–5161

19. Hurwitz BL, Deng L, Poulos BT, Sullivan MB (2013) Evaluation of methods to concentrate and purify ocean virus communities through comparative, replicated metagenomics. Environ Microbiol 15:1428–1440

20. Duhaime MB, Sullivan MB (2012) Ocean viruses: rigorously evaluating the metagenomics sample-to-sequence pipeline. Virology 434:181–186

21. Hurwitz BL, Sullivan MB (2013) The Pacific Ocean Virome (POV): a marine viral metagenomic dataset and associated protein clusters for quantitative viral ecology. PLoS One 8: e57355

22. Brum JR, Ignacio-Espinoza JC, Roux S, Doulcier G, Acinas SG, Alberti A, Chaffron S, Cruaud C, de Vargas C, Gasol JM, Gorsky G, Gregory AC, Guidi L, Hingamp P, Iudicone D, Not F, Ogata H, Pesant S, Poulos BT, Schwenck SM, Speich S, Dimier C, Kandels-Lewis S, Picheral M, Searson S, Tara Oceans Coordinators Bork P, Bowler C, Sunagawa S, Wincker P, Karsenti E, Sullivan MB (2015) Patterns and ecological drivers of ocean viral communities. Science 348:1261498

23. Brum JR, Sullivan MB (2015) Rising to the challenge: accelerated pace of discovery transforms marine virology. Nat Rev Microbiol 13:147–159

Chapter 5

Purification of Bacteriophages Using Anion-Exchange Chromatography

Dieter Vandenheuvel, Sofie Rombouts, and Evelien M. Adriaenssens

Abstract

In bacteriophage research and therapy, most applications ask for highly purified phage suspensions. The standard technique for this is ultracentrifugation using cesium chloride gradients. This technique is cumbersome, elaborate and expensive. Moreover, it is unsuitable for the purification of large quantities of phage suspensions.

The protocol described here, uses anion-exchange chromatography to bind phages to a stationary phase. This is done using an FLPC system, combined with Convective Interaction Media (CIM®) monoliths. Afterward, the column is washed to remove impurities from the CIM® disk. By using a buffer solution with a high ionic strength, the phages are subsequently eluted from the column and collected. In this way phages can be efficiently purified and concentrated.

This protocol can be used to determine the optimal buffers, stationary phase chemistry and elution conditions, as well as the maximal capacity and recovery of the columns.

Key words Bacteriophage, Concentration, CsCl, Ion-exchange chromatography, Purification

1 Introduction

Bacteriophage research and applications (e.g., phage display, proteomics, genomics, crystallography, and phage therapy) require purified phage suspensions. Nowadays, the standard procedure still involves polyethylene glycol (PEG) precipitation and/or CsCl-density gradient ultracentrifugation [1]. While PEG precipitation is an easy and inexpensive way to concentrate phages, it often fails at obtaining sufficiently pure phage preparations. CsCl-density ultracentrifugation on the other hand, results in highly purified phage particle suspensions, but it is very elaborate, time-consuming, expensive, and is not suitable for purifying industrial-size volumes. Furthermore, some phages cannot be purified via this method because of their instability in the high osmotic environment of the dense CsCl-gradients [2–4]. Thus, the need for a high-throughput method for bacteriophage purification still exists.

Martha R.J. Clokie et al. (eds.), *Bacteriophages: Methods and Protocols, Volume 3*, Methods in Molecular Biology, vol. 1681,
https://doi.org/10.1007/978-1-4939-7343-9_5, © Springer Science+Business Media LLC 2018

Chromatographic methods can be an alternative method of bacteriophage purification. Anion-exchange chromatography was already proposed in 1953 as a useful method to purify and concentrate phages [5]. Creaser and Taussig set up a reliable and easy protocol for the anion-exchange chromatography of phages using a resin [6]. Recently, monolithic ion-exchange columns were developed for the purification of biomolecules (e.g., plasmid DNA, viruses, proteins). The highly interconnected network of hollow channels with large and well-defined pore sizes, creates a strong and robust sponge-like structure [7]. This enables a laminar flow profile of the mobile phase, ensuring a large contact surface with limited pressure drops, even at high flow rates. These conditions favor a high binding capacity for large molecules, like phages [8, 9]. Furthermore, the resolution and capacity are not influenced by the flow rate. These features make this technique suitable for processing large volumes and is thus relevant not only on laboratory scale, but also for industrial and biomedical purposes.

Ion-exchange chromatography has been proven useful for the purification of phages, with more and more studies being published over the last few years describing the purification of a wide variety of bacteriophages [3, 8–11]. The amount of phages recovered from the purified fraction after the process is generally high, typically ranging from 35–70% of the original number of bacteriophages loaded, sometimes reaching as high as 99.99% [3]. This technique is specific enough to separate different phages from a mixture, given that the elution conditions are sufficiently different for each phage [9]. Lock and colleagues achieved a separation of the elution peaks of DNA packed and empty viral capsids [12]. Although the resolution was high enough for relative quantification of the different fractions by comparing the difference in peak surface, a higher resolution will be necessary to completely separate these particles into different fractions. Even though the research was done with viruses quite different from bacteriophages, the fact that it was possible to differentiate between empty and DNA-containing particles shows the high purification possibilities of ion-exchange chromatography.

The available column sizes range from 0.1 mL for small-scale research purposes to 8000 mL suitable for pilot tests and industrial purification of phage lysate at flow rates between 2 and 10 L/min. The binding and elution conditions can be optimized using a small-scale column, and easily transferred to larger columns, without the need for further optimization [13]. Although ion-exchange chromatographic purification of phages demands for an elaborate optimization, afterward, the protocol is fast, reproducible, can handle high-throughput volumes, is easily upscalable to industrial size, and can be almost entirely automated [3, 7, 10]. These advantages make this technique an alternative method for the classical CsCl-density gradient purification.

Here, we describe a step-by-step protocol for the optimization of phage purification applicable to almost any phage, using the Convective Interaction Media (CIM®) anion-exchange monolith columns from BIA Separations (Slovenia; http://www.biaseparations.com/) on an ÄKTA FPLC system (or the more recent ÄKTA pure system, GE Healthcare, UK; http://www3.gehealthcare.co.uk/) running UNICORN™ software.

2   Materials

2.1   Media Preparation

1. Degassed ultrapure water.

2. Loading buffer: 20 mM Tris–HCl, pH 7.5, 0.22 μm filter sterilized (see **Note 1**).

3. Elution buffer: 20 mM Tris–HCl, pH 7.5, 2 M NaCl, 0.22 μm filter sterilized (see **Note 1**).

4. Cleaning buffer: 1 M NaOH, 0.22 μm filter sterilized.

5. Storage buffer: 20% (v/v) ethanol, 0.22 μm filter sterilized.

6. Bacteriophage suspension, 0.22 μm filter sterilized (see **Note 2**).

2.2   FPLC and Accessories

1. CIM® Anion-exchange column DEAE or QA (BIA Separations, Slovenia).

2. HPLC or FPLC system with analyzing software (see **Note 3**).

3   Methods

Purification of phages using anion-exchange chromatography requires optimization for each phage isolate, after which only the stepwise gradient purification method of this methods section is needed for the purification of additional lysates. The optimization is a step-by-step process involving different parameters such as buffer composition, binding and elution conditions, and column chemistry. Prior to this, stability tests with the phages in the used buffers and pH should be performed.

Before proceeding with the protocol, the filtered phage suspensions should be diluted (1/1 v/v) in loading buffer in order to lower the ionic strength of the loaded sample (see **Note 4**).

Definitions and abbreviations	
CV	Column volume, the volume of a column as defined by the manufacturer
FT	Flow through, the loaded sample that ran over the column and is collected afterward
E	Eluate, the collected volume at the elution step
$W_{0\%}$	Wash using 0% elution buffer to wash unbound particles off the column
$W_{100\%}$	Wash using 100% elution buffer to wash all the bound particles off the column

3.1 Preparing the FPLC for Phage Purification

It is advised to keep the FPLC machine and monolithic columns in a 20% ethanol storage solution when not in use. Since the presence of ethanol or other chemicals can be harmful to phages, they need to be removed before phages are applied. The first step is removing the storage solution from the whole system.

1. Place the selected CIM® disk (QA or DEAE) in the column housing and attach it in line with the flow path.

2. Rinse the pumps, the sample loop, the column and the FPLC system's flow path with degassed ultrapure water to remove any traces of ethanol from the storage buffer and to create optimal binding conditions in the column.

3. Place pump A in a bottle with loading buffer and pump B in a bottle with elution buffer. Make sure an adequate amount of buffer is present so that the pumps always remain immersed. Fill both pumps with their respective buffer solution.

4. Wash the system with elution buffer, until the UV absorbance and conductivity signals stay stable.

5. Create the binding conditions by flowing loading buffer through the system. Do this until the UV absorbance and conductivity signals reached a stable base line.

6. Load the phage suspension onto the sample loop and attach the sample loop with phage suspension onto the injection valve in the correct position.

This protocol results in a clean machine with a loaded sample, ready for use. The following optimization protocols require these steps to be performed first and will start from this point.

In between two phage purification protocols, the right binding conditions have to be recreated. This is done by repeating **steps 4–6** of this protocol after the finalization of one purification protocol and before the start of the next. When the last phage purification is preformed, the column and machine need to be cleaned (*see* Cleaning protocol).

3.2 Optimization Protocol for Binding Conditions: One-Step Gradient

This protocol is the first step in the optimization of chromatographic purification of phages. The binding conditions for phages can depend on column chemistry, buffer composition and pH, or ionic strength of the phage suspension.

1. Load a small volume of phage suspension on the column at 0% elution buffer concentration. The FPLC software will show a peak in UV absorbance on the chromatogram when the phage suspension is loaded. This can be used to determine how long it takes for the phages to be loaded onto the column.

2. Wash the column with loading buffer until the baseline for the UV absorbance and conductivity is reached and stable.

3. Collect all of the flow through (**FT**) and the wash (**W$_{0\%}$**) of the column in one tube.

4. Elute the bound particles from the column using 100% elution buffer.

5. Collect the eluate (**E**) in one tube. A peak in UV absorbance indicates elution of particles bound to the column.

6. Titrate $FT + W_{0\%}$ and **E** fractions using the double agar layer method (under conditions appropriate for the purified phage). Ideal binding conditions are those with which no phages are found in the **FT + W$_{0\%}$** fraction and the majority of phages is present in **E**. Optimization of this step can be done by further diluting the phage suspension, by changing the loading buffer chemistry or pH, or by changing column chemistries.

3.3 Optimization Protocol for Elution Conditions: Linear Gradient

The linear gradient helps determining the conditions of phage elution. The percentage of elution buffer at which the bound phages will elute from the column can be defined precisely. The goal is to find the exact conditions at which the phages elute from the column, while leaving unwanted impurities bound to the column, in order to separate the phages from impurities.

1. Load a small volume of phage suspension on the column at 0% elution buffer concentration. Wash the column with loading buffer until the UV absorbance and conductivity signals reach their respective baselines.

2. Collect **FT** and **W$_{0\%}$** in one tube.

3. Set the system to linearly increase the ratio of elution buffer from 0 to 100%. Make sure this increase is done over a sufficient volume to derive the optimal elution conditions. An example of the chromatographic output of such a gradient is found in Fig. 1.

4. Collect the eluate periodically in fractions of 1–2 mL in separate tubes.

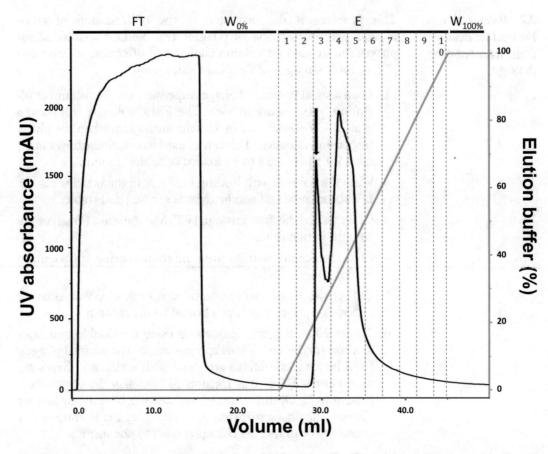

Fig. 1 Linear gradient of phage purification using a DEAE column. The UV absorbance (*black curve*) is given at a corresponding elution buffer concentration (*grey curve*). On top of the figure, the different phases in the process are indicated: Flow through (**FT**), Washing the column after loading (**W$_{0\%}$**), Elution (**E**) with the different collected fractions indicated by numbers, and the final wash to regenerate the column (**W$_{100\%}$**). The *arrow* indicates the peak that corresponds with the presence of phages

5. Wash the column with 100% elution buffer (**W$_{100\%}$**) until the UV absorbance signal shows no more peaks and has reached the baseline and the conductivity signal stays stable. Collect this **W$_{100\%}$** fraction.

6. Titrate all collected fractions using the double agar layer method (under conditions appropriate for the purified phage).

7. Match the titer of each **E** fraction with the UV absorbance peaks seen on the chromatogram (Fig. 1). This will show which of the peaks are caused by phage elution and which by elution of impurities. Based on this data, the concentration of the elution buffer at which impurities and phages elute can be calculated. If the titer of the **W$_{100\%}$** fraction is higher than 1% of the total phage content loaded on the column, binding conditions are too stringent for the chosen elution buffer used and the ionic strength of this buffer needs to be increased.

3.4 Optimization Protocol for Capacity Determination

Once the optimal binding conditions are set, the capacity of the column can be defined. This allows one to determine the maximum of phage particles that can be bound to the column, and subsequently purified (*see* **Note 5**). In short, this is done by loading an excess of phages onto the column. The moment the phages start to run off the column, the maximum amount of bound phages is reached. By subtracting the number of phages in the **FT** from the total number of phages loaded onto the column, the exact number of phages bound to the column can be derived.

1. Load an excess of phages. Literature shows a high variety in capacity for different phages, ranging from 10^9 up to over 10^{12} pfu/mL column volume for the CIM monolith with a CV equal to 0.34 mL [3]. For correct determination of the capacity, the number of loaded phages should exceed the maximal capacity.

2. Collect **FT** in one tube, discard $W_{0\%}$, **E** and $W_{100\%}$.

3. Titrate the **FT** fraction using the double agar layer method (under conditions appropriate for the purified phage). Phages found in the **FT** are phages which could not bind anymore to the column. Determine the capacity as follows:

$$Capacity = Total\ number\ of\ phages\ loaded$$
$$- Number\ of\ phages\ in\ \mathbf{FT}$$

When no phages are found in the **FT**, this indicates that the maximum capacity was not reached. The capacity cannot be derived correctly.

3.5 Purification of Phages Using Monolithic Anion-Exchange Columns: Stepwise Gradient

Once all the above parameters are set, this information can be used to optimally purify and concentrate the phages. The amount of phages should not exceed the capacity of the column, and optimal binding and elution conditions as determined in previous steps should be chosen (*see* **Note 6**). The elution will now happen in three distinct phases (Fig. 2).

1. Load a volume of the phage suspension from the sample loop onto the column at 0% elution buffer concentration so that the capacity of the column is reached, but not exceeded. Discard the **FT** (*see* **Note 7**).

2. Wash the column with loading buffer until UV absorbance and conductivity signals reach a stable baseline to remove all impurities. Discard the $W_{0\%}$.

3. Raise the percentage of elution buffer to 5–10% lower than the optimal concentration for phage elution as determined by the linear gradient. This will wash all the weaker bound impurities of the column. Discard this fraction.

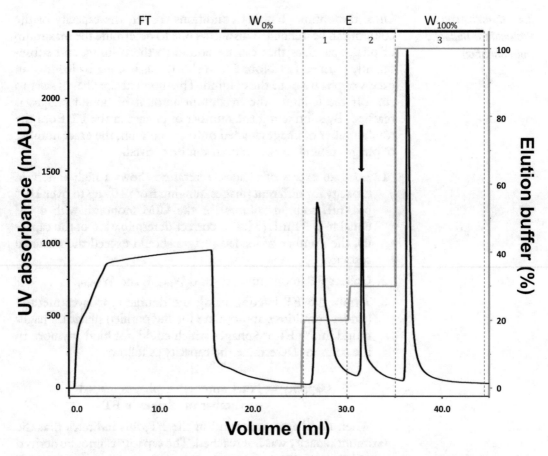

Fig. 2 Stepwise gradient of phage purification using a DEAE column. The UV absorbance (*black curve*) is given at a corresponding elution buffer concentration (*grey curve*). On *top* of the figure, the different phases in the process are indicated: Flow through (**FT**), Washing the column after loading (**W$_{0\%}$**), Elution (**E**), and the final wash to regenerate the column (**W$_{100\%}$**). The first fraction of **E** contains the impurities; the second is the purified phage fraction. The *arrow* indicates the peak that corresponds with the presence of phages

4. Raise the percentage of elution buffer to 5–10% above the optimal concentration for phage elution to elute phage particles. Collect this eluate, containing the purified phages.

5. Raise the percentage of elution buffer to 100% to remove all remaining particles from the column and regenerate it. Discard this **W$_{100\%}$**.

6. Titrate the fraction with the eluted phages using the double agar layer method (under conditions appropriate for the purified phage). The recovery of the phages from the column when used at capacity can now be calculated as follows (*see* **Note 8**):

$$\text{Recovery} = \frac{\text{Total number of phages in pure fraction}}{\text{Total number of phages loaded}}$$

The recovery reflects the efficiency of the column.

This final protocol results in a fraction containing the purified phages, which is collected in **step 4**.

The conditions of this final protocol can now be used for every additional purification of the same phage.

3.6 Cleaning Protocol After Phage Purification Using FPLC

To avoid cross-contamination with subsequent experiments and purifications, the FPLC, sample loop and column need to be extensively cleaned of all remaining phages. This cleaning protocol should be performed after every run.

1. Fill the sample loop with 1 M NaOH cleaning buffer.

2. Fill pump A of the FPLC with the cleaning buffer. Let the cleaning buffer flow through the whole flow path to clean the system. Make sure the column is in line with the pathway. Pass at least ten CVs through the column.

3. Remove the cleaning buffer from the sample loop, pumps, machine flow path and column by rinsing it with an extensive amount of degassed ultrapure water until the pH returns to neutral.

4. Regenerate the ion-exchange column by passing ten CVs of 100% elution buffer over the column. After regeneration the column is ready to be used again.

5. For long-term storage, rinse both pumps, the flow path and the column with storage buffer until the UV absorbance and conductivity signals reach a stable baseline. Remove the column from its housing and store in a closed vial, submersed in sterile storage buffer at 4 °C for optimal preservation.

4   Notes

1. Other compositions of loading and elution buffer are possible depending on the phage used. Examples used in recent publications include phosphate and HEPES buffers, although the latter was used for a eukaryotic enterovirus not a phage [3, 14]. Prior to the phage chromatography, stability testing of the phage in different buffers and at different pH should be performed. Adjust the buffer composition and pH accordingly.

2. Phage amplification is often a nonoptimized step, limiting the usability of other purification and concentration processes. This protocol allows one to purify phages from highly concentrated suspensions as well as from low phage-containing suspensions. For an optimal yield of purified phages, it is advised to adapt the loaded volumes of phage suspensions to reach the column's capacity. When the limitations of the FPLC system do not allow the loading of an adequate volume of phage lysate,

this might result in a lower yield, although the quality of the purified phage suspension should not be affected.

3. This protocol was optimized using an ÄKTA™ FPLC system, equipped with a P900 pump and UNICORN™ control and analysis software (GE Healthcare) and a CIM® disk with a total column volume of 0.34 mL (BIA Separations). A constant flow rate of 2 mL/min was used, according to the manufacturer's instructions (*see* **Note 9**). With few adaptations, this protocol is applicable to comparable systems and setups.

4. High ionic strength will result in the inability of the phages to properly bind the column. Phage lysates with a high ionic strength may require greater dilutions, while low ionic strength suspensions do not need dilution and can be used as is.

5. The maximum capacity of a column is different for each phage. For laboratory scale purifications using a 0.34 mL column, it was noted that the maximum capacity was often not reached. The main limiting steps are the finite amount of phage suspension that could be loaded, combined with a limited efficiency of phage amplification.

6. To prevent unnecessary loss of phages, it is advised to work at approximately 70% of the maximal capacity. Since the capacity can be influences by different factors (e.g., temperature, buffer conditions, column age), working with a slight sub-optimal amount of phages ensures that all phages can be bound to the column.

7. When performing the stepwise gradient elution for the first time, it can be useful to collect all different fractions. In case of unexpectedly low phage titers in the purified fraction, titration of all the collected fractions can give an indication of where the problem is situated.

8. This formula is only valid when the capacity of the column was not exceeded.

9. Read the column manufacturer's instructions carefully. Verify the optimal working conditions (e.g., flow rate, chemical resistance to acidic or basic solutions, maximal back pressure). Never neglect these important notes and adjust the protocol accordingly to avoid any damage to the FPLC system or ion-exchange columns.

References

1. Boulanger P (2009) Chapter 13: purification of bacteriophages and SDS-PAGE analysis of phage structural proteins from ghost particles. In: Clokie MRJ, Kropinski AM (eds) Bacteriophages: methods and protocols, volume 2: molecular and applied aspects. Humana Press, New York, pp 227–238

2. Sillankorva S, Pleteneva E, Shaburova O, Santos S, Carvalho C, Azeredo J, Krylov V (2010) Salmonella enteritidis bacteriophage

candidates for phage therapy of poultry. J Appl Microbiol 108(4):1175–1186

3. Adriaenssens EM, Lehman SM, Vandersteegen K, Vandenheuvel D, Philippe DL, Cornelissen A, Clokie MRJ, García AJ, De Proft M, Maes M, Lavigne R (2012) CIM® monolithic anion-exchange chromatography as a useful alternative to CsCl gradient purification of bacteriophage particles. Virology 434(2):265–270

4. Carlson K (2005) Appendix: working with bacteriophages: common techniques and methodological approaches. In: Kutter E, Sulakvelidze A (eds) Bacteriophages: biology and applications. CRC Press, Boca Raton, pp 437–494

5. Puck T, Sagik B (1953) Virus and cell interaction with ion exchangers. J Exp Med 97 (6):807–820

6. Creaser EH, Taussig A (1957) The purification and chromatography of bacteriophages on anion-exchange cellulose. Virology 4 (2):200–208

7. Oksanen HM, Domanska A, Bamford DH (2012) Monolithic ion exchange chromatographic methods for virus purification. Virology 434(2):271–277

8. Kramberger P, Honour RC, Herman RE, Smrekar F, Peterka M (2010) Purification of the Staphylococcus Aureus bacteriophages VDX-10 on methacrylate monoliths. J Virol Methods 166(1–2):60–64

9. Smrekar F, Ciringer M, Štrancar A, Podgornik A (2011) Characterisation of methacrylate monoliths for bacteriophage purification. J Chromatogr A 1218(17):2438–2444

10. Smrekar F, Ciringer M, Peterka M, Podgornik A, Štrancar A (2008) Purification and concentration of bacteriophage T4 using monolithic chromatographic supports. J Chromatogr B 861(2):177–180

11. Monjezi R, Tey BT, Sieo CC, Tan WS (2010) Purification of bacteriophage M13 by anion exchange chromatography. J Chromatogr B 878(21):1855–1859

12. Lock M, Alvira MR, Wilson JM (2012) Analysis of particle content of recombinant adeno-associated virus serotype 8 vectors by ion-exchange chromatography. Hum Gene Ther Methods 23(1):56–64

13. Liu K, Wen Z, Li N, Yang W, Hu L, Wang J, Yin Z, Dong X, Li J (2012) Purification and concentration of mycobacteriophage D29 using monolithic chromatographic columns. J Virol Methods 186(1–2):7–13

14. Kattur Venkatachalam A, Szyporta M, Kiener T, Balraj P, Kwang J (2014) Concentration and purification of enterovirus 71 using a weak anion-exchange monolithic column. Virol J 11(1):99

Chapter 6

Encapsulation Strategies of Bacteriophage (Felix 01) for Oral Therapeutic Application

Golam S. Islam, Qi Wang, and Parviz M. Sabour

Abstract

Due to emerging antibiotic-resistant strains among the pathogens, a variety of strategies, including therapeutic application of bacteriophages, have been suggested as a possible alternative to antibiotics in food animal production. As pathogen-specific biocontrol agents, bacteriophages are being studied intensively. Primarily their applications in the food industry and animal production have been recognized in the USA and Europe, for pathogens including *Salmonella*, *Campylobacter*, *Escherichia coli*, and *Listeria*. However, the viability of orally administered phage may rapidly reduce under the harsh acidic conditions of the stomach, presence of enzymes and bile. It is evident that bacteriophages, intended for phage therapy by oral administration, require efficient protection from the acidic environment of the stomach and should remain active in the animal's gastrointestinal tract where pathogen colonizes. Encapsulation of phages by spray drying or extrusion methods can protect phages from the simulated hostile gut conditions and help controlled release of phages to the digestive system when appropriate formulation strategy is implemented.

Key words Encapsulation, Microencapsulation by spray drying, Alginate–whey microsphere, Acid resistance of phages, Phage therapy, Bacteriophage

1 Introduction

Antibiotics are extensively used in agriculture (both plants and animals) and specifically in food animal production (poultry, swine, and cattle) as growth promoters for feed efficiency. This practice has been linked to antibiotic resistance in food pathogens and a decrease in their efficiency in clinical applications [1]. In response, European Union has banned use of antibiotics as growth promoters in animal feed and this issue is under consideration by many other countries [2]. It is expected that the international trade restriction and consumer demand would limit their nonclinical use.

Amongst a variety of strategies that are being suggested to replace antibiotics in food animal production, bacteriophages as a pathogen specific biocontrol is being intensively tried [3]. Currently use of bacteriophages against some pathogens such as

Martha R.J. Clokie et al. (eds.), *Bacteriophages: Methods and Protocols, Volume 3*, Methods in Molecular Biology, vol. 1681, https://doi.org/10.1007/978-1-4939-7343-9_6, © Springer Science+Business Media LLC 2018

Salmonella spp., *E. coli* O157:H7, and *Listeria monocytogenes*, in foods has been approved by FDA [4].

Although there are some published reports of successful phage therapy in food animals, the extent of the efficiency is variable, which may be related to the amount of active phage that makes it through to the pathogen site [3]. It is known that a variety of factors such as gastric acidity affect survival of some phages [5–7]. Therefore for successful phage therapy in animal production it is important to identify the limiting factors in the GI tract of each animal species and develop appropriate remedies to protect phage from damage during therapy and increase its therapeutic efficiency. To meet this challenge researchers have been developing encapsulation formulation to protect phage from gastric acid and control the release in simulated intestinal condition for efficient pathogen control [6, 8].

In this chapter, we present two protocols for encapsulation of phage where Felix O1 is used as a model example. We also include the protocols for testing the protection and release characteristics along with the infection efficiency of the encapsulated phage against the host pathogen in simulated gastric and intestinal conditions.

2   Materials

A microbiological laboratory equipped with laminar air flow chamber, chemical safety hood, and standard equipment such as balances, pH meter, and magnetic mixer.

2.1 Equipment and Supplies

1. Equipment, media, and supplies for preparing solid culture media, broth, buffers.

2. Micropipettes of variable and fixed volume (10–1000 μL) and sterile pipette tips.

3. Millex® 0.22 μm, 33 mm syringe filters, Corning® Spin-X® centrifuge tube filters (0.22 μm), Steriflip® centrifuge tube filter unit (Millipore), Corning disposable vacuum filtration system, (0.22 μm).

4. Centrifuge tubes (15 and 50 mL), 10 mL dilution tubes, Eppendorf tubes (1 and 2 mL).

5. Disposable petri dishes (100 mm), disposable square petri plates with grids.

6. Incubator shaker (such as Innova™ 4200, New Brunswick Scientific, NJ, USA).

7. Beckman preparative ultracentrifuge with Beckman SW28 rotor.

8. Thermo Scientific™ Sorvall™ RC 6 PLus Centrifuge with Fiberlite F14-6 × 250y rotor.

9. Bench-top microcentrifuge.

10. Laboratory homogenizer—Polytron 2500 E stand dispersion unit or similar.

11. Encapsulator—Inotech Encapsulator IER-50 (Inotech Biosystems Intl. Inc.) or similar with nozzle size 300 µm (*see* **Note 1**).

12. Spray dryer: Laboratory-scale spray dryer, such as Yamato spray dryer ADL 310 (Yamato Scientific America, Inc., CA, USA) or similar, nozzle orifice diameter 720 µm.

2.2 Chemicals

1. Eudragit S100 (Evonik Industries).

2. Skim milk powder (SMP), sorbitan monopalmitate (Span 40).

3. Gelatin from porcine skin (Type A).

4. Sodium citrate tribasic dehydrate, monobasic potassium phosphate (KH_2PO_4), sodium phosphate dibasic (Na_2HPO_4).

5. Maltose, maltodextrin (dextrose equivalent 16.5–19.5).

6. Whey protein (BiPro®, Eden Prairie, MN, USA).

7. Sodium alginate (medium viscosity, Sigma-Aldrich).

8. Pepsin (3200–4500 units/mg protein).

9. Pancreatin ($4\times$ USP specification, from porcine pancreas).

10. Sodium chloride (NaCl), calcium chloride, hexahydrate ($CaCl_2 6\ H_2O$), sodium bicarbonate ($NaHCO_3$), hydrochloric acid (HCl), cassium chloride (CsCl), ammonium hydroxide (NH_3OH).

11. Magnesium sulfate ($MgSO_4$), magnesium chloride ($MgCl_2$), Tris base ($NH_2C(CH_2OH)_3$).

12. DNase, RNase, polyethylene glycol (PEG 8000).

2.3 Buffers

1. Tris–HCl buffer, 50 mM (pH 8.0):

   Dissolve 6.057 g of Tris powder into 0.4 L of distilled deionized water. Add 1 M HCl to the Tris solution until the pH reaches 8.0, make up to a final volume of 1 L with distilled deionized water.

2. Microsphere breaking solution (MBS):

   50 mM of sodium citrate tribasic dehydrate and 200 mM of $NaHCO_3$ in 50 mM Tris–HCl buffer, pH 7.5.

3. SM Buffer:

   0.1% gelatin (w/v), 100 mM NaCl, 8 mM $MgSO_4$, 50 mM Tris–HCl, pH 7.5.

4. Hanks' balanced salt solution (HBSS):

   Mix 37 mM NaCl, 5.4 mM KCl, 4.2 mM $NaHCO_3$, 1.3 mM $CaCl_2$, 0.6 mM $MgSO_4$, 0.5 mM $MgCl_2$, 0.4 mM KH_2PO_4, 0.3 mM Na_2HPO_4 with distilled water, autoclave and refrigerate at 4 °C.

**2.4   Simulated
Gastrointestinal Fluids**

1 or 2% (w/v) porcine bile extract

*2.4.1   Simulated Bile
Fluid*

*2.4.2   Simulated Gastric
Fluid (SGF)*

1. Dissolve 3.2 mg of pepsin and 2.0 g of NaCl in 800 mL water.
2. Adjust the solution with 0.2 M HCl to a desired pH, such as, 2.0 and make up to 1 L with water (*see* **Note 2**).

*2.4.3   Simulated
Intestinal Fluid (SIF)*

1. Dissolve 6.8 g of KH_2PO_4 in 750 mL of water,
2. Add 10 g pancreatin (4× USP specification),
3. Adjust pH to 6.8 with either 0.2 M NaOH or 0.2 M HCl solutions, and make up to 1 L with water.

**2.5   Stock Solutions
for Spray Drying**

1. In a 250 mL glass bottle disperse 4 g Eudragit S100 in 63 mL of distilled water and stir for 30 min.
2. Inside a chemical safety hood, place the bottle on a magnetic stirrer and set it at 300 rpm.

*2.5.1   Eudragit S100
Solution*

3. Add 12 mL of 1 N NH_3OH drop wise and stir continuously till Eudragit completely dissolves.

*2.5.2   Phage Felix O1
Lysate Stock*

1. In a sterile glass vial dispense 10 mL Felix O1 liquid lysate containing desired titer of phage (*see* **Note 3**).
2. Add 1% (w/v) SMP and 1% (w/v) maltose to the phage lysate and shake gently until dissolved completely.

2.5.3   Gelatin Solution

Dissolve gelatin (1% w/v) in 5 mL distilled water at 40 °C under constant stirring.

*2.5.4   Nonionic
Surfactant Solution*

1. Take 10 mL of distilled water in a small glass vial and place on a water bath set at 60 °C.
2. Add Span 40 (1% w/v) to the hot water and shake to melt completely.
3. Using a laboratory homogenizer blend the molten Span 40 at high speed (~12,000 rpm) for 1 min. At this point no further creaming occurs. (Do not place the vial into hot water bath again.)

**2.6   Stock Solutions
for Encapsulation by
Extrusion**

1. Dissolve whey protein isolate (6–10% w/v) in deionized water in a screw capped bottle with continues stirring on a magnetic stirring plate for 1 h at room temperature (22 °C).
2. Adjust the protein solution to pH 8.0 using 1 M NaOH solution.

*2.6.1   Whey Protein
Solution*

3. Heat the protein solution at 80 °C for 30 min to denature the protein, and then cool to room temperature.

2.6.2 Sodium Alginate Solution

Dissolve sodium alginate (1.5–3% w/v) in 50 mM Tris–HCl buffer (pH 8.0) at 80 °C for 2 h under continues stirring in a water bath on a magnetic stirring plate.

2.7 Media

Prepare the following media according to manufacture's instruction: Tryptic Soy Agar (TSA), Brilliant Green Sulfa Agar (BGS Agar), Tryptic Soy Broth (TSB), Brain Heart Infusion Broth and Agar (BHI broth and agar).

2.7.1 Phage Agar (PA-Nal)

For enumeration of phages by direct plating plaque assay.

8.4 g NaCl, 13 g bacteriological agar, 10 g Nutrient Broth, water 1000 mL, autoclave. Add 20–40 µg/mL nalidixic acid (Nal) filtered using 0.2 µm syringe filter and mix prior to pouring into petri dishes.

2.8 Culture

Salmonella Typhimurium DT104 Nal^R (Nalidixic acid resistance) strains, Bacteriophage Felix O1 (Felix d'Herelle reference center, Université Laval, Quebec, Canada).

---

3 Methods

3.1 Bacteria Culturing

1. Use *Salmonella* Typhimurium DT104 Nal^R (nalidixic acid resistant, from Guelph Research and Development Center collection) for propagating and enumerating phage Felix O1.

2. Transfer 10 µL of culture into 5 mL BHI broth containing 20–40 µg/mL nalidixic acid (BHI-Nal). Incubate overnight (O/N) at 37 °C with vigorous shaking at 180 rpm.

3. Purify the O/N culture by streaking a small amount of inoculum on a TSA-Nal plate and incubate for 18–20 h at 37 °C.

4. Using a sterile loop pick a single colony and transfer to 5 mL BHI-Nal broth and incubate O/N at 37 °C.

5. Subculture these O/N cultures at 10^{-2} dilution into fresh TSB and grow until an optical density of 0.2 at 600 nm (OD_{600}) is reached.

3.2 Bacteriophage Propagation and Purification

Propagate phage Felix O1 on Typhimurium strain DT104 according to [9], then purify the bacteriophage Felix O1 using the method of [10] with some modification. A brief description of the method is given below.

1. Prepare crude phage Felix O1 lysate in TSB liquid culture, filter-sterilize by a vacuum-driven disposable filter (0.22 µm), and store at 4 °C.

2. Clarify the crude TSB lysate by centrifugation at $13,000 \times g$ for 20 min at 4 °C.

3. Digest the collected supernatant with 1 µg/mL each DNase and RNase. Centrifuge at $13,000 \times g$ for 10 min at 4 °C to further clarify the lysate.

4. Precipitate the supernatant in the presence of 10% polyethylene glycol (PEG) 8000 and 1 M NaCl at 4 °C O/N.

5. Centrifuge at $11,000 \times g$ for 10 min at 4 °C, and discard the supernatant.

6. Reconstitute the pellet in SM buffer and store at 4 °C.

7. Extract the PEG from the reconstituted phage suspension by mixing with equal volume of chloroform and centrifuge at $11,000 \times g$ for 20 min at 4 °C.

8. Purify the phage suspension by CsCl density gradient centrifugation (0.75 g/mL CsCl, run at $25,000 \times g$ at 4 °C for 24 h) in a Beckman SW28 rotor.

9. Resuspend the phage pellet in SM buffer, dialyze extensively against Hanks' balanced salt solution (HBSS).

10. Sterilize the purified phage suspension using 0.22 µm filters, and store as a high-titer stock at 4 °C until use.

3.3 Phage Titer— Rapid Plaque Assay

Bacteriophages can be enumerated by double agar overlay plaque assay [11], small drop plaque assay [12] or direct plating plaque assay [13]. However, we have used rapid plaque assay where there is no requirement of top agar overlay which makes the set up faster and simpler. The procedure is as follows:

1. Inoculate 5 mL of BHI broth with a colony of *S. typhimurium* DT104 Nal^R and incubate O/N at 37 °C at 180 rpm.

2. Transfer 200 µL of an O/N culture of *S. typhimurium* DT104 to 20 mL of BHI.

3. Incubate at 37 °C with shaking (220 rpm) for 5–6 h.

4. Transfer 900 µL of SM buffer to the appropriate number of sterile, disposable microtiter tubes.

5. Label the tubes according to the samples being titrated and the following dilutions: 10^{-1}–10^{-8}.

6. Using filter-tipped 200 µL pipette tips, transfer 100 ul of stock phage lysate to the 10^{-1} tube, discard the tip.

7. Using a new filter-tip, mix the contents of the tube by gently pipetting up and down, 20 times.

8. Transfer 100 µL of this dilution to the 10^{-2} tube, discard the tip and repeat **step 7**.

9. Repeat **steps 7–8** until all dilutions have been prepared.

10. Label the PA-Nal plates with sample name and dilutions.

11. In the biological safety cabinet, dilute the fresh log-phase culture of *S. typhimurium* DT104 1/10 in BHI, vortex gently (prepare as much of the 1/10 dilution as you will need to seed the appropriate number of PA-Nal plates).

12. Seed a PA-Nal plate by flooding the surface of a plate with 3–4 mL of the *S. typhimurium* DT104 1/10 dilution. Gently rock the plate to ensure that the entire surface has been wetted by the culture.

13. Tilt the plate, and using a sterile pipette, aspirate the excess culture into a sterile flask (waste container).

14. Seed as many plates as required.

15. Keep the lids of the plates open to allow the seeded agar surface to dry.

16. Spot 10 µL of each dilution of phage, in duplicate, to the appropriate squares on the PA-Nal plates.

17. Allow spots to dry. Cover the plates with lids and incubate O/N at 37 °C.

18. Count plaques and calculate plaque forming units (PFU)/mL or gram. Using the following formula

PFU/mL or g = Number of plaques × 100 × reciprocal of counted dilution

4 Encapsulation of Phage Felix O1 into Alginate–Whey Protein Microspheres

4.1 Preparation of Encapsulation Mixture

1. Mix sodium alginate and whey protein stock solutions at desired concentration ratio (*see* **Note 4**).

2. Add 1 mL of phage Felix O1 to 20 mL of alginate–whey protein solution, mix for 10 min at room temperature with a magnetic stir bar at slow speed.

4.2 Preparation of Alginate–Whey Microspheres (Wet)

1. Set the encapsulator at 1000 Hz, 1.2 kV.

2. Place the encapsulation mixture on a magnetic stirrer and stir at low speed.

3. Dip the feeding tube connected to the encapsulator into the bottom of the encapsulation mixture.

4. Adjust the flow speed of the feeding solution until the stream coming out of the nozzle splits into many fine streams of droplets.

5. Collect the droplets in a gelation bath containing 0.1 M $CaCl_2$ solution with slow agitation by a magnetic bar at room temperature.

6. Leave the microspheres in the gelation bath for 30 min to harden.

7. Drain the CaCl$_2$ solution and collect the microspheres using a sieve, rinse with distilled water, and then remove the excess water by shaking gently.

8. Weigh the microspheres and place into capped 50-mL sterile polypropylene tubes and store at 4 °C.

9. Perform plaque assay to determine the phage load and encapsulation efficiency.

4.3 Preparation of Alginate–Whey Microspheres (Dry)

1. Immerse 5 mL of the freshly prepared wet microspheres in 5 mL of 40% (w/v) maltodextrin solution. To allow sufficient diffusion of the solutes, keep the microspheres in this protective solution for 12 h at 4 °C (*see* **Note 5**).

2. Drain the excess liquid from the microspheres, then spread them on a filter paper and leave them in a running laminar air flow chamber at 22 °C for 30 h.

3. Immediately after drying, determine the surviving phage titer.

4. Store the dried microspheres in sealed bottles at 4 °C or 23 °C to study the storage stability of the phages inside the microspheres (*see* **Note 6**).

5. Perform plaque assay to determine the titer (survival rate) of encapsulated phage upon storage for desired time period.

5   Encapsulation of Phage Felix 01 by Spray Drying

Spray drying encapsulation technique is ideal for producing bulk materials at an economical manner. The formulation of the micro-encapsulated phages by spray drying method is different from the alginate–whey-based encapsulation. We have prepared the spray drying feed solution using an enteric polymer which dissolves at around pH 7 (*see* **Note 7**).

5.1 Encapsulation Mixture Preparation

To prepare the encapsulation mixture for spray drying the components need to be mixed together in an orderly manner.

1. Mix the phage solution with gelatin solution; pour this phage-gelatin mixture into the surfactant solution followed by homogenizing at a speed of 6000 rpm for 1 min.

2. Pour the homogenized solution into eudragit S100 solution with constant stirring at 150 rpm on a magnetic stirrer plate for 2 h.

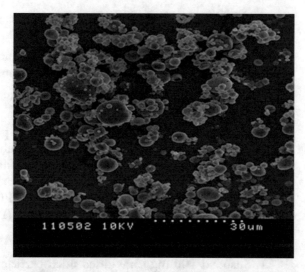

Fig. 1 Scanning electron microscopic image of spray-dried encapsulated phage Felix O1 powders. Feed solution containing Felix O1 was spray-dried at inlet temperature of 75 °C and 0.1 Mpa atomization air pressure

5.2 Spray Drying of Bacteriophages

The spray drying process needs to be optimized depending on the type and model of the spray dryer. For this encapsulation protocol a laboratory scale Yamato spray dryer model ADL 310 was used.

1. Start up the spray dryer according to the manufacturer's instruction.

2. Set inlet temperature, atomization, drying air flow, and feeding pump flow rate at 75 °C, 0.1 Mpa, 0.8 m³/min, and 4 mL/min respectively.

3. Submerge the silicone tubing connected to the feeding pump into the encapsulation mixture in order to deliver the feed solution to the spray nozzle for atomization. The atomized particles will be carried to the cyclone by the drying air and get deposited in to the collection vessel. Turn off the spray dryer when the process is finished and harvest the spray-dried particles. (Fig. 1 and *see* **Note 8**).

4. Determine the surviving phage titer by plaque assay and store rest of the particles in sealed vials at 4 °C (*see* **Note 9**).

6 Phage Survival and Encapsulation Efficiency Determination

6.1 Alginate–Whey Microsphere

We have used direct plating plaque assay to determine phage titer per gram of microspheres.

1. Weigh 200 mg of wet or 100 mg of dry microspheres into a vial.

2. Add MBS solution to dissolve the beads into a volume corresponding to a 10^{-1} initial dilution (*see* **Note 10**).

3. Enumerate phage by rapid plaque assay and calculate plaque forming units (PFU) per gram of microspheres (*see* **Note 11**).

4. Calculate encapsulation efficiency of phage Felix O1 using the following formula:

   Encapsulate efficiency (%) = (quantity of phage released from the microspheres/quantity of phage initially added to the mixture for preparing the microspheres) × 100.

5. Determine the survival rate of phage after drying as follows:

   Survival rate (%) = (number of phage (PFU per g wet microsphere) after drying/number of phage (PFU/g wet microsphere) before drying) × 100.

6.2 Spray-Dried Powder

1. Suspend 100 mg spray-dried microparticles in 10 mL of pre-warmed (39 °C) SM buffer adjusted to pH 7.5 and mix gently by inverting the tube for 30 s.

2. Incubate the sample at 39 °C for 120 min shaking at 150 rpm. Pipette 100 μL of samples at the end of 120 min followed by decimal dilutions up to 10^{-7} to determine surviving phage titer by rapid plaque assay.

3. Use the following formula to determine encapsulation efficiency of spray-dried particles (Fig. 1 and *see* **Note 11**).

   Encapsulation efficiency (%) = (quantity of phage released from the dissolved dried particles/quantity of phage added to the feed solution) × 100.

7 Resistance to Bile Salts, Simulated Gastric Fluid, and Release in Simulated Intestinal Fluid

7.1 Stability in Bile Salts

We have found that titer of free Felix O1 significantly decreases in 2% bile salt solution upon 1–3 h exposure but remained viable when protected by encapsulation. Hence, stability of the phages in the wet, dried microspheres or spray-dried powder was evaluated in simulated bile solution.

7.1.1 Free Phages

For free phage, dispense 100 μL of phage suspension (about 10^8 PFU/mL) into a culture tube with 9.9 mL of simulated bile fluid and incubate at 39 °C or 41.4 °C for 1 h and 3 h. Serially dilute the samples in SM buffer and determine phage survival by direct plating plaque count.

7.1.2 Encapsulated in Microspheres

1. Place 100 mg of microspheres into a tube with 9.9 mL of simulated bile fluid and incubate at 39 °C for 1 h and 3 h (*see* **Note 12**).

2. Afterward, collect the microspheres and wash with SM buffer. Dissolve the microspheres in 10 mL MBS and immediately assay the survival of released phage by plaque assay.

7.1.3 Encapsulated in Spray-Dried Powder

1. Take 100 mg spray-dried powder and add to 15 mL centrifuge tubes containing 9.9 mL of prewarmed (39 °C) bile solution (1 or 2%).

2. Incubate at 39 °C for 1 and 3 h while shaking gently.

3. At 1 and 3 h take the respective sample tubes and centrifuge at $9,820 \times g$ for 5–10 min. Discard the supernatant and replace with 9.9 mL of SM buffer pH 7.5. Incubate the tubes again at 39 °C for 1 h while shaking gently to dissolve the particles.

4. Draw 100 µL aliquots from each tube, serially dilute for enumeration of surviving phages by plaque assay (*see* **Note 13**).

7.2 Resistance to Simulated Gastric Fluid (SGF)

7.2.1 Free Phages

1. Add 100 µL of phage suspension (~10^8 PFU/mL) to 9.9 mL of prewarmed (39 °C or 41.1 °C) SGF solution (final pH of 3.8 and 4.4) (*see* **Note 14**).

2. Incubate for 5, 15, 30, 60, 90, and 120 min at 39 °C or 41.1 °C.

3. Pipette out 100 µL aliquots, serially dilute and immediately assay for phage survival.

7.2.2 Encapsulated Phages in Microspheres

1. Weigh approximately 250 mg of microspheres and add it to a test tube containing 9.75 mL SGF (final pH 2.0 and 2.4 or pH 3.8 and 4.4) prewarmed to desired temperature (41.1 °C).

2. Incubate at the same temperature for up to 120 min.

3. At predetermined time intervals, such as 30, 60, and 90 min, drain the SGF solution; dissolve the microspheres in 2.25 mL MBS solution.

4. Transfer 100 µL aliquots and immediately conduct plaque assay to enumerate viable phages (*see* **Note 15**).

5. Record the phage titer and compare with the initial phage titer in the microspheres before the SGF treatment (Fig. 2).

7.2.3 Encapsulated Phages in Spray-Dried Powder Particles

1. Add 100 mg spray-dried powder to a 15-mL centrifuge tube containing prewarmed (at 39 °C) SGF adjusted to pH 2.0 and/ or 2.4.

2. Incubate the tube at the same temperature for 2 h.

3. At 5, 15, 30, 60, 90, and 120 min collect 100 µL aliquots in Eppendorf tubes, centrifuge the tubes at $9,820 \times g$ for 3–5 min, drain the SGF solution and replace with 9.9 mL of SM buffer pH 7.5. Incubate the tubes again at 39 °C while

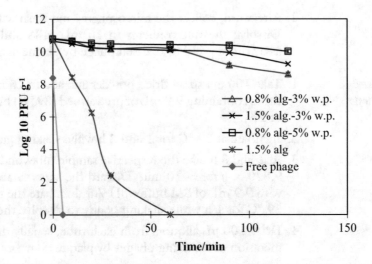

Fig. 2 Survival of phage Felix 01 encapsulated in alginate (alg) and whey protein (w.p.) microspheres during incubation in simulated gastric fluid at pH 2.0. Each value in the figure represents the mean $\pm$ SE ($n = 3$)

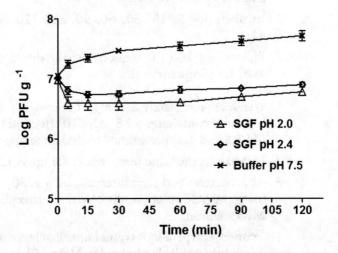

Fig. 3 Survival of phage Felix 01 encapsulated in spray-dried powders during incubation in simulated gastric fluid at pH 2.0 and pH 2.4. SM buffer at pH 7.5 was used as a control

shaking gently at 150 rpm to dissolve the particles for 1 h. (*see* **Note 16**).

4. Withdraw 100 µL aliquots, serially dilute to enumerate the number of viable phages by plaque assay (Fig. 3).

7.3 Release of Phages in Simulated Intestinal Fluid

Phage encapsulation strategies are intended to protect them from adverse gastrointestinal condition especially gastric acid and delivering the phages in the intestinal tract by dissolution at neutral pH. Hence it is very important to determine release of the phages from the encapsulation matrices in simulated intestinal condition.

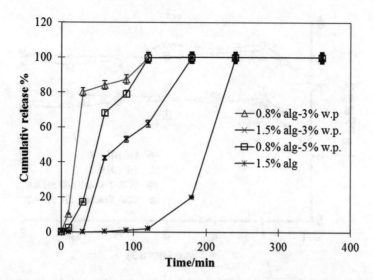

Fig. 4 Release profile of biologically active phage Felix O1 from alginate (alg) and whey protein (w.p.) microspheres in simulated intestinal fluid at pH 6.8. Each value in the figure represents the mean $\pm$ SE ($n = 3$)

7.3.1 From Alginate–Whey Microspheres

1. Weigh 250 mg of microspheres and add it to a test tube containing 9.75 mL SGF prewarmed to desired temperature (41.1 °C) and incubate at the same temperature for 2 h.

2. Drain the microspheres, add 19.75 mL SIF pH 6.8, and incubate at the same temperature while shaking at 100 rpm.

3. At predetermined time points, such as 0.5, 1, 2, 3, and 4 h, remove 100 µL of this solution and immediately assay for phage titer. Replace the same volume of the fresh SIF withdrawn at each time point.

4. Plot the cumulative amount of released phage against time in order to get phage release profile (Fig. 4 and *see* **Note 17**).

7.3.2 From Spray-Dried Powder Particles

1. Add 100 mg of the spray-dried powder to a 15 mL centrifuge tube containing 10 mL prewarmed SGF pH 2.0 or 2.4 (at 39 °C) and incubate at the same temperature while shaking at 150 rpm.

2. After 2 h centrifuge the tube at 9,820 × *g* for 5–10 min and discard the supernatant.

3. Add 9.9 mL of prewarmed (at 39 °C) SIF adjusted to pH 6.8 or pH 7.2 and incubate for 6 h at 39 °C with shaking at 150 rpm.

4. Every 1 h, transfer 100 µL of samples into Spin-X centrifuge filter tubes and centrifuge at 11,000 × *g* for 1–2 min. Replace the same volume of the fresh SIF withdrawn at each time point (*see* **Note 18**).

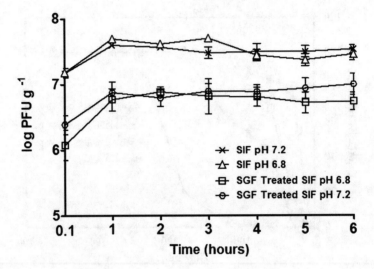

Fig. 5 Release profile of Felix 01 encapsulated in spray dried powders incubated in simulated intestinal fluid (SIF) at pH 6.8 and 7.2. Samples were treated with simulated gastric fluid at pH 2.0 for 2 h prior to subjecting in SIF. Samples which were not treated with SGF were used for comparison

5. To the filtrate, add 900 μL of SM buffer and serially dilute to carry out plaque assay. Plot phage titer against each time point to plot release profile (Fig. 5).

8   Notes

1. The instrument equipped with a series of nozzles in the size range of 0.08–1.0 mm. As a rule of thumb, the size of microspheres is approximately two times the nozzle size used to produce them; it decreases with the frequency of electrical current applied. The sizes of microspheres are also affected by the total polymer concentration in the encapsulation formula. In general, the higher of the polymer concentration, the larger of the microsphere will be. Repeated adjustment of these parameters will lead to the desired microsphere size.

2. According to United States Pharmacopeia 33-28NF (2010), the pH of SGF is 1.2, which is intended to represent stomach acid in a fasted state of healthy humans. The pH of gastric fluid of a monogastric animal can range from 1.2 to 5.0 depending on the fed state.

3. There is no specific requirement for phage titer to be encapsulated. It is mainly determined by the target applications. High titer of 10^{11} pfu/mL was encapsulated with negligible loss of phage using this protocol [14].

4. We have initially tested various ratios and concentrations of alginate and whey protein in the encapsulation mixture

[8, 14]. The two polymer concentrations can be selected to suit for intended applications. Both the protection effect and the release rate of phage from the microspheres are influenced by varying the formulation, as shown in Subheadings 7.2.2 and 7.3.1 of this chapter.

5. We have also tested other protectants, among which maltose, which also provides good protection to phage Felix O1, whereas sucrose, maltodextrin and skim milk showed good protection to phage K [15].

6. No loss of viability is observed for phage Felix O1 in such dried microspheres when stored at 4 °C for 2 weeks.

7. It is possible to choose different enteric polymers that release the encapsulated phage at different pH levels according to the applications.

8. Adjust the spray parameters if outlet temperature increases to higher than 57–60 °C as the dried particles loaded with phages remain exposed in the outlet temperature till the spray drying process is completed. Prolonged exposure to higher temperature may reduce survival of certain phages. SEM micrographs (Fig. 1) revealed spray-dried powder particle of varying sizes ranging 350 nm to 15 μm.

9. Plaque assay can also be performed at desired time periods in order to determine the titer (survival rate) of encapsulated phage upon storage. Felix O1 encapsulated in spray-dried particles did not show loss of viability for 6 months when stored at 4 °C.

10. Use 5× or lower volume of MBS for the samples subjected to storage stability assessment and where low phage survival is expected.

11. The final beads contained up to ~1 × 10^{11} PFU/(g of wet beads).

12. Determine incubation temperature based on the end application. Normal human body temperature is 37 °C while average body temperature of a chicken is 41–45 °C and 38–40 °C for a pig. Our application was focused on delivering the phages in chicken and pig digestive system; hence incubation temperature was chosen around the range of body temperature of the test animals (41.1 and 39 °C, respectively).

13. Survival of phages per g of spray-dried powder found to be ~7.71 ± 0.16 log pfu while the initial loading of Felix O1 was ~8 log pfu / g in the encapsulation mixture; approximately 96.3% encapsulation efficiency was achieved.

14. Our initial results indicated that pH 2.0 and 2.5 is not conducive for free phage Felix O1 survival, hence for the comparative

study, experiments were also conducted at higher pH 3.8 and 4.4.

15. A minimum of three replicates are required for each sample at each time interval. When encapsulated in the current formulation, less than 0.5 log pfu reduction is observed after 2 h incubation in SGF at pH 2.0, whereas at pH 2.5, the viability of phage was fully maintained. The protection effect varies with the formulation. In general, higher total polymer concentration favors a better protection as shown in Fig. 2. However, a further increase in polymer concentration is restricted by the increasing viscosity of the solution.

16. A small amount of residual SGF remaining at the bottom of the tube can lower the pH of the SM buffer which is added to dissolve the spray-dried powder. Prior to adding the SM buffer solution, make sure to pipette out as much SGF as possible without agitating the pelleted particles. After adding SM buffer, check the pH of the solution and, if required, adjust to pH 7.5. Subjecting the spray-dried particles to either SGF pH 2.0 or pH 2.4 for 2 h resulted in ~0.29 and 0.23 log pfu reduction in phage viability respectively [16] (Fig. 3).

17. The release rate of phage from the microsphere is determined by the formulation and size of the microsphere (Fig. 4). Studies have clearly demonstrated that high polymer concentration tends to retard the release of encapsulated component from alginate based microspheres [8, 14].

18. This step is important to ensure that only the phages released from the spray-dried particles during the incubation period are plated while preventing any undissolved particles that were unintentionally pipetted out.

References

1. Boerlin P (2010) Implications of antimicrobial agents as therapeutics and growth promoters in food animal production. In: Sabour PM, Griffiths MW (eds) Bacteriophages in the control of food- and waterborne pathogens. ASM Press, Washington, DC, pp 1–9

2. Antibiotic resistance - Milestones, http://www.efsa.europa.eu/en/topics/topic/amr.htm

3. Goodridge LD (2010) Designing phage therapeutics. Curr Pharm Biotechnol 11(1):15–27

4. Goodridge LD, Bisha B (2011) Phage-based biocontrol strategies to reduce foodborne pathogens in foods. Bacteriophage 1(3):130–137

5. Lee N, Harris DL (2001) The effect of bacteriophage treatment as a Preharvest intervention strategy to reduce the rapid dissemination of salmonella typhimurium in pigs. American Association of Swine Veterinarians (AASV), Perry, IA: AASV, pp 555–557

6. Ma Y, Pacan JC, Wang Q, Xu Y, Huang X, Korenevsky A, Sabour PM et al (2008) Microencapsulation of bacteriophage Felix O1 into chitosan-alginate microspheres for oral delivery. Appl Environ Microbiol 74(15):4799–4805

7. Wang Q, Sabour PM (2010) Encapsulation and controlled release of bacteriophages for food animal production. In: Sabour PM, Griffiths MW (eds) Bacteriophages in the control of food- and waterborne pathogens. ASM Press, Washington, DC, pp 237–255

8. Tang ZX, Huang X, Sabour PM, Chambers JR, Wang Q et al (2015) Preparation and characterization of dry powder bacteriophage K for intestinal delivery through oral administration. Food Sci Tech – LWT 60(1):263–270

9. Whichard JM, Sriranganathan N, Pierson FW et al (2003) Suppression of salmonella growth by wild-type and large-plaque variants of bacteriophage Felix O1 in liquid culture and on chicken frankfurters. J Food Prot 66:220–225

10. Sambrook J, Fritsch EF, Maniatis T (1989) Molecular cloning: a laboratory manual, 2nd edn. Cold Spring Harbor Laboratory Press, Cold Spring Harbor, NY

11. Kropinski AM, Mazzocco A, Waddell TE, Lingohr E, Johnson RP (2009) Enumeration of bacteriophages by double agar overlay plaque assay. In: Clokie M RJ, Kropinski AM (eds) Bacteriophages: methods and protocols, vol. 1: isolation, characterization, and interactions. Humana Press, New York, NY, p 69

12. Kropinski AM, Mazzocco A, Waddell TE, Lingohr E, Johnson RP (2009) Enumeration of bacteriophages by double agar overlay plaque assay. In: RJ CM, Kropinski AM (eds) Bacteriophages: methods and protocols, vol. 1: isolation, characterization, and interactions. Humana Press, New York, NY, p 69

13. Kropinski AM, Mazzocco A, Waddell TE, Lingohr E, Johnson RP (2009) Enumeration of bacteriophages using the small drop plaque assay system. In: RJ CM, Kropinski AM (eds) Bacteriophages: methods and protocols, vol. 1: isolation, characterization, and interactions. Humana Press, New York, NY, p 81

14. Tang ZX, Huang X, Baxi S, Chambers JR, Sabour PM, Wang Q et al (2013) Whey protein improves survival and release characteristics of bacteriophage Felix O1 encapsulated in alginate microspheres. Food Res Int 52:460–466

15. Ma Y, Pacan JC, Wang Q, Sabour PM, Huang X, Xu Y et al (2012) Enhanced alginate microspheres as means of oral delivery of bacteriophage for reducing Staphylococcus aureus intestinal carriage. Food Hydrocoll 26 (2):434–440

16. Islam GS, Wang Q, Sabour PM, Warriner K et al (2012) Microencapsulation of bio-control bacteriophage by spray drying method. In: 20th international conference on bioencapsulation, Orillia, ON, Canada, 21–24 Sept 2012

Chapter 7

Encapsulation of *Listeria* Phage A511 by Alginate to Improve Its Thermal Stability

Hanie Ahmadi, Qi Wang, Loong-Tak Lim, and S. Balamurugan

Abstract

Microencapsulation is a versatile method for enhancing the stability of bacteriophages under harsh conditions, such as those which occur during thermal processing. For food applications, encapsulation in food-grade polymer matrices is desirable owing to their nontoxicity and low cost. Here, we describe the encapsulation of *Listeria* phage A511 using sodium alginate, gum arabic, and gelatin to maximize its viability during thermal processing.

Key words Gelatin, Alginate, Gum arabic, Microencapsulation, Bacteriophages, *Listeria* phage A511, Thermotolerance

1 Introduction

Bacteriophages are generally sensitive to temperatures greater than 60 °C. In order to successfully apply them as a preservative agent in foods, especially those that are heat-treated, the viruses must be stabilized against harsh thermal processes. Microencapsulation (ME) is a versatile technique for stabilizing bioactives within polymeric solid, liquid or semisolid miniature particles [1]. This technique is effective for protecting bioactives from deleterious effects of food-processing, as well as reducing their undesirable interactions with the environment [2]. Various ME techniques have been used to encapsulate bioactive compounds. Spray-drying has been commonly used for the encapsulation of oil soluble vitamins and fatty acids [3, 4]. Emulsion and electrospining have also shown the potential for encapsulation of omega-3 fatty acids, yeasts, and enzymes [5–7]. However, ME technologies applied to live microorganisms are generally limited [8].

Microencapsulation using gel particles is promising for protecting viable microorganism. Among the various polymers available as encapsulant material, alginate is by far the most useful for the encapsulation of viable cells [9]. For example, alginate has been

Martha R.J. Clokie et al. (eds.), *Bacteriophages: Methods and Protocols, Volume 3*, Methods in Molecular Biology, vol. 1681, https://doi.org/10.1007/978-1-4939-7343-9_7, © Springer Science+Business Media LLC 2018

extruded into a calcium chloride solution to encapsulate live micro-organisms. Starch and whey proteins have been blended with algi-nate in order to improve the encapsulation efficacy [10, 11]. ME with alginate has been applied to enhance the viability of probiotic bacteria for oral delivery and in simulated gastric juices and a bile salt solution [12, 13].

In several studies, bacteriophages have been encapsulated for effective delivery to the lower intestines of animals for control of bacterial pathogen [14–17]. Tully et al. proposed the use of encap-sulation of bovine herpesvirus-1 with anionic polymer-amine mem-brane for delivering vaccine orally [18]. Ma et al. reported a method to microencapsulate bacteriophage Felix O1 for oral deliv-ery [14]. In another study, T4 bacteriophage was incorporated into core/shell electrospun fibers made from poly ethylene oxide (PEO), cellulose diacetate (CDA), and their blends for potential uses as food packaging materials [19]. However, encapsulation of phage for enhancing viability during thermal processing has not been reported in literature.

Here, we report a process for the microencapsulation of *Lis-teria* phage A511 in sodium alginate gel particles by extruding the phage-containing alginate solution into calcium chloride solution. In this approach, gum arabic and gelatin are blended with alginate in order to improve the thermal stability of the *Listeria* phage. Encapsulating *Listeria* phage A511 by this method significantly improves the thermal stability of the phage compared to free phage.

2   Materials

2.1 Personal Protective Equipment

1. Disposable gloves.
2. Lab coat.
3. Biological safety cabinet (BSC).
4. Appropriate footwear.

2.2 Equipment

1. Refrigerators set at 4 °C.
2. Incubator and shaking incubator set at 37°C.
3. Magnetic stir bar heater.
4. Pipettor and pipettes.
5. Micropipettors and pipette tips.
6. Stomacher® 80 *micro*Biomaster lab blender (Seward Labora-tory Systems, Inc. Bohemia, NY, USA) and Stomacher® 80 bags.
7. 10 mL sterile, disposable syringe.
8. Disposable needles.
9. Laboratory film (Parafilm®, Oshkosh, USA).

10. Sterile disposable loops.

11. Water baths for tempering media (50 °C).

12. Vortex.

2.3 Reagents

1. *Listeria* phage A511 (*see* **Note 4.1**).

2. *Listeria monocytogenes* 08-5578 (serotype 1/2a) (*see* **Note 4.2**).

3. Tryptic Soy Broth (TSB; BD Biosciences, San Jose, CA, U.S. A.) prepared according to manufacturer's directions.

4. Tryptic Soy Agar (TSA; BD Biosciences) prepared according to manufacturer's directions.

5. Overlay TSA agar (*see* **Note 4.3**).

6. Saline-Magnesium (SM) buffer (10 mM NaCl, 10 mM $MgSO_4$, 50 mM Tris–HCl, pH 7.5).

7. 1 M Calcium chloride ($CaCl_2$).

8. 0.1 M Calcium chloride ($CaCl_2$).

9. 0.1 M Phosphate buffer solution (pH 7).

10. Sodium alginate, from brown algae (Sigma-Aldrich, Steinheim, Germany), Viscosity of 2% solution at 25 °C, 250 cps.

11. Gelatin, Type B: from bovine skin, Approximately 75 bloom (Sigma-Aldrich).

12. Gum arabic, from acacia tree (Sigma-Aldrich).

13. Distilled deionized (Nanopure®) water.

3 Methods

3.1 Encapsulation Solution: Alginate 3% (w/w), gum Arabic 1% (w/w), Gelatin 1% (w/w) Solution (100 g)

1. Weight 3 g of alginate in a 300 mL beaker (*see* **Note 4.4**).

2. Add 1 g of gum Arabic and 1 g of gelatin to the same beaker.

3. Adjust the weight to 100 g by adding distilled water.

4. Seal the top of beaker using Parafilm®.

5. Dissolve the ingredients in water by mixing on a magnetic stir bar heater (80 °C).

6. When completely dissolved turn the heater off to cool to room temperature with stirring.

3.2 Encapsulation of Phage (Fig. 1)

1. Transfer 49 mL of the encapsulation solution to a 300 mL beaker using 50 mL pipette.

2. Add 1 mL of A511 phage (10^9 PFU/mL).

3. Mix suspension using magnetic stirrer for 10 min (*see* **Note 4.5**).

4. Transfer the suspension into 10 mL syringe with a 16-gauge needle.

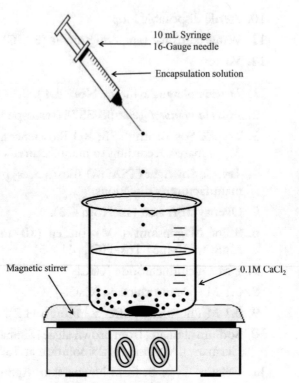

10 mL Syringe
16-Gauge needle

Encapsulation solution

Magnetic stirrer

0.1M CaCl$_2$

Fig. 1 Encapsulating phage A511 by alginate beads in CaCl$_2$

5. Place a 300 mL beaker on magnetic stir plate and add 200 mL of sterile 0.1 M CaCl$_2$.

6. Inject the phage suspension through the needle by applying gently pressing the plunger as to generate droplets into the beaker with 0.1 M CaCl$_2$ to form beads (*see* **Note 4.6**).

7. Allow to stand for 1 h for solidification.

8. Decant the supernatant and rinse the beads with sterile water three times.

9. Collect the beads by draining or filtration.

10. Add 5 mL SM buffer to keep the beads moist.

11. Seal and store at 4 °C.

3.3 Phage Release and Enumeration

1. Suspend 1 g of beads in 9 mL of sterile phosphate buffer solution in stomacher bag.

2. Homogenize suspension using a Stomacher at normal setting for 10 min (*see* **Note 4.7**).

3. Perform appropriate serial dilution of homogenate in SM buffer.

4. Perform the overlay assay as described by Kropinski et al. [20] (*see* **Note 4.8**).

5. Incubate the plated samples at 30 °C overnight (*see* **Note 4.9**).

6. Count the plaques on plates with 30–300 plaques.

---

4 Notes

1. Phage Propagation:

   Phage A511 was propagated with the bacterial host; *Listeria monocytogenes* 08-5578 (serotype 1/2a) as described by Radford et al. [21].

   (a) Inoculate 5 mL of tryptic soy broth (TSB) with a single colony of *L. monocytogenes* and incubate at 37 °C for 18 h at 120 rpm.

   (b) Mix 200 µL of above culture (10^9 CFU/mL) and 100 µL of phage A511 (10^9 PFU/mL) with 4 mL of molten top agar, supplemented with $CaCl_2$ (TSB, 5% granulated agar, 10 mM $CaCl_2$).

   (c) Gently mix the solution to avoid generation of air bubbles and pour onto sterile tryptic soy agar (TSA) plates.

   (d) Incubate the plates at 30 °C for 18 h to form a top agar layer of phage-host coculture.

   (e) Add 5 mL of SM buffer to the plates after the 18 h of incubation to cover the surface of top agar entirely and refrigerated at 4 °C overnight.

   (f) Extract all the liquid after refrigeration time using a micropipette and then filter using a 0.2 µm sterile syringe filter (VWR International, Texas, USA)

   (g) Retain the filtrate at 4 °C until required.

   (h) Determine the titer of propagated phages by overlay assay as described by Kropinski et al. [20].

2. Bacterial culture

   (a) Pipet 5.0 mL of sterile TSB into 15 mL screw cap, sterile falcon tubes.

   (b) Inoculate with a single colony of *L. monocytogenes* 08-5578 (serotype 1/2a).

   (c) Incubate the tube at 37 °C and 120 rpm overnight.

3. Top agar

   (a) Dissolve 30 g of TSB in 1 L of distilled water (as per manufacturer's direction).

   (b) Add 5.0 g of granulated agar.

   (c) Dissolve ingredients by heating on a magnetic stir plate heater to mix.

      (d) Add 10 mL of 1 M $CaCl_2$ solution (10 mM final concentration).

      (e) Autoclave for 20 min at 121 °C.

      (f) Aliquot 200 mL and cool.

      (g) Store at 4 °C.

4. Using sodium alginate with higher viscosity or higher concentration gives harder beads.

5. Mix on low setting (but well) to prevent phage damage.

6. To better control the bead size, consider using an encapsulator. Parameters that affect the beads size include polymer concentration, spray voltage, vibration frequency, and flow rate. Higher vibration frequency can help to produce smaller beads. To obtain beads with good morphology (I.E., round in shape and uniform in size), the alginate concentration must be >1.5% (w/w).

7. If particles are visible in stomacher bag, continue to stomach the content for two or more minutes.

8. Phage enumeration by overlay assay technique.

      (a) Label TSA underlay plates.

      (b) Melt the overlay agar.

      (c) Transfer 3 mL of overlay agar to glass tube.

      (d) Place tubes in 50 °C water bath.

      (e) Transfer 100 μL of host bacteria and 100 μL of homogenate dilution to glass tube by sterile pipette.

      (f) Vortex.

      (g) Transfer content over the surface of TSA plate.

      (h) Incubate overnight at 30 °C.

9. Retrieve plate from incubator overnight. Incubating plates for longer times will result in larger plaques, which may be more difficult to enumerate.

Acknowledgment

This work was supported by Agriculture and Agri-Food Canada.

References

1. Lim L-T (2015) Encapsulation of bioactive compounds using electrospinning and electrospraying technologies, Nanotechnology and functional foods. Wiley, Hoboken, NJ, pp 297–317

2. Hussain MA, Liu H, Wang Q, Zhong F, Guo Q, Balamurugan S (2017) Use of encapsulated bacteriophages to enhance farm to fork food safety. Crit Rev Food Sci Nutr 57:2801–2810

3. Kolanowski W, Laufenberg G, Kunz B (2004) Fish oil stabilisation by microencapsulation with modified cellulose. Int J Food Sci Nutr 55(4):333–343

4. Hategekimana J, Masamba KG, Ma J, Zhong F (2015) Encapsulation of vitamin E: effect of physicochemical properties of wall material on retention and stability. Carbohydr Polym 124:172–179

5. Chowdhuri S, Cole CM, Devaraj NK (2016) Encapsulation of living cells within giant phospholipid liposomes formed by the inverse-emulsion technique. Chembiochem 17 (10):886–889

6. Zhou Y, Lim LT (2009) Activation of lactoperoxidase system in milk by glucose oxidase immobilized in electrospun polylactide microfibers. J Food Sci 74(2):C170–C176

7. Moomand K, Lim L-T (2015) Properties of encapsulated fish oil in electrospun zein fibres under simulated in vitro conditions. Food Bioprocess Technol 8(2):431–444

8. Kailasapathy K (2002) Microencapsulation of probiotic bacteria: technology and potential applications. Curr Issues Intest Microbiol 3 (2):39–48

9. Gasperini L, Mano JF, Reis RL (2014) Natural polymers for the microencapsulation of cells. J R Soc Interface 11(100):20140817

10. Dembczynski R, Jankowski T (2002) Growth characteristics and acidifying activity of *Lactobacillus rhamnosus* in alginate/starch liquid-core capsules. Enzym Microb Technol 31 (1–2):111–115

11. Gerez CL, Font de Valdez G, Gigante ML, Grosso CR (2012) Whey protein coating bead improves the survival of the probiotic *Lactobacillus rhamnosus* CRL 1505 to low pH. Lett Appl Microbiol 54(6):552–556

12. Jiang T, Singh B, Maharjan S, Li H-S, Kang S-K, Bok J-D et al (2014) Oral delivery of probiotic expressing M cell homing peptide conjugated BmpB vaccine encapsulated into alginate/chitosan/alginate microcapsules. Eur J Pharm Biopharm 88(3):768–777

13. Lee KY, Heo TR (2000) Survival of *Bifidobacterium longum* immobilized in calcium alginate beads in simulated gastric juices and bile salt solution. Appl Environ Microbiol 66 (2):869–873

14. Ma Y, Pacan JC, Wang Q, Xu Y, Huang X, Korenevsky A et al (2008) Microencapsulation of bacteriophage felix O1 into chitosan-alginate microspheres for oral delivery. Appl Environ Microbiol 74(15):4799–4805

15. Annan NT, Borza AD, Hansen LT (2008) Encapsulation in alginate-coated gelatin microspheres improves survival of the probiotic *Bifidobacterium adolescentis* 15703T during exposure to simulated gastro-intestinal conditions. Food Res Int 41(2):184–193

16. Stanford K, McAllister TA, Niu YD, Stephens TP, Mazzocco A, Waddell TE et al (2010) Oral delivery systems for encapsulated bacteriophages targeted at *Escherichia coli* O157:H7 in feedlot cattle. J Food Prot 73(7):1304–1312

17. Tang Z, Huang X, Baxi S, Chambers JR, Sabour PM, Wang Q (2013) Whey protein improves survival and release characteristics of bacteriophage Felix O1 encapsulated in alginate microspheres. Food Res Int 52 (2):460–466

18. Speaker TJ, Clark HF, Moser CA, Offit PA, Campos M, Frenchick PJ (2001) Aqueous solvent based encapsulation of a bovine herpes virus type-1 subunit vaccine. Patent EP0873752B1, European Patent register.

19. Korehei R, Kadla JF (2014) Encapsulation of T4 bacteriophage in electrospun poly(ethylene oxide)/cellulose diacetate fibers. Carbohydr Polym 100:150–157

20. Kropinski AM, Mazzocco A, Waddell TE, Lingohr E, Johnson RP (2009) Enumeration of bacteriophages by double agar overlay plaque assay. Methods Mol Biol 501:69–76

21. Radford DR, Ahmadi H, Leon-Velarde CG, Balamurugan S (2016) Propagation method for persistent high yield of diverse *Listeria* phages on permissive hosts at refrigeration temperatures. Res Microbiol 167(8):685–691

Chapter 8

Application of a Virucidal Agent to Avoid Overestimation of Phage Kill During Phage Decontamination Assays on Ready-to-Eat Meats

Andrew Chibeu and S. Balamurugan

Abstract

We describe a method for assessing the effectiveness of tea extract based virucide (TeaF) application to remove phage LISTEX™ P100 not bound to *Listeria monocytogenes* from stomached rinses prior to direct plating and bacterial enumeration, where the phage is being used as a decontaminant to reduce *L. monocytogenes* levels on ready-to-eat meat.

Key words Bacteriophage, Virucide, Tea extract, LISTEX™ P100, *Listeria monocytogenes*, Ready-to-eat meat

1 Introduction

Bacteriophages are effective in the control of *Listeria monocytogenes* on food [1–7]. Two bacteriophage preparations (LISTEX™ P100 and LMP-102) were approved by the United States Food and Drug Administration to control *L. monocytogenes* contamination in a select number of foods [8–10]. Health Canada issued a letter of no objection for application of LISTEX™ P100 as a processing aid against *L. monocytogenes* in deli meat and poultry products, cold-smoked fish, vegetable prepared dishes, soft cheeses, and/or other dairy products [11]. The Food Standards Australia/New Zealand (FSANZ) also approved LISTEX™ P100 as a processing aid to reduce contamination of *L. monocytogenes* in a variety of foods [12].

Guenther et al. [4] pointed out that phage-based pathogen intervention in ready-to-eat (RTE) products greatly depends on the phage dose applied, the chemical composition of the food and its specific matrix. The study further suggested that there is a need to individually optimize protocols for the application of phages with respect to the viruses and the target bacteria as well as considering the food matrix. In addition to these considerations,

Martha R.J. Clokie et al. (eds.), *Bacteriophages: Methods and Protocols, Volume 3*, Methods in Molecular Biology, vol. 1681, https://doi.org/10.1007/978-1-4939-7343-9_8, © Springer Science+Business Media LLC 2018

researchers should avoid common sources of errors that negate their study assays depicting actual phage effect on bacterial cell numbers on RTE product in a real life setting. One common source of error is the overestimation of the bacteriophage killing effect due to determining bacterial viable counts without removal of unbound phages from stomached rinses prior to direct plating to enumerate the number of remaining viable bacteria. The rationale of including this additional phage removal step is to prevent the phages which had not come into contact with the bacterial cells on the meat surface from interacting with the bacteria during stomaching and plating, thus causing additional lysis to cells more than what occurred on the food surface. Chibeu et al. [6] put into consideration this source of error and applied the use of the virucidal agent to remove unbound phage particles from stomached rinses prior to plating in a study on phage use in decontamination of RTE meats.

Jassim et al. [13] developed a virucidal agent which comprised of a mixture of a ferrous sulfate and a plant extract of *Punica granatum* (pomegranate) rind, *Viburnum plicatum* (Japanese snowball) leaves or flowers, *Camellia sinensis* (tea) leaves, or *Acer saccharum* (sugar maple) leaves in aqueous solution. This virucidal agent destroyed phages of various species without damaging metabolically active bacterial cells, by the activation of Fe^{2+} ions with the plant extracts. The tea with ferrous sulfate (TeaF) virucidal agent has been used previously as specific inhibitor of phage during phage amplification assays [14] and, to decrease phage populations in soils to determine how bottom-up and top-down controls differentially affected microbial respiration in situ [15]. More recently, Helsley et al. [16] studied TeaF as a virucidal agent to assess the role of phage predation in soils.

The effectiveness of each virucide in removal of external phages in phage amplification assays is variable based on the phage type [14, 17]. Therefore, in phage decontamination of ready-to-eat meats, it is necessary to ascertain the efficacy of the virucidal agent before use to avoid overestimation of phage kill.

2   Materials

2.1 Personal Protective Equipment

1. Disposable gloves.
2. Lab coat.
3. Disposable solid-front gown.
4. Biological safety cabinet (BSC).
5. Appropriate footwear.

2.2 Equipment

1. Refrigerators set at 4 °C and 10 °C.
2. Incubator and shaking incubator set at 37 °C.

3. Styrofoam meat trays (Dyne-A-Pak Inc., Laval, QC Canada; http://www.dyneapak.com/en/).

4. 8″ × 6″ commercial barrier bags [oxygen transmission rate: 40–50 cc m^{-2} 24 h^{-1}; (Winpak Ltd., Winnipeg, MB, Canada; http://www.winpak.com/)].

5. Chamber machine C200 (MULTIVAC AGI, Knud Simonsen Industries Ltd., Rexdale, ON, Canada).

6. Sterile disposable spreaders (Arben Bioscience Inc., Rochester, NY, USA; Catalogue number KG-5P; http://www.arbenbio.com/).

7. Pipettor and pipettes.

8. Micropipettors and pipette tips.

9. Meat core cutter (Custom made to cut RTE meat slices into 10 cm^2 cores; alternatively can use a biscuit cutter, e.g., Endurance®, RSVP International, Inc. Seattle, WA, USA; http://www.rsvp-intl.com/).

10. Steel plate work surfaces precooled to 4 °C (Custom made).

11. Stomacher® 80 *micro*Biomaster lab blender (Seward Laboratory Systems, Inc. Bohemia, NY, USA; http://seward.co.uk/) and Stomacher® 80 bags.

12. Benchtop centrifuge with swinging bucket rotor (Eppendorf 5804 R; Westbury, NY, USA).

13. Sealable Tupperware® containers.

14. Autoclavable plastic Nalgene™ buckets (Thermo Fisher Scientific Inc., Waltham, MA, USA).

15. 0.45 μm syringe filter.

16. 10 mL sterile, disposable syringe.

17. Whatman™ filter paper (Grade 4; 20–25 μm).

18. Spectrophotometer.

19. Disposable cuvettes with lids.

20. Water baths for tempering media (42 and 50 °C).

21. Vortex.

2.3 Reagents

1. LISTEX™ P100 (Micreos Food Safety B.V., Wageningen, Netherlands; http://www.micreos.com/) (*see* **Note 1**).

2. *Listeria monocytogenes* cocktail (must consist of 1/2a, 1/2b, and 4b serotypes) (*see* **Notes 2** and **3**).

3. Phosphate Buffered Saline (PBS; 100 mM NaCl, 20 mM Na$_2$HPO$_4$, pH 7.4).

4. Tryptic Soy Broth (TSB; BD Biosciences, San Jose, CA, USA; http://m.bdbiosciences.com/) prepared according to manufacturer's directions.

5. Sliced, ready-to-eat meat (roast beef or cooked turkey).

6. 5 M HCl.

7. Tryptic Soy Agar (TSA; BD Biosciences, San Jose, CA, USA) prepared according to manufacturer's directions.

8. Oxford Agar (EMD Chemicals Inc., Gibbstown, NJ, USA; http://www.emdmillipore.com) prepared according to manufacturer's directions.

9. Black loose-leaf tea (Kenyan Tinderet from Davids Tea, Montreal, QC, Canada or equivalent).

10. SM buffer (10 mM NaCl, 10 mM $MgSO_4$, 50 mM Tris–HCl, pH 7.5).

11. Ferrous sulfate ($FeSO_4 \cdot 7H_2O$).

12. Distilled deionized (Nanopure®) water.

3   Methods

3.1 Virucidal Solution Preparation

Prepared 1–2 min prior to use and protected from sunlight.
Preparation of tea infusion (can be done in advance)

1. Add 7.5% w/v loose-leaf black tea to Nanopure® water and boil for 10 min.

2. Filter the infusion using Whatman™ paper (Whatman™ International Ltd., Ipswich, UK).

3. Sterilize the filtered solution by autoclaving at 121 °C for 15 min, cool and store at 4°C.

4. **Preparation of 4.3 mM $FeSO_4$ in SM buffer:** Freshly prepare 0.53% $FeSO_4$ stock solution. (0.053 g $FeSO_4 \cdot 7H_2O$ in 10 mL SM buffer) and sterilize by membrane filtration (0.45 μm). Transfer 4.1 mL of 0.53% sterile $FeSO_4$ stock solution to a sterile test tube containing 14 mL sterile SM buffer to prepare 4.3 mM $FeSO_4$ solution.

5. Prepare virucidal solution by mixing 330 μl of sterile tea infusion with 700 μl of freshly prepared 4.3 mM sterile $FeSO_4$.

6. Any unused 0.53% stock solution cannot be stored for future use and must be discarded.

3.2 Ready-to-Eat Meat Sample Preparation (Fig. 1)

1. Obtain freshly sliced meat products direct from the processing facility and store in tightly sealed Tupperware® containers at 4 °C until ready to use.

2. Place fresh, refrigerated samples on the precooled steel block work surfaces (wrapped with clean aluminum foil and refrigerated to 4 °C).

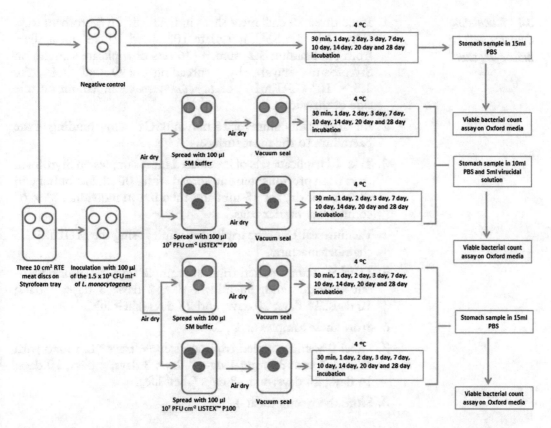

Fig. 1 Sample preparation flowchart

3. Using autoclave sterilized meat core cutter or stainless steel cookie cutters, cut 135 uniform slices of meat with 10 cm² top-surface area.

4. Discard the remaining meat remnants in a biohazard waste bag.

5. Return all sliced meat samples to the Tupperware® container at 4 °C while not in use.

3.3 Preparation of Negative Controls

1. Place three 10 cm² meat slices individually on Styrofoam trays in the BSC and put in individual 8″ × 6″ commercial barrier bags (Fig. 1).

2. Vacuum seal 9 of the bags using the MULTIVAC chamber machine.

3. Store 9 of the vacuum sealed triplicate samples at 4°C and label them "negative control" and: 30 min; 1 day; 2 days; 3 days; 7 days; 10 days; 14 days; 20 days; and 28 days (shelf life) (Fig. 1).

3.4 Inoculating Samples with L. monocytogenes

1. Place three 10 cm^2 meat slices individually on Styrofoam trays (Fig. 1) in the BSC, inoculate 108 sliced 10 cm^2 meat slices from Subheading 3.2, **step 3** (36 sets of triplicate samples on Styrofoam trays) by spreading 100 μl of the 1.5×10^3 CFU mL^{-1} of *L. monocytogenes* inoculum on one side of the slice.

2. Air-dry the inoculum for 15 min in BSC to allow binding of the bacterium to the meat surface.

3. Take 18 triplicate sets of inoculated meat samples on Styrofoam trays from previous step and spread with 100 μL SM buffer and allow to air dry for 15 min and then put in individual 8″ × 6″ commercial barrier bags.

4. Vacuum seal 18 of the triplicate samples using the MULTIVAC chamber machine.

5. Label 9 vacuum sealed triplicate sample trays "L. mono without virucide" and: 30 min; 1 day; 2 days; 3 days; 7 days; 10 days; 14 days; 20 days; and 28 days (shelf life).

6. Store these samples at 4 °C.

7. Label 9 vacuum sealed triplicate sample trays "L. mono with virucide" and: 30 min; 1 day; 2 days; 3 days; 7 days; 10 days; 14 days; 20 days; and 28 days (shelf life).

8. Store these samples at 4 °C.

3.5 Inoculating Samples with Phage

1. Determine the volume of phage dilution to be spread over the meat slice to ensure application of 10^7 PFU cm^{-2}.

   - If phage was accurately diluted to ~10^9 PFU mL^{-1} expected plating volume will be 100 μl.

2. Take the remaining 18 triplicate sets of inoculated meat samples on Styrofoam trays from Subheading 3.3, **step 2** and spread the appropriate volume of phage preparation on the same surface as the *L. monocytogenes* inoculation.

3. Vacuum seal 18 of the samples using the MULTIVAC chamber machine.

4. Label 9 of triplicate sample trays "L. mono + phage without virucide" and: 30 min; 1 day; 2 days; 3 days; 7 days; 10 days; 14 days; 20 days; and 28 days (shelf life).

5. Store these samples at 4 °C.

6. Label 9 sets of triplicate sample trays "L. mono + phage with virucide" and: 30 min; 1 day; 2 days; 3 days; 7 days; 10 days; 14 days; 20 days; and 28 days (shelf life).

7. Store these samples at 4 °C.

3.6 Enumerating Viable Bacteria in the Samples

1. Use dissecting scissors to aseptically open the vacuum sealed meat samples.

2. Using sterile forceps aseptically transfer each meat sample to an appropriately labeled Stomacher® 80 bag.

   - Double-bag each sample to minimize the risk of infectious material leaking from the bags.

3. For all samples labeled "negative control" and "without virucide", use a sterile pipette to add 15 mL of sterile PBS to the bag.

4. Place the bag in an autoclavable Nalgene™ bucket.

5. Place the bag into the Stomacher® lab blender, taking care to leave the top 3–4 inches of the bag above the paddles.

6. Blend the sample for 2 min (use a timer) at medium setting.

7. Transfer the bag containing the homogenized sample to another autoclavable Nalgene™ bucket.

8. For all samples labeled "with virucide", use a sterile pipette to add 10 mL of sterile PBS to the bag.

9. Using a sterile pipette, add 5 mL TeaF virucidal solution (Subheading 3.1) to the bag to inactivate the remaining phage on the samples.

10. Repeat Subheadings 3.6, **step 5** and 3.6, **step 7**.

11. Serially dilute the homogenate, tenfold, in sterile PBS to yield 1000 μL each of 10^{-1} and 10^{-2} dilutions.

12. For each sample plate 100 μL of the 10^{-1} and 10^{-2} dilutions on 90 mm Oxford agar plates in triplicate.

13. If no colonies are observed on any of the plates, plate 1000 μL of undiluted homogenate (spread plate four 250 μL aliquots of the undiluted homogenate on four 90 mm Oxford agar plate).

14. Incubate the plates for 48 h at 37 °C and enumerate typical *Listeria* colonies.

15. *L. monocytogenes* appears on Oxford Agar as green colonies surrounded by a black halo.

4 Notes

4.1. Phage preparation
Fresh LISTEX™ P100 should be prepared and employed in the amount recommended by the manufacturer. Phage stock should be serially diluted in sterile SM buffer to a working stock of 2×10^9 PFU/mL. Standard soft agar overlay method can be employed to confirm the phage titers.

Titration plates must be incubated at 30 °C. Plated volumes should be adjusted to ensure plating of 10^7 PFU cm^{-2}.

4.2. Handling of *Listeria monocytogenes*: Aseptic Precautions

1. All manipulations of pathogen will be performed in a BSC.

2. All disposable plastic ware will be disposed in the autoclave waste bucket in the hood, and autoclave sterilized prior to disposal.

3. All glassware will be decontaminated by autoclaving prior to washing and reuse.

4. All work areas, and laboratory equipment used should be labeled with signs indicating the use of *Listeria monocytogenes*.

4.3. *Listeria monocytogenes* Inoculum Preparation

1. Using a sterile, disposable inoculating loop, transfer a single colony of *L. monocytogenes* from a fresh plate (not more than 3 days old) to a labeled culture tube containing 5 mL of tryptic soy broth (TSB).

2. Incubate for 24 h at 37 °C with shaking, at 160 rpm, to obtain a concentration of approximately 10^9 CFU mL^{-1} (equivalent to an optical density at 600 nm [OD600] ~1.2).

3. Confirm the optical density, OD, by transferring 600–1000 μL of culture to a cuvette and measuring absorbance at $\lambda = 600$ nm.

4. Transfer the remaining culture to a sterile centrifuge tube.

5. Harvest the cells by centrifuging at $7000 \times g$ for 10 min.

6. Use a sterile pipette to aspirate the supernatant from the tube.

7. Resuspend the cell pellet in 5 mL of PBS.

8. Repeat **steps 5–7** two times to wash the cells twice.

9. Prepare 10 mL of serial tenfold dilutions of the *L. monocytogenes* cell suspensions, in sterile PBS, to obtain the desired cell concentrations (The target counts on spiked food are 10^3 CFU cm^{-2}).

10. Mix equal volumes (e.g., 10 mL) of prepared *L. monocytogenes* cell suspensions of isolates belonging to serotypes 1/2a, 1/2b, and 4b and one representative outbreak strain.

References

1. Leverentz B, Conway WS, Camp MJ, Janisiewicz WJ, Abuladze T, Yang M et al (2003) Biocontrol of *Listeria monocytogenes* on fresh-cut produce by treatment with lytic bacteriophages and a bacteriocin. Appl Environ Microbiol 69:4519–4526

2. Carlton RM, Noordman WH, Biswas B, de Meester ED, Loessner MJ (2005) Bacteriophage P100 for control of *Listeria monocytogenes* in foods: genome sequence, bioinformatic analyses, oral toxicity study, and application. Regul Toxicol Pharmacol 43:301–312

3. Schellekens MM, Woutersi J, Hagens S, Hugenholtz J (2007) Bacteriophage P100 application to control *Listeria monocytogenes* on smeared cheese. Milchwissenschaft 62:284–287

4. Guenther S, Huwyler D, Richard S, Loessner MJ (2009) Virulent bacteriophage for efficient biocontrol of *Listeria monocytogenes* in ready-to-eat foods. Appl Environ Microbiol 75:93–100

5. Soni KA, Nannapaneni R (2010) Bacteriophage significantly reduces *Listeria monocytogenes* on raw salmon fillet tissue. J Food Prot 73:32–38

6. Chibeu A, Agius L, Gao A, Sabour PM, Kropinski AM, Balamurugan S (2013) Efficacy of bacteriophage LISTEX™P100 combined with chemical antimicrobials in reducing *Listeria monocytogenes* in cooked turkey and roast beef. Int J Food Microbiol 167:208–214

7. Soni KA, Nannapaneni R, Hagens S (2010) Reduction of *Listeria monocytogenes* on the surface of fresh channel catfish fillets by bacteriophage Listex P100. Foodborne Pathog Dis 7:427–434

8. U.S. Food and Drug Administration (2006) Agency response letter GRAS notice no. GRN 000198. http://www.fda.gov/Food/Ingred ientsPackagingLabeling/GRAS/NoticeInven tory/ucm154675.htm

9. U.S. Food and Drug Administration (2006; Revised 2013) Food additive permitted for direct addition to food for human consumption: Listeriaspecific bacteriophage preparation. http://www.accessdata.fda.gov/scripts/cdrh/cfdocs/cfCFR/CFRSearch.cfm?fr=172.785

10. U.S. Food and Drug Administration (2007) Agency response letter GRAS notice no. GRN 000218. http://www.fda.gov/Food/Ingred ientsPackagingLabeling/GRAS/NoticeInvent ory/ucm153865.htm

11. Micreos Food Safety (2010) Use of the bacteriophage preparation "Listex P100" as an antimicrobial intervention method against *Listeria monocytogenes*. In or on variety of meat, pultry, seafood and cheese. http://microeosfoodsafety.com/images/Processing%20Aid%20Canada.pdf. Accessed 10 Dec 2014

12. Food Standards Australia New Zealand (2012) Application A1045 - bacteriophage preparation as a processing aid. http://www.foods tandards.gov.au/code/applications/pages /applicationa1045bact4797.aspx. Accessed 10 Dec 2014

13. Jassim SAA, Denyer SP, Stewart GSAB (2001) Antiviral and antifungal composition and method. US Patent 6:187,316

14. de Siqueira RS, Dodd CER, Rees CED (2006) Evaluation of the natural virucidal activity of teas for use in the phage amplification assay. Int J Food Microbiol 111:259–262

15. Allen B, Willner D, Oechel WC, Lipson D (2010) Top-down control of microbial activity and biomass in an Arctic soil ecosystem. Environ Microbiol 12:642–648

16. Helsley KR, Brown TM, Furlong K, Williamson KE (2014) Applications and limitations of tea extract as a virucidal agent to assess the role of phage predation in soils. Biol Fertil Soils 50:263–274

17. Stewart GS, Jassim SA, Denyer SP, Newby P, Linley K, Dhir VK (1998) The specific and sensitive detection of bacterial pathogens within 4 h using bacteriophage amplification. J Appl Microbiol 84:777–783

Part II

Sequencing Analysis of Bacteriophages

Chapter 9

Sequencing, Assembling, and Finishing Complete Bacteriophage Genomes

Daniel A. Russell

Abstract

Next-generation DNA sequencing (NGS) technologies have made generating genomic sequence for organisms of interest affordable and commonplace. However, NGS platforms and analysis software are generally tuned to be used on large and complex genomes or metagenomic samples. Determining the complete genome sequence of a single bacteriophage requires a somewhat different perspective, workflow, and sensitivity to the nature of phages. Because phage genomes consist of mostly coding regions (*see* Pope/Jacobs-Sera chapter), a very high standard should be adopted when completing these genomes so that the subsequent steps of annotation and analysis are not sabotaged by sequencing errors. While read quality and assembly algorithms have continued to improve, achieving this standard still requires a significant amount of human oversight and expertise. This chapter describes our workflow for sequencing, assembling, and finishing phage genomes to a high standard by the NGS platforms Illumina, Ion Torrent, and 454.

Key words DNA sequencing, Library preparation, Coverage, Illumina, Ion Torrent, 454, Sanger, Newbler, PhagesDB, Consed, PAUSE, Genome termini, Galaxy, AceUtil

1 Introduction

In some ways, sequencing a phage genome is a simple proposition. Phage genomes are relatively small (~15 kb to ~500 kb) and only very rarely contain the long repetitive elements that hinder assembly of larger genomes. Yet phage genomes can present their own specific challenges, such as determining the type and location of genome ends, or planning the optimal number of genomes that can be multiplexed in a single high-throughput NGS run. In addition, the modern emphasis in sequence quality control has been toward higher coverage and away from finishing work, such as careful attention to questionable areas or targeted Sanger sequencing. Experience has shown that finishing phage genomes in the modern area of sequencing is both manageable and important; a bit of time and attention to the quality of an assembly can substantially reduce

Martha R.J. Clokie et al. (eds.), *Bacteriophages: Methods and Protocols, Volume 3*, Methods in Molecular Biology, vol. 1681, https://doi.org/10.1007/978-1-4939-7343-9_9, © Springer Science+Business Media LLC 2018

sequencing errors and provide a firm foundation for later bioinformatic or comparative genomics findings.

The broad strokes of our workflow are below.

(a) Plan an NGS run so that each multiplexed phage gets enough coverage.

(b) Prepare libraries from phage DNA for sequencing.

(c) Select a subset of the resulting reads for assembly (recommended).

(d) Assemble reads with Newbler.

(e) Open and view the assembly using consed.

(f) Verify that the entire genome has been sequenced.

(g) Run PAUSE to locate genome ends, if present.

(h) Run AceUtil to locate potential problems with the consensus sequence.

(i) Run targeted Sanger sequences, if necessary, to resolve problem areas.

(j) Generate a final sequence (.fasta) file.

The workflow described in this chapter has several features that should make it useful to most people doing phage genome sequencing. First, all the software is free for academic use. Second, it is not limited to a single sequencing platform, and can be used with data from Illumina, Ion Torrent, or 454 sequencers. Third, with experience and good data, it can take less than 1 h to complete the steps from assembly through publishable sequence. Finally, this method has been extensively tested, having been used on hundreds of genomes of phages of dozens of hosts that have been sequenced and/or quality-controlled at the Pittsburgh Bacteriophage Institute.

A complete guide to every possible situation that may be encountered when sequencing, assembling, and finishing phage genomes would require far more space than a single chapter, as would complete user manuals to UNIX, Newbler, consed, PAUSE, and AceUtil. Instead, I have attempted to provide an overview of the entire sequencing process that emphasizes the questions and concerns that should occupy the phage genome researcher's mind, while still giving detailed walkthroughs at some critical steps. The workflow described here will be enough to produce high quality sequences in many circumstances, but protocols for marginal situations and further details are available at PhagesDB (http://phagesdb.org/).

2 Library Preparation

Most commercially available library preparation kits and workflows are satisfactory for sequencing phage genomes, with one important exception. "Tagmentation" kits—those that rely on transposon-mediated shearing and adapter ligation, such as Nextera kits—should often be avoided. Though these kits are fast and easy to use, the libraries they generate will rarely have the phage's genome ends represented, resulting in a petering out of coverage near the termini, and no buildups of read starts to indicate ends. Because end determination is an important part of phage genome sequencing, it is best to avoid these kits unless the type and location of ends are already known, or if the phage is known to have a circularly permuted genome.

Because the yield from NGS platforms is so high, any library preparation workflow will need to include a DNA barcoding step so that multiple genomes can be sequenced together in a single run (multiplexed) and the resulting reads can subsequently be separated bioinformatically. Fortunately, all NGS platforms have DNA tagging available in their library-prep kits. See the Coverage section below for details on how many samples can be multiplexed on different platforms.

3 Coverage

Coverage, or read depth, is the number of reads underlying a given position in a consensus sequence after assembly. It can be thought of as the number of times a particular base in the genome has been sampled; the more reads that are generated for a genome, the higher the average coverage will be. Generally, higher average coverage is better because random errors in individual reads will be overcome by a multitude of correct reads, and because it is more likely that the entire genome will be represented. Too much coverage, however, can cause problems with assembly, represents a waste of money, and doesn't overcome platform-specific errors. (*See* **Note 3** for some information on systematic errors). Sequencing phage genomes on NGS platforms thus requires a careful balancing act. On one hand, you want enough reads to make sure your entire genome is represented and high quality. At the same time, it is quite easy to have superfluous reads.

So what is the right amount? For phage genomes, 100-fold average coverage with NGS reads is a good minimum number to shoot for. The median phage genome in GenBank is ~65 kb long, so to get 100-fold coverage for this representative genome you would need $65,000 \times 100 = 6.5$ Mb (megabases, or million bp) of shotgun sequence. When sequencing novel phage genomes, however, the expected genome length may be unknown, so we aim to acquire ≥10 Mb of sequence per phage.

Using 10 Mb per phage as a guideline, you can decide how many genomes can be multiplexed on a single NGS run. For example, an Ion Torrent run using 400 bp reads and an Ion 314™ Chip v2 is expected to yield 60–100 Mb of sequence [1], and therefore 6–10 phage genomes could be multiplexed in this run. Similarly, a 454 GS FLX+ run yields 700 Mb of sequence [2] and could accommodate up to 70 multiplexed phage genomes. An Illumina MiSeq run using a 150-cycle Reagent Kit v3 yields ~3.3 Gb (gigabases, or billion bp) of data [3], meaning as many as 330 phage genomes could be combined on this run, though the number of available indexing sequences may not allow for such a large number. Knowing how much data your sequencer will generate, and how much data is needed for each genome, is essential to getting the best value for your sequencing dollar. *See* **Note 6** for more details on multiplexing.

4   Assembly

Assembly is the process whereby shotgun sequencing reads are aligned and contigs (contiguous assemblies of reads) are generated. Though assembly can be a difficult process for complex genomes, it is usually fairly simple for phage genomes provided the input DNA sample was pure. An ideal assembly would contain a single contig representing the complete phage genome.

There are many software options for assembling phage genomes de novo from sequencing reads, but Newbler (formally known as GS De Novo Assembler) is our program of choice. It handles reads from 454, Illumina, or Ion Torrent in .fastq or .sff format; has an easy-to-use graphical user interface; is provided for free upon request; and produces consed-ready output for use in downstream quality control. Newbler must be installed on a Linux system, but you can request a Windows- or Mac-compatible virtual disk image with Newbler preinstalled from the author.

One common pitfall occurs when attempting to assemble a phage genome using *all* available reads, particularly when the average coverage is very high. Though counterintuitive, often an excess of reads can be detrimental to the quality and length of the assembly process. We recommend assembling with no more than 100,000 reads, and often use 50,000 reads for our initial assemblies. Reads not included in the initial assembly need not be disregarded or discarded, however, as they can be incorporated into an assembly project at a later point (*see* **Note 13**).

The protocol below describes how to assemble a phage genome with Newbler, including limiting the number of input reads if necessary. It assumes that you are working within a system that has Newbler installed, and that reads are in .fastq format. If they are .sff reads, they can be converted using the tool sff2fastq, available at: https://github.com/indraniel/sff2fastq.

4.1 Assembling a Phage Genome with Newbler

1. Prepare your fastq file.

   (a) Open a terminal and navigate to the location of your fastq file.

   (b) Determine the number of lines in your fastq file using the "wc" command. Replace the sample name below with your file name.

   ```
   wc my_phage_reads.fastq
   ```

   (c) Divide the first number shown (the number of lines in the file) by 4, and the result is the number of reads present in the file.

   (d) If your fastq file contains more than 100,000 reads, select a subset of reads to be used for assembly, using the "head" command. For example, the command below selects 50,000 reads (200,000 lines) from the file my_phage_-reads.fastq and stores them in a new file called my_phage_50k.fastq.

   ```
   head -n 200000 my_phage_reads.fastq > my_phage_50k.fastq
   ```

   (e) The number of reads used can drastically affect assembly quality.

2. Create a Newbler project.

   (a) Launch Newbler and select "New Assembly Project."

   (b) Enter a name for your project and click "OK".

3. Add reads to your project.

   (a) Select the "Project" tab.

   (b) Select the "FASTA and FASTQ reads" sub-tab.

   (c) Click the "+" icon on the left.

   (d) Navigate to the directory containing your prepared fastq file, select it, and click "OK."

4. Set parameters.

   (a) Select the "Parameters" tab.

   (b) Select the "Output" sub-tab.

   (c) In the "ACE Format:" field, select the "Complete Consed folder" option.

   (d) Other parameters may also be set as desired, though we most commonly use the defaults.

5. Run the assembly.

   (a) Click the "Start" button on the right side of the window.

   (b) If prompted, select "Yes."

(c) Wait for "Ready for Analysis" to display in the progress window. Assemblies using 50,000 reads on modern computers should take less than 10 min.

6. Check results.

   (a) Select the "Alignment results" tab to see a list of assembled contigs and their sizes.

   (b) Select the "Project" tab to view the numbers and percentages of reads used.

   (c) Select the "Result files" tab to view some of the files output by Newbler.

7. Quit and save.

   (a) Click the "Exit" button on the right side.

   (b) When prompted, select "Yes" to save the assembly.

4.2 Consed

There are several popular and commercially available software suites that can be used for bioinformatics purposes, but none is tailored to quality-control of assemblies like consed. Consed was developed during the Human Genome Project to do just the type of jobs necessary for genome finishing, such as assessing quality values, closely inspecting reads, tagging of consensus positions of interest, locating sequence matches between or within contigs, editing the consensus, joining contigs, designing primers, incorporating new reads, and exporting .fasta files [4]. Importantly, consed is also is free for academic use.

Though consed was developed before NGS technologies were available, it has been consistently updated to handle new read types and higher coverage levels while retaining the suite of features mentioned above. Therefore, it is the basis for all of our post-assembly finishing work. As described in the subsequent sections, we use other tools such as PAUSE and AceUtil, but ultimately they point us to regions we should look at in consed.

4.3 Verifying the Sequence is Complete

In most cases, when the complete phage genome sequence is present in an assembly, the genome will "circularize," meaning the reads at one end of the contig will overlap with those at the other end of the contig. This is true for circularly permuted genomes, sticky-overhang (5′ or 3′ cohesive end) genomes, and direct terminal repeat genomes. To determine whether a sequence is complete, its assembly can be opened in consed, and then the reads at one end of the contig of interest can be compared to reads at the other end (*see* Fig. 1). If they match, the genome has been circularized, and all of the sequence is present. If not, PCR across the contig ends followed by Sanger sequencing may be required.

Left end of Contig 1

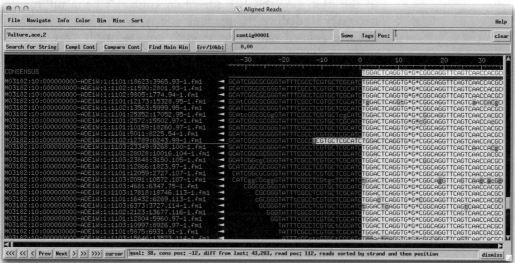

Right end of Contig 1

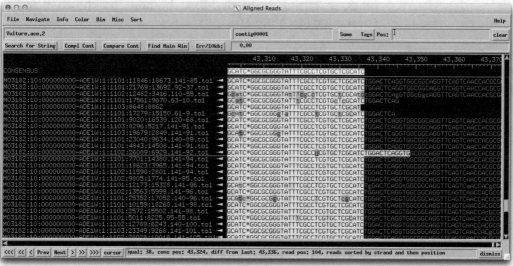

Fig. 1 Checking for genome circularization. Reads from the *left* and *right* end of the same contig in consed are shown. The bases highlighted in *yellow* match, indicating that the contig has been circularized and the entire genome is present

5   Determining the Genome Ends

An important but sometimes overlooked step in phage genome sequencing is determining the type and location of biological genome ends. Many types of genome ends have been identified, and experiments have been devised to ascertain the nature of the termini for a given phage [5].

While the process of end determination may have been onerous in the past, now NGS reads often contain all the information necessary to deduce the genome end types and locations. This is because defined chromosome ends are essentially "pre-sheared" positions in the input DNA sample, so those ends will be over-represented in library fragments, and thus reads will disproportionately begin on the precise base of a genome end. By searching for large buildups of read starts and changes in coverage, ends can often be identified. The protocol below describes how to use a program called PAUSE (Pileup Analysis Using Starts and Ends) to obtain a useful diagram showing coverage levels and read pileups that will draw attention quickly to the locations of potential ends. PAUSE requires a draft sequence file in fasta format, and a .fastq file of reads; the former can be exported from consed, and the latter is simply the file of reads output by the sequencer.

Once ends have been determined, a final sequence (.fasta) file should faithfully represent the DNA molecule packaged in each phage capsid (*see* **Note 15**). Note that some genomes are circularly permuted and will not have defined genome ends. For these genomes, a starting point or "Base 1" must be chosen based on convention (*see* **Note 16**).

5.1 Running PAUSE (Pileup Analysis Using Starts and Ends)

1. Create an account on the CPT (Center for Phage Technology) public Galaxy instance.

   (a) Go to: https://cpt.tamu.edu/galaxy-public/

   (b) Under the "User" dropdown menu select "Register" and complete the necessary steps.

2. Import the PAUSE workflow.

   (a) Select the "Shared Data" dropdown menu, then click on "Published Workflows."

   (b) Click on an appropriate PAUSE workflow (for example, "PAUSE v4.0 (Single End)") and select "Import."

   (c) This import only needs to be done once per user account.

3. Upload your .fastq and .fasta files.

   (a) Select "Get Data" from the left column, then click "Upload File."

   (b) In the window that opens, select "Choose local file" and select your .fasta file. Manually set the "Type" column to **fasta**.

   (c) Select "Choose local file" again and select your .fastq file. Manually set the "Type" column to **fastqsanger**.

   (d) Click "Start" and wait for the upload to complete.

4. Run the PAUSE workflow.

   (a) From the left column, click on "All workflows."

(b) Click on your PAUSE workflow (for example, "imported: PAUSE v4.0 (Single-End)").

(c) In the center pane, make sure your .fasta file is selected under the "Library to re-format" field.

(d) In the center pane, make sure your .fastq file is selected under the "FASTQ file" field.

(e) Scroll to the bottom of the center pane and click "Run workflow."

5. Check/download results.

(a) Once all boxes in the right column have turned green, click on the Eyeball icon in the "Aligned BAM PAUSE Plotter..." box. Your PAUSE image will display in the center pane.

(b) Right-click on the image to save it.

6. Analyze results.

(a) Scan your coverage/buildup graph for spikes in read starts and abrupt coverage changes. View these positions in consed to determine whether they are true ends.

(b) *See* Fig. 2 for examples.

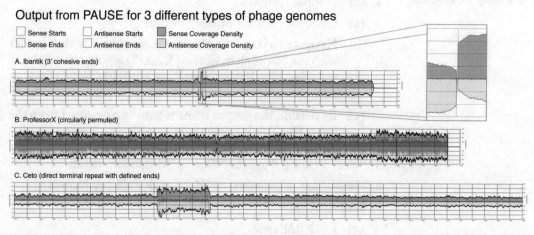

Output from PAUSE for 3 different types of phage genomes

☐ Sense Starts   ☐ Antisense Starts   ■ Sense Coverage Density
☐ Sense Ends   ☐ Antisense Ends   ☐ Antisense Coverage Density

A. Ibantik (3′ cohesive ends)

B. ProfessorX (circularly permuted)

C. Ceto (direct terminal repeat with defined ends)

Fig. 2 PAUSE Output Examples. Output from PAUSE shows *forward-* and *reverse*-strand read coverage in *gray above* and *below* the *center line*. Buildups of read starts and stops are shown in *red, green, purple,* or *blue*. (**a**) Ibantik, a phage with 3′ cohesive ends, shows even coverage throughout the genome except for a sharp drop over a small region, the 10 bp overhang (magnified area). The overhang is flanked by a read buildup on each strand. (**b**) ProfessorX, a circularly permuted genome, shows fairly even coverage throughout with no obvious read buildups. (**c**) Ceto, a phage with direct terminal repeats, shows a region with approximately double coverage flanked by a sharp read buildup on either side

6   Identifying and Resolving Weak Areas

Assembling reads into a single contig is a critical step, but it is not sufficient for producing high-quality complete genomes. Sequencers and assemblers make errors, and so a human review of assembly quality is essential to resolving potential weak areas. But scanning an entire assembly by eye to search for potential problems is impractical in today's high-throughput world, so we use a program called AceUtil to efficiently draw our attention to only the consensus positions that may be problematic.

AceUtil takes an .ace file as input, scans the assembly for positions that have a high number of discrepancies or low coverage, and outputs a new .ace file with those positions tagged. The output .ace file can be opened in consed, and the tagged positions quickly reviewed and, if necessary, changed. AceUtil can scan one contig at a time, so to tag multiple contigs from the same assembly, simply run AceUtil again and select a different contig. Through experience, we have tweaked AceUtil's default parameters to catch all positions of an assembly that may be problematic without overwhelming the user with false positives, but the parameters can be adjusted to perform a more or less stringent inspection when desired (*see* **Note 14**). AceUtil is written in Java, and is thus cross-platform. More details and installation instructions are available here: http://phagesdb.org/AceUtil/.

6.1 Using AceUtil to Identify Weak Areas

1. Launch AceUtil.
2. Set the input/output.
   (a) Click the "Set Input File" button.
   (b) Navigate to your .ace file. If you assembled with Newbler with the "Complete Consed folder" option, this will be in a folder like ProjectName → assembly → consed → edit_dir.
   (c) By default, the output name is auto-incremented from the input file. This is perfectly acceptable, but if a new name is desired, click the "Set Output File" button.
3. Select a contig from the "Contig" dropdown menu.
4. Use default parameters, or adjust as desired (*see* **Note 14**).
5. Run AceUtil.
   (a) Click "Analyze."
   (b) A short report of the number of tagged regions is shown in the terminal window.
   (c) A new ace file is created with regions of interest tagged.
6. Review tags in consed.

(a) Open a terminal and navigate to your project's edit_dir.

(b) Type "consed" and press enter.

(c) Select the output .ace file generated by AceUtil, and click "Open".

(d) Double-click on the contig of interest to open its Aligned Reads view.

(e) In the Aligned Reads view, select the "Navigate" menu, then "Tags," then "comment" tags.

(f) Review the list of positions tagged by AceUtil that opens by using the "next" button in the bottom left corner of the Aligned Reads window. Many will be simple misalignments or sequencing errors, but some will require the consensus sequence to be changed, or targeted Sanger reads to resolve.

6.2 Targeted Sanger Sequencing

Ideally, after shotgun sequencing, no additional wet-lab experiments will be necessary to produce a finished genome sequence; indeed, this is the case in over 90% of the genomes we sequence with our Illumina MiSeq. At times, however, even close inspection of the existing shotgun reads is not enough to give confidence that a consensus sequence is correct, and targeted Sanger reads may be required to definitively resolve a region. This may happen, for example, when discrepancies are caused by systematic errors in NGS platforms (*see* **Note 3**). Fortunately, consed makes both primer design and Sanger read incorporation simple.

Details on primer design are available in the consed documentation, but the process can be as simple as opening a contig's Aligned Reads view and right-clicking on the consensus sequence at a position of interest, then selecting "Pick primer from Clone." Consed creates a list of acceptable primers from which the user can select one. We have used thousands of primers designed by consed in both direct sequencing and PCR reactions, and have found consed's primer design algorithm to be excellent.

When resolving low-coverage or poor quality regions, Sanger reads can follow a PCR amplification of the region of interest. In this case, you must be mindful to use polymerases with proofreading activity, to limit mutations that may impact the sequencing result. If Sanger reads are needed to help determine ends, they must be done using phage genomic DNA as a direct template, as only reads generated in this manner will have the characteristic abrupt signal drops that confirm chromosome ends (*see* Fig. 3). Though this is more difficult than acquiring a high-quality read from a PCR product or plasmid, we regularly have success performing Sanger reactions with genomic DNA as the template using the following recipe. We run these reactions with standard cycling conditions except that we use 35–50 cycles instead of 25.

A. Babsiella (3′ cohesive ends)

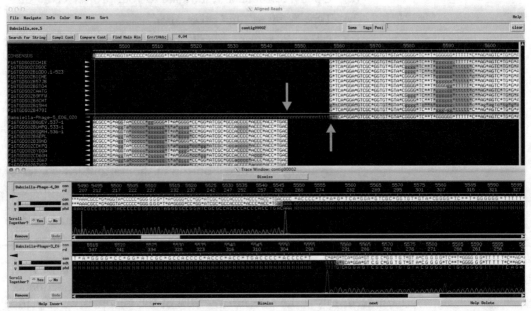

B. Monty (direct terminal repeats)

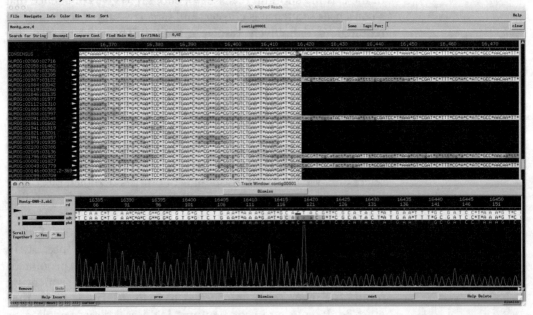

Fig. 3 Shotgun and Sanger reads at genome ends. The precise end of a genome can be determined in consed using read buildups and/or Sanger reads. (**a**) Babsiella, a genome with 3′ cohesive ends, shows read start buildups on each strand in close proximity (*orange arrows, top*). Sanger reads from either direction terminate suddenly, confirming physical genome ends (*bottom*). (**b**) Monty, a genome with direct terminal repeats, shows a read start buildup at one end of the terminal repeat (*top*). A Sanger read toward this end has an abrupt drop to half the intensity because those primers that annealed within the left copy of the terminal repeat can continue extension, but those primers that annealed within the right copy of the terminal repeat hit a physical end

Reactions can be cleaned up with any standard sequencing-reaction cleanup product, then run on a Sanger sequencer.

Amount	Component
x μL	Clean phage genomic DNA (300–500 ng)
1 μL	BigDye® terminator v3.1 mix
1.5 μL	5× BigDye sequencing buffer
2 μL	5 M Betaine (particularly helpful for GC-rich genomes)
0.5 μL	5 μM primer
Up to 10 μL	HPLC water

Once the Sanger read has been generated, it can be incorporated into the existing assembly using steps outlined in the consed documentation. Be sure to collect the .ab1 file as this is the file you need to add a Sanger read to a consed project. After incorporation, the consensus sequence can be adjusted or confirmed based on the results of the Sanger read, and the Sanger trace can be viewed within consed (*see* Fig. 3).

7 Generating a Final Fasta File

After ends have been determined, presence of the entire genome has been confirmed, and all weak areas have been resolved, the remaining step is to generate a final sequence file for annotation and publication. From the Aligned Reads view in consed, the "File" menu has two useful options for this purpose. The first, "Export consensus sequence", writes the current contig's entire consensus sequence to a fasta-formatted text file. This file can then be edited if necessary, for example to relocate genome ends to the start and end of the file or remove duplicated sequence. The second menu option, "Export consensus sequence (with options)", allows you to select only a portion of the consensus sequence to write to a file using the contig coordinates. In this case, a final genome can be exported in two parts and the text files combined. Care should be taken that the final .fasta file is of the appropriate length, orientation, and has ends in the correct places (*see* Note 15).

8 Notes

1. I've created some video tutorials to elaborate upon the techniques described here. They are available at: http://phagesdb.org/workflow/Sequencing/

2. Our current preferred sequencing method is Illumina due to the ease of finishing genomes thus sequenced, so long as tagmentation library preps are avoided. Illumina reads are high quality, have few systematic errors that cause problems, and show clear read buildups at genome ends. We typically multiplex 32–48 phage genome libraries on a single MiSeq run, using a 150-cycle Reagent Kit v3, and the total reagent cost with such a setup is less than $100 per phage genome.

3. Each NGS platform has nonrandom errors that may occur. Both 454 and Ion Torrent have intrinsic difficulties with runs of a single nucleotide longer than six bases [6]. Ion Torrent reads have an issue with strand-specific deletions [7, 8]. Illumina reads are more likely to contain errors following certain sequence motifs, such as GGCGGG [9]. Because these errors are systematic, increasing coverage is unlikely to fix the problems they cause. Awareness of these types of errors is important when checking assemblies for weak areas and deciding whether or not targeted Sanger reads are needed to resolve problematic consensus regions.

4. Since tagmentation library preps should be avoided, and because some labs may be preparing libraries in-house without access to a Covaris or other recommended shearing option, we suggest using a low-cost enzymatic shearing option if necessary. We use dsDNA Shearase™ from Zymo Research. A 20-min, 10 µL reaction with 250 ng of input phage genomic DNA and 0.75 µL of Shearase produces fragments that are ready for an Illumina TruSeq library prep kit. The amount of Shearase can be adjusted to increase or decrease fragment size as needed.

5. Assembly using Newbler works best with read lengths of at least 100 bp. Longer reads may assist in assembly, but even 75 bp Illumina reads will generally assemble into a single contig, provided a pure sample and good sequencing run, so having the longest available reads is much less of a concern than it is when sequencing bacterial or eukaryotic genomes.

6. When considering the number of genomes to include in a single NGS run, remember that the barcode distribution will not be precisely even; there may be an order of magnitude difference between the most- and least-represented samples. Thus, it is often advisable to err on the side of including fewer genomes to ensure that even the least-represented sample gets enough coverage. For example, if you set up a run so that each phage will get on average 30 Mb of data, one might get 80 Mb of coverage, while another will get 8 Mb.

7. Because phage genomes are generally simple to assemble, the information contained in paired-end reads is superfluous.

Single-end reads are sufficient, though no harm will come from having paired-end reads.

8. The number of reads to use in an initial assembly can vary from platform to platform or even sample to sample. Because running an assembly with Newbler often takes only a few minutes, it is possible to experiment with different numbers of input reads to obtain the best possible assembly. We usually begin with 50,000 reads, and increase or decrease the number if necessary.

9. When Newbler assembles a circularized genome, it must select a break point to display the contig. By default, it cleaves all underlying reads precisely at this break point, and puts half the read at one end of the contig, and the other half at the other end. This is convenient, as there is no duplicated sequence at the contig ends to remove. The left end picks up precisely on the base where the right end leaves off (*see* Fig. 1).

10. Contig ends should not be confused with genome ends. Often, genome ends will appear within the interior of the main contig because the genome has circularized. This can happen because cohesive ends ligate during library preparation, or direct terminal repeats assemble on top of one another.

11. Assemblies that result in multiple contigs may or may not be problematic. In some cases, a single genome may be split into several contigs and can be rejoined. In other cases, small contigs may be contamination, and can be disregarded. BLAST and/or relative coverage can be useful for determining whether or not small contigs are part of the phage genome.

12. Sometimes, an input DNA sample is inadvertently or deliberately mixed, meaning it contains two or more phage genomes. If the genomes in the mixed sample contain little sequence similarity, it is likely they will assemble independently and can be quality-controlled separately; this is sometimes a strategy for getting two library preps for the price of one. If, however, the two genomes share sequence similarity, it is likely that Newbler will break the contigs at the questionable regions, and the project may have to be scrapped.

13. A video tutorial for incorporating additional reads into an existing assembly is available. This is useful if some reads were excluded during the initial assembly. It can be found here: http://phagesdb.org/workflow/Sequencing/

14. AceUtil can scan for four types of assembly issues, with default values shown in parentheses: high discrepancies ($>18\%$ of reads disagree with the consensus), low coverage (<20-fold coverage at the position), high strand-specific discrepancies ($>70\%$), and low strand-specific coverage (<4-fold). We've used these

parameters with success for all types of NGS output, though they can be adjusted.

15. Final fasta sequence files should have the following characteristics.

    (a) Oriented so the terminase is transcribed in the forward direction. (This is a convention.)

    (b) Defined genome ends, if present, at the start and end of the sequence.

    (c) If a 3′ cohesive overhang is present it should be at the right end of the genome. If a 5′ cohesive overhang is present it should be at the left end of the genome.

    (d) If a direct terminal repeat is present, both copies should be included (one at the left end and one at the right end).

16. To choose "Base 1" of a circularly permuted genome, we use the following criteria.

    (a) If a highly similar genome is present in GenBank, select the same Base 1 as that genome.

    (b) Using DNA Master (*see* Pope/Jacobs-Sera chapter), locate the terminase gene in the genome, and orient the genome so that the terminase is transcribed in the forward direction.

    (c) Begin at the terminase, and look upstream for a logical break point. This can be a noncoding gap or a flip between forward and reverse genes.

    (d) Choose Base 1 so that it is upstream of, but close to, the terminase, and is at a logical break point. We recommend using the first base of the start codon of the first gene downstream of the break point.

Acknowledgments

I would like to thank Eric Rasche from the Center for Phage Technology at Texas A & M University for the development of PAUSE, and Charlie Bowman at the University of Pittsburgh for the development of AceUtil. Thanks to both for making these programs easy-to-use and freely available, and for their prompt responses to feature requests and bug fixes. Thanks to the PHIRE and SEA-PHAGES students who continue to isolate and prepare DNA from a multitude of interesting phages so we have fodder for our sequencers. I am grateful to Debbie Jacobs-Sera and Welkin Pope for helpful discussions about phage biology, and to Ching-Chung Ko for teaching me almost everything I know about sequencing and for being a continual sounding board for new ideas. And a special thank you to Graham Hatfull for his leadership, vision, and support.

References

1. LifeTechnologies (2013) Ion PGM system specification sheet. https://tools.lifetechnologies.com/content/sfs/brochures/PGM-Specification-Sheet.pdf Accessed 15 Feb 2015

2. Roche Diagnostics GmbH (2011) GS FLX+ System. http://454.com/downloads/GSFLXApplicationFlyer_FINALv2.pdf Accessed 15 Feb 2015

3. Illumina (2015) MiSeq specifications. http://www.illumina.com/systems/miseq/performance_specifications.html. Accessed 15 Feb 2015

4. Gordon D, Abajian C, Green P (1998) Consed: a graphical tool for sequence finishing. Genome Res 8:195–202

5. Casjens SR, Gilcrease EB (2009) Determining DNA packaging strategy by analysis of the termini of the chromosomes in tailed-bacteriophage virions. Methods Mol Biol 502:91–111

6. Loman NJ, Misra RV, Dallman TJ et al (2012) Performance comparison of benchtop high-throughput sequencing platforms. Nat Biotechnol 30:434–439

7. Quail MA, Smith M, Coupland P et al (2012) A tale of three next generation sequencing platforms: comparison of ion torrent, Pacific biosciences, and Illumina MiSeq sequencers. BMC Genomics 13:341

8. Russell DA (2013) Quirks of ion torrent PGM data: on the other strand… http://phagesdb.Org/blog/posts/12/ Accessed 10 Jan 2015

9. Nakamura K, Oshima T, Morimoto T et al (2011) Sequence-specific error profile of Illumina sequencers. Nucleic Acids Res 39(13):e90

Chapter 10

Identification of DNA Base Modifications by Means of Pacific Biosciences RS Sequencing Technology

Philip Kelleher, James Murphy, Jennifer Mahony, and Douwe van Sinderen

Abstract

Whole phage genomes can be sequenced readily using one or a combination of next generation sequencing (NGS) technologies. One of the most recently developed NGS platforms, the so-called Single-Molecule Real-Time (SMRT) sequencing approach provided by the PacBio RS platform, is particularly useful in providing complete (i.e., un-gapped) genome sequences, but differs from other technologies in that the platform also allows for downstream analysis to identify nucleotides that have been modified by DNA methylation. Here, we describe the methodological approach for the detection of genomic methylation motifs by means of SMRT sequencing.

Key words SMRT sequencing, PacBio, Bacteriophage, Methylase, Methylome

1 Introduction

Epigenetic modification of genomes by methylation plays an important role in expanding the functionality of the four traditional DNA bases, and is a process that is carried out by DNA methyltransferases (MTases). MTases are encoded by eukaryotes, prokaryotes, viruses, and (bacterio)phages [1], and their modifications play a variety of important roles including regulation of the cell cycle, DNA repair, and pathogenesis, although they are most frequently linked to their cognate restriction endonucleases (REase) to form restriction-modification (R–M) systems [2–5]. Three methylase classes exist in prokaryotes, all functioning by methyl transfer from S-adenosyl-L-methionine (SAM) to a target nucleotide base. Methylase classes I and II target the exocyclic nitrogens at position $N6$ in adenine and position $N4$ in cytosine, forming N^6-methyladenine (m6A) and N^4-methylcytosine (m4C), respectively. Class III MTases target the C^5 position in cytosine yielding C^5-methylcytosine (m5C) [6]. Akin to their bacterial hosts, many

Martha R.J. Clokie et al. (eds.), *Bacteriophages: Methods and Protocols, Volume 3*, Methods in Molecular Biology, vol. 1681, https://doi.org/10.1007/978-1-4939-7343-9_10, © Springer Science+Business Media LLC 2018

phage genomes harbor genes encoding (predicted) MTases with several different functions [7], including the regulation of progeny phage particle release as exemplified by the *Escherichia coli* phage P1 which encodes a DAM MTase [8]. Failure to methylate the GATC sites within the phage packaging site results in a reduction of the number of released phage particles [9]. The majority of phage-encoded MTases are, however, believed to have been acquired by a given phage as a defense mechanism against host-encoded R-M systems. Several lactococcal phages harbor MTases thereby blocking the activity of restriction endonucleases, a finding echoed in *Bacillus* phages [10, 11]. Nonetheless, as the number of available phage genomes increases, it is clear that many encode one or more putative MTases. The development of Single-Molecule Real-Time (SMRT) DNA sequencing provides an experimental platform to couple whole genome sequencing to the characterization of (phage-encoded) MTases.

SMRT DNA sequencing is a single molecule, sequence-by-synthesis approach developed by Pacific Biosciences (through sequence analysis employing a PacBio RS instrument). The SMRT technology utilizes a single molecule polymerase enzyme immobilized on a zero-mode waveguide (ZMW) nanostructure to incorporate fluorescently labeled nucleotides complementary to a DNA template strand [12, 13]. The incorporation of a new nucleotide results in the cleavage of the phosphate backbone, to which the base-specific fluorophores are linked, generating a nucleotide-specific fluorescent signal which is captured in a real-time movie format [14].

SMRT sequencing is the first high-throughput approach that can directly detect DNA base modifications concomitantly with the acquisition of primary DNA sequence information [15]. The signal generated during fluorophore cleavage, called a "pulse," can be exploited to detect m6A, m4C and m5C base modifications on the template strand. This method exploits two parameters of the kinetic rate of incorporation of each nucleotide: (1) the pulse width (PW), which reflects the duration of time the polymerase is bound to a particular nucleotide and (2) the interpulse duration (IPD), which is the time it takes for the polymerase to move from one nucleotide to the next. Distinct variations in the PW and IPD are observed when the polymerase encounters a modified base in the DNA template and this signature is used to identify individual modifications [16]. However, there are some issues which affect the detection of these signatures. The m5C kinetic signature is difficult to detect accurately and (at low sequence coverage) requires the DNA to be treated with 10–11 translocation methylcytosine dioxygenase 1 (Tet1) enzyme prior to sequencing. Tet1 treatment converts 5-methylcytosine to 5-carboxylcytosine which has a stronger pulse signature and thus makes m5C modifications easier to detect [17]. The other dependent factor is the

Table 1
Minimum sequence fold-coverage for accurate detection of base modifications[a]

Base Modification	Fold-coverage[b]
4-methylcytosine	25×
5-methylcytosine	250×
5-hydroxymethylcytosine	250×
Glucosylated 5-hydroxymethylcytosine [18]	25×
Hydroxymethylcytosine[c]	25×
6-methyladenine	25×
8-oxoguanine	25×

[a] Data sourced from (http://www.pacb.com/applications/base_modification/)
[b] SMRT sequencing follows a Poisson distribution across a genome, to ensure the entire genome meets the 25× threshold, Pacific Biosciences recommend a total coverage of 100×
[c] Enriched with the Hydroxymethyl Collector™ Kit available from (http://www.activemotif.com/catalog/775/hydroxy methyl-collectortrade)

fold-coverage of the sequencing data. Pacific Biosciences recommends individual minimum fold-coverage of a particular analysed sequence for each type of base modification (Table 1). The fold-coverage obtained from a sequencing run will depend on the run chemistry and genome size with larger genomes requiring the use of multiple SMRT cells to achieve higher fold-coverage.

2 Materials

The following protocol describes the steps necessary to conduct base modification analysis on SMRT sequencing projects. This protocol utilizes the Pacific Biosciences SMRTanalysis v2.3.0 software, which supports analysis of DNA sequence data from the PacBio *RS* sequencing platform. All software used is open source and available from the Pacific Biosciences web page (http://www.pacb.com/devnet/), along with detailed installation guides. The installation of the SMRTanalysis software package may be technically challenging and will require a reasonable level of proficiency with Linux-based systems. System administrator privileges are required for installation and the set-up of user accounts to access the portal. Installation and running SMRTanalysis requires a 64-bit Linux machine with a minimum of 8 cores with 2GB RAM per core and 250 GB of disk base. An installation summary with complete system requirements is available from (https://github.com/PacificBiosciences/SMRT-Analysis/wiki/SMRT-Analysis-Software-Installation-v2.3.0). In this study, SMRTanalysis v2.3.0 was

Table 2
Software used for PacBio *RS* SMRT assemblies

Software package	Download source
SMRTanalysis v2.3.0	http://www.pacb.com/devnet/
SSH secure Shell	http://en.kioskea.net/download/download-1423-ssh-secure-shell
Xming X server	http://sourceforge.net/projects/xming/
Bio-Linux	http://environmentalomics.org/bio-linux/

installed on a Bio-Linux machine and accessed using SSH Secure Shell Client. Xming X server for Windows was used to tunnel Mozilla Firefox browser to connect to the SMRT portal (Table 2). A number of commercial sequencing providers are currently providing PacBio sequencing services and an exhaustive list of PacBio NGS providers is available from (http://www.pacb.com/support/sequencing_provider/).

3 Methods

3.1 Protocol Overview

There are a number of steps involved in performing the base modification analysis protocol detailed below (schematically outlined in Fig. 1). The starting point of the analysis will depend on the input data, particularly the availability of a good quality reference genome.

3.1.1 To Perform the Analysis with Raw SMRT Cell Data Only

In this case all steps in the protocol will be necessary. Subheading 3.2 "Importing Data" describes how to import the raw SMRT cell data from the sequence service provider into the SMRT portal. The base modification analysis protocol requires a reference genome to detect modifications. This reference genome is generated from SMRT cell data using the protocol described in Subheading 3.3 "RS_HGAP_Assembly.2 Protocol". Following this protocol, the reference sequence is generated from an initial assembly of the SMRT data, and Subheading 3.4 "HGAP Assembly Results" describes how to check and download this reference. The final step in this section of the protocol is to import the reference genome sequence to SMRT portal; this is described in Subheading 3.5 "Import HGAP Reference." To initiate the base modification analysis, follow the steps described in Subheading 3.6 "RS_Motif_and_Modification.1 Protocol". On completion of this protocol analysis of the results is described in Subheading 3.7 "Base Modification Results."

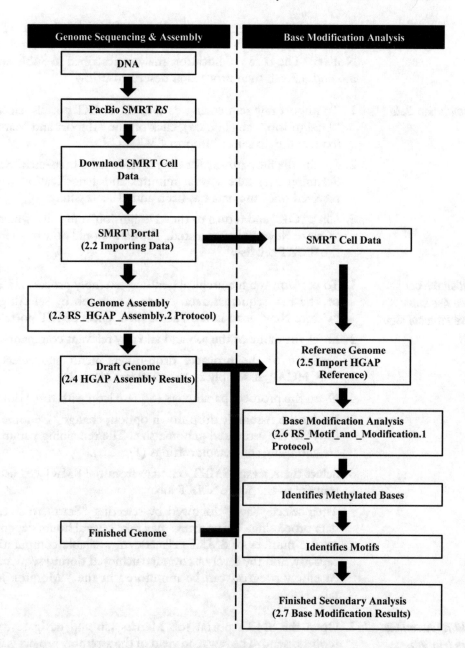

Fig. 1 Flowchart overview of SMRT RS sequencing, assembly, and base modification analysis

3.1.2 To Perform the Analysis with Raw SMRT Cell Data and an Available Reference Genome

In this case it is not necessary to generate a new reference genome from the SMRT cell data and some steps can be removed from the protocol. To begin, follow the steps described in Subheading 3.2 "Importing Data," this allows the user to import the raw SMRT cell data from the sequence service provider into SMRT portal. Subheadings 3.3 and 3.4 are no longer required and the user can proceed to Subheading 3.5 "Import HGAP Reference." Here, the available reference genome can be supplemented for the HGAP assembly sequence that previously came from Subheading 3.4 (*see* **Note 6**). The base modification analysis described in Subheadings 3.6 and 3.7 can then proceed as described below.

3.2 Importing Data

1. To import raw sequencing data to the SMRT portal, select the "Design Job" tab (Fig. **2a**), click on the "Import and Manage" icon, and then select "Import SMRT Cells."

2. Specify the file pathway for the required data, then click "Scan." Scanning may take several minutes and a notification will be received once the data has been added successfully.

3. Click "OK" and return to the "Design Job" tab, after which the "Create New" icon is selected. The data should now be visible in the SMRT cell list.

3.3 Hierarchical Genome Assembly Process Protocol (See Note 1)

1. To perform the hierarchical genome assembly process (HGAP) on the raw sequence data, create a new job by selecting the "Create New" icon in the Pacific Biosciences SMRT portal.

2. Enter the name of the job and add any relevant comments.

3. Navigate to the protocol drop-down menu and select the "RS_HGAP_Assembly.2" protocol (*see* **Note 2**).

4. Open the protocol parameters (Square icon with three dots).

5. Under the Assembly drop-down option, change "Genome Size (bp)" to the estimated genome size. The remaining parameters are run using the default settings (Fig. **2b**).

6. Select the relevant SMRT cell data from the SMRT cell list and transfer to the active SMRT job.

7. Then select "Save," followed by selecting "Start" to start the data processing. This process may take several hours depending on the number of SMRT cells added, available computational capacity, and the level of coverage achieved during sequencing. Assembly progress can be monitored in the '"Monitor Jobs" tab.

3.4 HGAP Assembly Results (Fig. 2c)

1. Open the SMRT portal Job Metrics tab and navigate to the reports menu. The raw read yield of the assembly project is listed under Assembly (*see* **Note 3**). The mapping and coverage data of the assembly is listed under Resequencing.

2. Under the polished assembly tab, the number of contigs in the final polished assembly of the phage genome is shown, along with the size of the contigs.

3. To download the fasta file of the assembled phage genome, which will be used as the reference file for the base modification protocol, navigate to the "Data" menu and under "Assembly" click on the polished assembly "FASTA" (*see* **Note 4**).

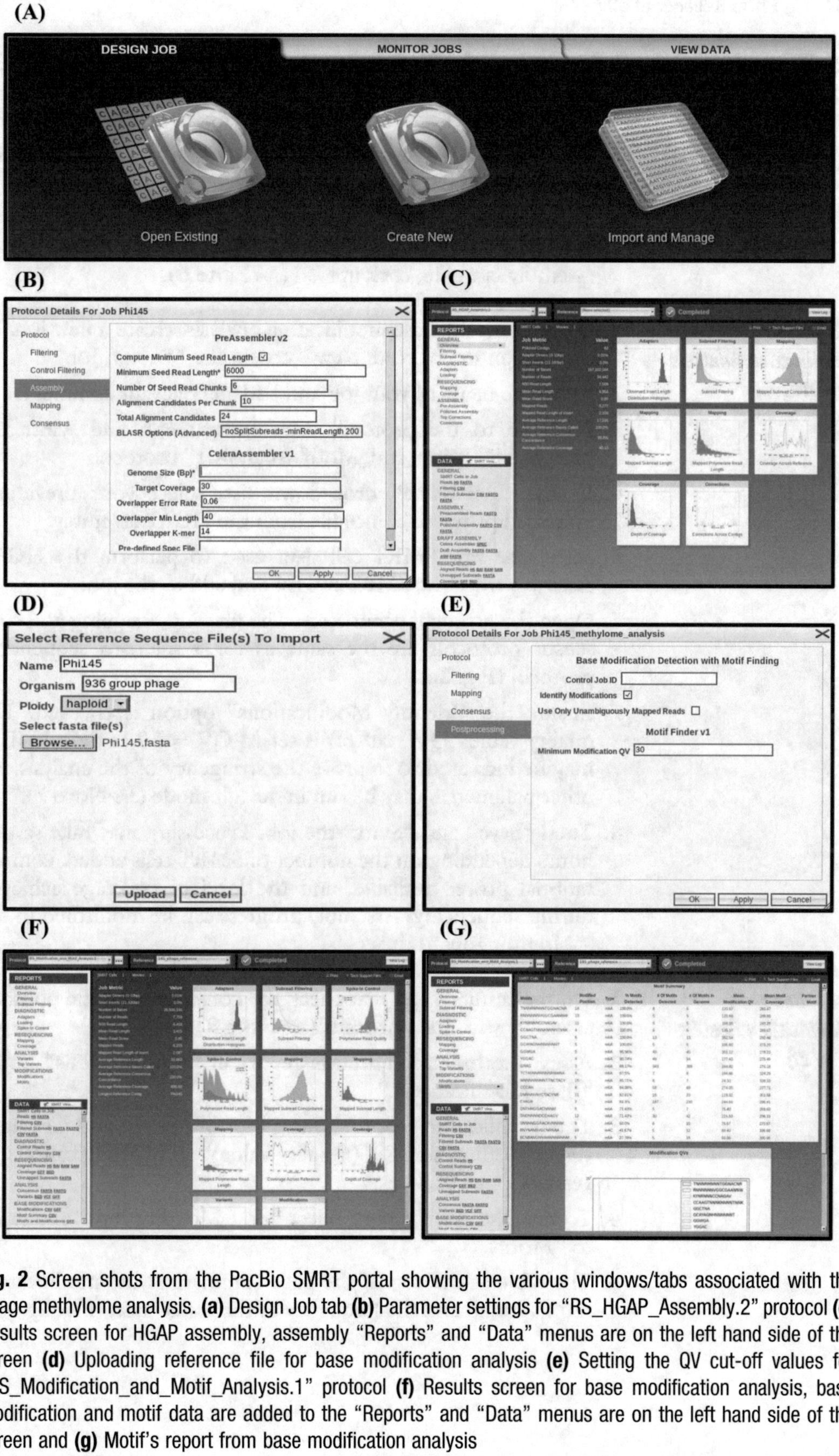

Fig. 2 Screen shots from the PacBio SMRT portal showing the various windows/tabs associated with the phage methylome analysis. (a) Design Job tab (b) Parameter settings for "RS_HGAP_Assembly.2" protocol (c) Results screen for HGAP assembly, assembly "Reports" and "Data" menus are on the left hand side of the screen (d) Uploading reference file for base modification analysis (e) Setting the QV cut-off values for "RS_Modification_and_Motif_Analysis.1" protocol (f) Results screen for base modification analysis, base modification and motif data are added to the "Reports" and "Data" menus are on the left hand side of the screen and (g) Motif's report from base modification analysis

3.5 Protocol for Importing HGAP Reference

1. To upload the phage genome reference file from the HGAP assembly, select the "Import and Manage" icon in the "Design Job" tab and click on the reference sequences link (*see* **Note 5**).

2. Select "New" and enter the assembly name and organism (Fig. 2d).

3. To select the reference file, use the "Select Fasta File(s)" option and browse to your previously downloaded polished phage assembly fasta file; click upload (*see* **Note 6**).

3.6 Base Modification Analysis Protocol

1. To perform the base modification analysis, create a new job, by clicking on the "Create New" icon in the "Design Job" tab.

2. Enter the name of your job and add any relevant comments.

3. Navigate to the protocol drop-down menu and select the "RS_Modification_and_Motif_Analysis.1" protocol.

4. In the "Reference" drop-down list, select your previously uploaded phage reference file from the HGAP assembly.

5. Select the raw SMRT cell data used to perform the HGAP assembly from the SMRT cell list and add to the job.

6. Open the protocol parameters. The filtering, mapping and consensus protocols are the same as for a standard sequencing protocol (Fig. 2e).

7. Ensure the "Identify Modifications" option is checked. The quality value "QV" cut-off is set to QV = 30 as default. This may be increased to improve the stringency of the analysis. All other parameters may be run in default mode (*see* **Note 7**).

8. Then "Save" and "Start" the job. Processing may take several hours depending on the number of SMRT cells added, computational power available, and the level of coverage achieved during sequencing. Assembly progress can be monitored in the '"Monitor Jobs" tab.

3.7 Base Modification Results (Fig. 2f)

1. On the results page, first check the consistency of the polished assembly with the raw reads (*see* **Note 8**).

2. Assess the base modification report in "Modifications" under "Base Modifications."

3. The modification report shows a plot of "Per-Strand Coverage versus Modification QV (Quality Value)" and "Modification QV versus Count of Bases."

4. To view the motif report, select "Base Modifications" and scroll to "Motifs."

5. The table at the top of the page summarizes the modification motifs found and the frequency of their detection (Fig. 2g).

6. The plot shows the Modifications QVs with the identified motifs in comparison to all other modifications not found to be within a motif.

7. The base modification data can be downloaded from the "Data" menu under "Base Modifications" to be used for further analysis.

4 Notes

1. To conduct base modification analysis on PacBio SMRT sequencing data, a reference genome sequence is required (i.e., an assembled [phage] genome). The reference sequence is obtained by performing a HGAP assembly on the raw sequence data.

2. "RS_HGAP_Assembly.2" in the protocol is specific to SMRTanalysis v2.3.0. When employing SMRTanalysis v2.0, use assembly protocol "RS_HGAP_Assembly.1", to perform the HGAP assembly.

3. The preassembly report provides the preassembled yield, which is the number of preassembled reads over the number of preassembled bases. This is a useful check of the raw read coverage of the sequencing project. If this value is low, the assembly may be limited by the raw-read coverage.

4. Files downloaded from the SMRT portal are stored in the Internet browsers' Downloads folder. The polished phage assembly fasta reference file will need to be transferred to a reference file drop box on the server, from where it can be uploaded directly to the SMRT portal for the base modification analysis protocol.

5. A directory should be created to store the reference sequences on the server running SMRT portal. This directory should be set up as a reference drop-box on the SMRT analysis user account.

6. It has been found in previous analyses [10] that a completed phage genome, if available, should be used as a reference, rather than an assembly fasta file. Once uploaded your reference file will be validated, which may take several minutes. An e-mail will be sent upon completion.

7. When using SMRTanalysis v2.0, it is required that the user specify in the parameters menu if the sample has been Tet1 treated. To reliably detect 5mc base modifications, samples should be treated with Tet1 enzyme prior to sequencing. SMRTanalysis v2.3.0 does not require the user to specify if the sample has been Tet1 treated.

8. During the initial HGAP assembly, polishing has been implemented with Quiver (consensus polisher program). This process is repeated during the base modification protocol. Looking at the variants analysis under "Reports" indicates whether this assembly polishing was successful or if the assembly can be polished further; if there are few variants shown it indicates that the data polishing was successful. Regions with insufficient coverage will have a PHRED quality score of zero (usually located at the edge of contigs), these regions should be removed from the final sequence. If there is a large number of variants shown, the analysis can be repeated with a lower "Minimum Subread Length" (default is 500) or a lower "Minimum Polymerase Read Quality" (default is 0.80). These parameters can be adjusted in the "Protocol parameters" filtering section, in the "Design Job" tab, when setting up the protocol. This is usually caused by highly repetitive targets, and will be unnecessary with most phage sequencing projects.

References

1. Korlach J, Turner SW (2012) Going beyond five bases in DNA sequencing. Curr Opin Struct Biol 22:251–261

2. Boye E, Lobner-Olesen A (1990) The role of dam methyltransferase in the control of DNA replication in E. coli. Cell 62:981–989

3. Julio SM, Heithoff DM, Provenzano D, Klose KE, Sinsheimer RL, Low DA, Mahan MJ (2001) DNA adenine methylase is essential for viability and plays a role in the pathogenesis of Yersinia pseudotuberculosis and Vibrio cholerae. Infect Immun 69:7610–7615

4. Mohapatra SS, Fioravanti A, Biondi EG (2014) DNA methylation in Caulobacter and other Alphaproteobacteria during cell cycle progression. Trends Microbiol 22:528–535

5. Marinus MG, Casadesus J (2009) Roles of DNA adenine methylation in host-pathogen interactions: mismatch repair, transcriptional regulation, and more. FEMS Microbiol Rev 33:488–503

6. Roberts RJ, Belfort M, Bestor T, Bhagwat AS, Bickle TA, Bitinaite J, Blumenthal RM, Degtyarev S, Dryden DT, Dybvig K, Firman K, Gromova ES, Gumport RI, Halford SE, Hattman S, Heitman J, Hornby DP, Janulaitis A, Jeltsch A, Josephsen J, Kiss A, Klaenhammer TR, Kobayashi I, Kong H, Kruger DH, Lacks S, Marinus MG, Miyahara M, Morgan RD, Murray NE, Nagaraja V, Piekarowicz A, Pingoud A, Raleigh E, Rao DN, Reich N, Repin VE, Selker EU, Shaw PC, Stein DC, Stoddard BL, Szybalski W, Trautner TA, Van Etten JL, Vitor JM, Wilson GG, Xu SY (2003) A nomenclature for restriction enzymes, DNA methyltransferases, homing endonucleases and their genes. Nucleic Acids Res 31:1805–1812

7. Murphy J, Mahony J, Ainsworth S, Nauta A, van Sinderen D (2013) Bacteriophage orphan DNA methyltransferases: insights from their bacterial origin, function, and occurrence. Appl Environ Microbiol 79:7547–7555

8. Coulby JN, Sternberg NL (1988) Characterization of the phage P1 dam gene. Gene 74:191

9. Sternberg N, Coulby J (1990) Cleavage of the bacteriophage P1 packaging site (pac) is regulated by adenine methylation. Proc Natl Acad Sci U S A 87:8070–8074

10. Murphy J, Klumpp J, Mahony J, O'Connell-Motherway M, Nauta A, van Sinderen D (2014) Methyltransferases acquired by lactococcal 936-type phage provide protection against restriction endonuclease activity. BMC Genomics 15:831

11. Trautner TA, Pawlek B, Gunthert U, Canosi U, Jentsch S, Freund M (1980) Restriction and modification in Bacillus subtilis: identification of a gene in the temperate phage SP beta coding for a BsuR specific modification methyltransferase. Mol Gen Genet 180:361–367

12. Korlach J, Bjornson KP, Chaudhuri BP, Cicero RL, Flusberg BA, Gray JJ, Holden D, Saxena R, Wegener J, Turner SW (2010) Real-time DNA sequencing from single

polymerase molecules. Methods Enzymol 472:431–455

13. Eid J, Fehr A, Gray J, Luong K, Lyle J, Otto G, Peluso P, Rank D, Baybayan P, Bettman B, Bibillo A, Bjornson K, Chaudhuri B, Christians F, Cicero R, Clark S, Dalal R, Dewinter A, Dixon J, Foquet M, Gaertner A, Hardenbol P, Heiner C, Hester K, Holden D, Kearns G, Kong X, Kuse R, Lacroix Y, Lin S, Lundquist P, Ma C, Marks P, Maxham M, Murphy D, Park I, Pham T, Phillips M, Roy J, Sebra R, Shen G, Sorenson J, Tomaney A, Travers K, Trulson M, Vieceli J, Wegener J, Wu D, Yang A, Zaccarin D et al (2009) Real-time DNA sequencing from single polymerase molecules. Science 323:133–138

14. Timp W, Mirsaidov UM, Wang D, Comer J, Aksimentiev A, Timp G (2010) Nanopore sequencing: electrical measurements of the code of life. IEEE Trans Nanotechnol 9:281–294

15. Flusberg BA, Webster DR, Lee JH, Travers KJ, Olivares EC, Clark TA, Korlach J, Turner SW (2010) Direct detection of DNA methylation during single-molecule, real-time sequencing. Nat Methods 7:461–465

16. Clark TA, Murray IA, Morgan RD, Kislyuk AO, Spittle KE, Boitano M, Fomenkov A, Roberts RJ, Korlach J (2012) Characterization of DNA methyltransferase specificities using single-molecule, real-time DNA sequencing. Nucleic Acids Res 40:e29

17. Clark T, Lu X, Luong K, Dai Q, Boitano M, Turner S, He C, Korlach J (2013) Enhanced 5-methylcytosine detection in single-molecule, real-time sequencing via Tet1 oxidation. BMC Biol 11:1–10

18. Josse J, Kornberg A (1962) Glucosylation of deoxyribonucleic acid: III. α- AND β-glucosyl transferases from T4-infected *escherichia coli*. J Biol Chem 237:1968–1967

Chapter 11

Analyzing Genome Termini of Bacteriophage Through High-Throughput Sequencing

Xianglilan Zhang, Yahui Wang, and Yigang Tong

Abstract

High-throughput sequencing (HTS) is an effective tool for bacteriophage genome and its termini analysis. HTS technology parallelizes the sequencing process, producing thousands to millions of reads concurrently. Terminal information of a bacteriophage genome is important and basic knowledge for understanding the biology of the bacteriophage. We have created a high-occurrence reads as termini theory and developed practical methods to determine the bacteriophage genome termini, which is based on the large data of HTS. With this method, the termini of the bacteriophage genome can be efficiently and reliably identified as a by-product of bacteriophage genome sequencing, by solely analyzing the sequence statistics of the raw sequencing data (reads), without any further lab experiments.

Key words High-throughput sequencing (HTS), Bacteriophage, Genome termini, High Occurrence Read Termini theory, High frequency read sequence (HFS)

1 Introduction

Bacteriophages play an important role in molecular biology research. For example in 1952, Hershey–Chase demonstrated that DNA is the hereditary material by using bacteriophage [1]. A variety of useful enzymes are encoded in bacteriophage genomes, such as T4 DNA ligase, T4 RNA ligase, and T4 polymerase.

Recently, a rise in the spread and severity of bacterial antibiotic-resistance has become a menacing health problem [2]. For example, strains of enterococci have emerged to resistant to vancomycin (vancomycin-resistant *Enterococcus*, VRE), with an increased incidence reported worldwide [3–5]. Vancomycin was thought to be one of the last antibiotics that are reliably effective against the multidrug-resistant superbug [6]. Furthermore, methicillin-resistant *Staphylococcus aureus* (MRSA) causes intractable infections [7]. Except VRE and MRSA, vancomycin-resistant *S. aureus* (VRSA), extended spectrum beta-lactamase (ESBL) producing bacteria and multidrug-resistant *Acinetobacter baumannii*

Martha R.J. Clokie et al. (eds.), *Bacteriophages: Methods and Protocols, Volume 3*, Methods in Molecular Biology, vol. 1681, https://doi.org/10.1007/978-1-4939-7343-9_11, © Springer Science+Business Media LLC 2018

(MRAB) are also common types of antibiotic-resistant bacteria [8]. Bacteriophage therapy has great potential to be developed to target bacteria that are resistant to antibiotics and therefore the identification and study of bacteriophages specifically infect antibiotic-resistant bacteria is likely to have a significant medical impact in the near future.

An important aspect of characterizing bacteriophages is their genome packaging, which plays an important role in the phage's life cycle, starting from initiation [9], to viral DNA replication [10], termination and regulation of transcription [11]. Genome termini identification is a crucial stage of the whole DNA packaging process. High-throughput sequencing (HTS) is an effective tool for bacteriophage genome sequence analysis [12–14], including genome termini analysis. HTS generates a large number of reads data. Determination of phage genome termini is usually a challenge to molecular biologists. Conventional methods first use these data to assemble the full sequence of a bacteriophage genome, and then carry out molecular biology experiments to identify its termini. Conventional termini analysis is complicated, time-consuming and expensive. Unlike these conventional methods, we propose a High Occurrence Read Termini theory to study the bacteriophages' genome termini, as well as genome packaging, directly using only the HTS reads data. Compared to conventional methods, our approach acquires relevant information of a bacteriophage's genome termini, including genome type, the position and sequence of termini and the length of gene repeated termini sequences, without any secondary lab experiments, largely shortens the analysis time and decreases the cost.

Experiments on the T4-like bacteriophages, the *Enterococcus* bacteriophages, the Twort-like bacteriophages, T7-like bacteriophage, Viuna-like bacteriophage, and some other bacteriophages (*see* Table 1) have proved the correctness and effectiveness of the High Occurrence Read Termini theory and the related termini determination method. Using this theory, we have already identified the genome termini and sequence characteristics along with replication of T4-like, *Enterococcus faecalis*, *Enterococcus faecium*, Twort-like *Staphylococcus aureus*, and λ-like bacteriophages. For this chapter we have summarized the different groups of bacteriophage termini and replication characteristics. For example the T3, T7-like and N4-like bacteriophages, have biospecific and repeat termini (the termini analysis of T3 and the replications of T3 and IME-11, are given in Table 2 and Fig. 1). The T4-like bacteriophages, have bioconsensus (nonspecific) and repeat termini (the results of IME-08 are given in Table 3 and Fig. 1). *Enterococcus* bacteriophages IME-EF4, IME-EFm1 and IME-EF3, have biospecific and nonrepeat termini with 9 bp 3′ protruding cohesive ends (we listed the results of IME-EF4 and IME-EFm1 here, *see* Table 4 and Fig. 1). The Twort-like *S. aureus* IME-SA1 and IME-SA2 have

Table 1
Bacteriophages analyzed in our laboratory

Bacteriphage	Family	Genus	Genome length (bp)	Host bacteria	Accession number
IME-EF4	Siphoviridae	N/A	40,713	*Enterococcus faecalis*	NC_023551.1
IME-EFm1	Siphoviridae	N/A	42,599	*Enterococcus faecium*	NC_024356.1
IME-SA1	Myoviridae	Twort-like virus	140,218	*Staphylococcus aureus*	
IME-SA2	Myoviridae	Twort-like virus	140,906	*Staphylococcus aureus*	
IME-08	Myoviridae	T4-like	172,253	*Escherichia coli 8099*	NC_014260
IME-09	Myoviridae	T4-like	166,499	*Escherichia coli 8099*	NC_019503
IME-EC1	Myoviridae	T4-like	170,335	*Escherichia coli*	
IME-EC2	N/A	N/A	41,510	*Escherichia coli*	
IME-EC16	Podoviridae	T7-like	38,870	*Escherichia coli*	
IME-EC17	Podoviridae	T7-like	38,870	*Escherichia coli*	
IME-11	Podoviridae	N4-like	72,570	*Escherichia coli*	NC_019423
T3	Podoviridae	T7-like	38,208	*Escherichia coli*	KC960671
IME-SF1	Podoviridae	T7-like	38,842	*Shigella flexneri*	
IME-SF2	Podoviridae	T7-like	40,387	*Shigella flexneri*	
IME-AB2	Myoviridae	N/A	43,665	*Acinetobacter baumannii*	JX976549
IME-AB3	Myoviridae	N/A	43,050	*Acinetobacter baumannii*	NC_023590.1
IME-SM1	N/A	N/A	149,960	*Serratia marcescens*	
IME-SL1	Myoviridae	Viuna-like virus	153,667	*Salmonella*	
Iridovirus W150	Iridoviridae	Irido virus	162,590	*Aedes albopictus C6/36 cells*	

Table 2
T3 terminal sequence occurrence frequency statistics

Bacteriophage	Strand	Ave. freq.	Ter. freq.	Occurrence frequency ratios	Terminal sequence
Tagged T3	Positive	1.35	890	659	TCTCATAGTTCAAGAACCCA
	Negative		709	525	AGGGACACATAGAGATGTAC
Un-tagged T3	Positive	5.12	1570	306	TCTCATAGTTCAAGAACCCA
	Negative		1262	246	AGGGACACATAGAGATGTAC

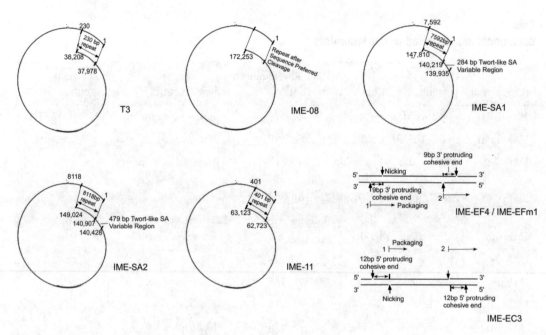

Fig. 1 Bacteriophage replication

bispecific and long repeat termini with an adjacent variable region (*see* Fig. 1). The λ-like bacteriophage IME-EC3 and IME-EC2, have biospecific and nonrepeat termini with about 10 bp 5′ protruding cohesive ends (*see* Fig. 1 about the replication of IME-EC3). Finally the *Acinetobacter baumannii* bacteriophages IME-AB1 and IME-AB2 have no special termini characteristics, and thus we hypothesize that they have completely random termini.

Theoretically, this method can be applied to analysis of other microbe genome termini, like that of plant and animal viruses.

2   Materials

2.1   Software and Website

Velvet, ABYSS, SOAPdenovo, CLC genomics workbench (Aarhus, Denmark), MEGA 5.10, in-house UNIX shell commands, NCBI website, RAST (Rapid Annotation using Subsystem Technology), Kodon (Applied Math, Sint-martens-Latem, Belgium), the bacteriophage genome database of the European Molecular Biology Laboratory (EMBL), tRNAscan-SE (v.1.21).

2.2   Reagents

DNase I and RNase A (Thermo Scientific, USA), 10% sodium dodecyl sulfonate, 500 mM EDTA, 1 mg/mL proteinase K, phenol, phenol–chloroform–isoamyl alcohol (25:24:1), isopropanol, ice 75% ethanol, distilled water, Ion Shear™ Plus Reagents, E-Gel® SizeSelect™ agarose gel, Fastx toolkit, AMPure beads (Beckman Coulter, California, USA), T4 DNA polymerase

Table 3
Top 20 reads with the highest occurrence frequencies

Read sequence	Occurrence frequencies				Genome presence		
	Rank	Total	1.fq	2.fq	Strand	Position	Seq. containing upstream
GCTCTTCCGGAAAGGTCAAAAACAGTTTGAG	1	828	427	401	Positive	30,641	TCTATTTGGAGCTCTTCGGA
GTTTACAGAATCGTACTCGGCCTTGTTCG	2	705	388	317	Positive	3272	AATTACTGGAGTTTACAGA
GTATAATGATTCATCAACAAACAAAAGACA	3	692	383	309	Negative	30,486	CCCTTTTGGAGTATAATGAT
GCGTAATTCCACCTTTTCTTCCCAATCTT	4	673	352	321	Positive	52,555	TCTTGTTGGAGCGTAATTCC
GGTATACATCATTAAATAACGATGTATATC	5	641	333	308	Positive	163,251	AGAAATTGGAGTATACATC
GTATTTCAAGAAACGTGATAAAGCCCAGGC	6	577	318	259	Positive	121,764	AACGTTTGGAGTATTTCAAG
GCGTAATTGCTTCAGGTAAGCCTTTAGGAT	7	505	256	249	Positive	73,140	AGAATATGGAGCGTAATTGC
GTGCATGATTGGTAACAGTTCGGCAACCCA	8	505	277	228	Positive	40,411	GGTCTTTGGAGTGCATGATT
GTTTTACAGACAACGCAAATCTTATCTGAC	9	496	253	243	Positive	115,803	ATCGATTGGAGTTTTACAGA
GCTGAAAAGGCAGCTGAAACTAAAGCCGCT	10	494	270	224	Negative	3702	TAAATTAGCAGCTGAAAAGG
GTATAATGTAAAAACAAACCTGAGGAAATT	11	490	274	216	Negative	32,654	CTCCCTTGGAGTATAATGTA
GTATTAACAAGATTCCAGAATTTCTCACCC	12	481	253	228	Negative	75,276	GTTTTCTTGGAGTATTAACAA
GTTTCTCAGCGATTTAATCGACCACTCTT	13	448	238	210	Positive	29,924	TCGTCTTGGAGTTTCTCAGC
GTTACATAAGCATCAGGAGCAGATGGTCCC	14	445	254	191	Negative	102,003	TTGCTTTGGAGTTACATAAG
GCTTTAATCTTAACAATAGTGCCGAGATAA	15	443	245	198	Positive	165,136	GTATTTACCTGCTTTAATCT
GCTGAACGTACCGAAGTTGCAGGTATGACT	16	440	266	174	Negative	28,799	GTTGTTCAGAGCTGAACGTA
GTATAATCTTTCTATCAACTTGAGGAGAAT	17	434	217	217	Negative	46,215	GATGGATGGAGTATAATCTT
GCTGCATCTTCAGATTGGTCTTCGTCTTCA	18	431	251	180	Positive	5448	TTCAGATGGAGCTGCATCTT
GTTATTACTAAACAAGTTTTTAACCGCACT	19	426	222	204	Negative	122,567	CTCCCTTGGAGTTATTACTA
GTTAACAAATGCCATACGACATTTAAGGGA	20	425	208	217	Positive	56,968	AACGTTTAGAGTTAACAAAT

Table 4
IME-EF4 and IME-EFm1 terminal sequence occurrence frequency statistics

Bacteriophage	Strand	Ave. freq.	Ter. freq.	Occurrence frequency ratios	Terminal sequence
IME-EF4	Positive	6.73	1322	196	ATTAGTTTCTTCAAAAAATT
	Negative	6.73	2318	344	CTTTCGCTTAAACGAATCTC
IME-EFm1	Positive	12.95	3194	246	ATTAATTCGTTATAAAAAGG
	Negative	12.95	4412	341	CTCTTCTTCGCACGAAATTC

2.3 Instruments

(IonTorrent, San Francisco, USA). All the primers were synthesized by the Sangon Company (Beijing, China). Life Technologies Ion Torrent Personal Genome Machine (PGM) Ion Torrent sequencer (IonTorrent), Solexa HiSeq2000 Genome Analyzer (Illumina, San Diego, USA).

2.4 Bacteriophages

The analyzed lytic *Enterococcus* bacteriophages IME-EF4 and IME-EFm1, Twort-like bacteriophages IME-SA1 and IME-SA2, T4-like bacteriophage IME-08 and IME-09, N4-like bacteriophage IME-11, and λ-like bacteriophage IME-EC3 and IME-EC2 were isolated from sewage in the Chinese PLA hospital 307 (Beijing, China),. The detail information of above bacteriophages and others analyzed in our laboratory is listed in Table 1. The information of bacteriophages from IME-08 to Iridovirus W150 is reproduced from [3], Fig. 2 displays their genera.

3 Methods

All bacteriophages were processed through isolation, purification, concentration (optional step), genomic DNA extraction and high-throughput sequencing. The termini and replication characteristics of abovementioned bacteriophages were discovered using High Occurrence Read Termini Theory theory.

3.1 Samples Preparation

3.1.1 Bacteriophage Isolation and Purification

Enrichment cultures [4] were used to isolate bacteriophages from sewage in the Chinese PLA hospital 307 (Beijing, China), while the host bacteria for these bacteriophages were isolated from clinical samples in the same hospital. Bacteriophage purification, concentration, and replication were carried out by standard methods as described previously [5]. The bacteriophage titer was assessed using the double layer agar technique according to methods described in [4].

3.1.2 Bacteriophage Genomic DNA Extraction

Bacteriophage DNA was extracted based on the previously published method [6]. In brief, DNase I and RNase A were added to

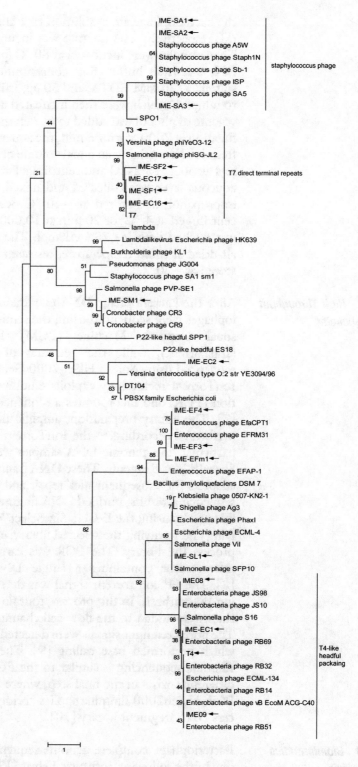

Fig. 2 Phylogenetic tree of bacteriophages. The bacteriophages analyzed in our laboratory are indicated by *arrows*. The tree was generated from an amino acid alignment (gap open cost—10; gap extension cost −1; end gap cost—free) using the Maximum Likelihood method with 1000 bootstrap replicates (MEGA 5.10)

the bacteriophage stock solution to a final concentration of 1 μg/ mL respectively. The mixture was incubated overnight at 37 °C. Then samples were incubated at 80 °C for 15 min to deactivate the DNase I. Lysis buffer (final concentration, 0.5% sodium dodecyl sulfonate, 20 mM EDTA, and 50 μg/mL proteinase K) was added to samples, which were then incubated at 56 °C for 1 h. An equal volume of phenol was added to extract the DNA. Following centrifugation at $7000 \times g$ for 5 min, the aqueous layer was removed to a fresh tube containing an equal volume of phenol–chloroform–isoamyl alcohol (25:24:1) and centrifuged at $7000 \times g$ for 5 min. The aqueous layer was collected and mixed with an equal volume of isopropanol, and stored at −20 °C overnight. The mixture was centrifuged at 4 °C for 20 min at $10,000 \times g$, and the DNA pellet was washed with ice 75% ethanol. The resulting DNA was then air-dried at room temperature, resuspended in distilled water, and stored at −20 °C.

3.2 High-Throughput Sequencing

After the bacteriophage DNA extraction, the genomes of bacteriophages were sequenced using the semiconductor sequencer Personal Genome Machine (PGM) Ion Torrent sequencer (IonTorrent), while the genomes of T3 and IME-11 were sequenced using Solexa HiSeq2000 Genome Analyzer. This PGM IonTorrent technology exploits emulsion polymerase chain reaction (PCR) and incorporates a sequencing-by-synthesis approach [7]. The library preparation, amplification, and sequencing were performed according to the IonTorrent sequencing protocols. In particular, the genome DNA samples were sheared using the Ion Shear™ Plus Reagents. These DNA fragments were then ligated to adapters for subsequent nick repair and purification. For the best sequencing results, purified DNA fragments of about 300 bp were selected by using the E-Gel® SizeSelect™ agarose gel. After amplifying and purifying the selected library, emulsion PCR was used to process the library. The PCR was carried out in a water-in-oil microreactor containing a single DNA molecule on a bead [8]. The H^+ ion torrent signal was detected during the sequencing-by-synthesis. In this process, four fluorescently labeled nucleotides were added to the flow-cell channels during DNA synthesis. The florescent light signals were detected by the Genome Analyzer, which performed base calling [9]. The protocol of HiSeq2000 Illumina sequencing is similar to the PGM IonTorrent. The only difference exists in the final step, where instead of using emulsion PCR, HiSeq2000 Illumina applies "bridging" amplification to process DNA fragment library [10].

3.3 Bioinformatics Analysis

Bacteriophage complete genome sequences were assembled using one of the following software: Velvet [12], ABYSS [13], SOAPdenovo [14], and CLC genomics workbench (Aarhus, Denmark). The adaptor sequences were removed using Fastx toolkit [15]. The occurrence frequency of each read was calculated using

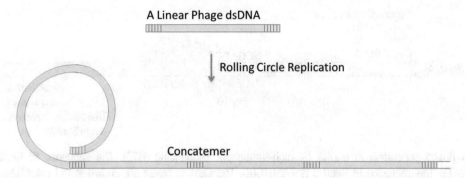

Fig. 3 Simplified sketch of the bacteriophage linear dsDNA. Terminal repetitions are located at 5′ and 3′ ends

in-house UNIX shell commands. Blast search was performed on NCBI website [16] to find each bacteriophage's similar sequences. Genome sequences without adaptors were mapped onto the assembled sequences using CLC genomics workbench. Genome annotation was implemented using RAST (Rapid Annotation using Subsystem Technology) [17]. Bacteriophage conserved coding DNA sequences (CDSs), such as large terminase subunit, were selected for phylogenetic analysis using MEGA 5.10. The potential coding regions of genome was predicted using the software Kodon (Applied Math, Sint-martens-Latem, Belgium) with a minimum open reading frame (ORF) size of 50 amino acids, and with the "Bacterial and Plant Plastid Code" as translation table. These putative coding regions were then aligned with the bacteriophage genome database of the European Molecular Biology Laboratory (EMBL), where the best matches were used to annotate and finally identify the number ORFs. tRNA genes were predicted using tRNAscan-SE (v.1.21) [18].

3.4 High Occurrence Read Termini Theory

As illustrated in Fig. 3, a bacteriophage with a linear double-stranded DNA (dsDNA) has terminal repetitions. These repetitions are used for homologous recombination during the bacteriophages DNA replication. The bacteriophages dsDNA can be circularized through the genome terminal repetitions; therefore, identifying the natural genome termini of bacteriophage, which are cleaved by the terminase, is difficult. In this study, we conceived the High Occurrence Read Termini theory, which can find the natural termini using the read occurrence frequency (*see* **Note 1**).

Suppose that there are m identical genomes, the length of each genome is L. All the genomes are divided into N_r short sequences. Each short sequence is called a read. The average length of the reads is L_{reads}.

Theorem 1: $R = \frac{Freq_{ter}}{Freq_{ave}} = 2 \times L_{reads}$.

Proof 1.: *There are* m *identical genomes. As illustrated in Fig. 4, the high-throughput sequencing (HTS) machine reads each read from 5′*

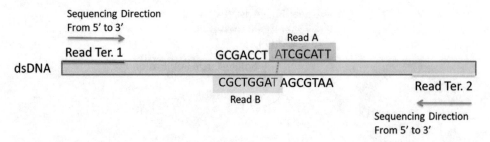

Fig. 4 Reads generation in dsDNA by high-throughput sequencing (HTS). The sequence of Read A is ATCGCATT. The sequence of Read B is TAGGTCGC. The starting bases are shown in red color. Read Ter. 1 and Read Ter. 2 are the two reads beginning with the natural termini

to 3′. Thus, each genome with dsDNA has two termini. The occurrence frequency of the reads starting with a natural terminus is:

$$\text{Freq}_{\text{ter}} = m \tag{1}$$

There are totally N_r reads for the m genomes. As shown in Fig. 4, two different reads, A and B, start with base A and base T. Thus, the average occurrence frequency of all the reads is:

$$\text{Freq}_{\text{ave}} = \frac{N_r}{2 \times L} \tag{2}$$

The ratio of Freq_{ter} to Freq_{ave} is

$$R = \frac{\text{Freq}_{\text{ter}}}{\text{Freq}_{\text{ave}}} = \frac{m}{\frac{N_r}{z \times L}} = \frac{2 \times m \times L}{N_r} = 2 \times L_{\text{reads}} \tag{3}$$

3.4.1 High Occurrence Read Termini Theory Verification (Optional Step)

To verify High Occurrence Read Termini theory, we firstly tag the termini of a bacteriophage genomic DNA. If the termini are identical with the HTS reads having the highest occurrence frequencies, the High Occurrence Read Termini theory works. In this chapter, the HTS reads having the highest occurrence frequencies is shortly named high frequency read sequences (HFS).

In particular, a pair of complementary oligonucleotides is designed as shown in Table 5. The two oligonucleotides were annealed together and formed a double stranded adaptor with a base T overhanging at the 3′ terminus. As shown in Fig. 5, this designed adaptor was used as a tag of bacteriophage termini. The termini of a bacteriophage genome were polished blunt with T4 DNA polymerase (IonTorrent, San Francisco, USA) and were phosphorylated at the 5′ end with T4 Polynucleotide Kinase (Ion Torrent). A base A was added at the 3′ end with Taq DNA polymerase (Ion Torrent) using standard protocols. The designed tag adaptors wehre then ligated to the modified bacteriophage termini in a reaction mixture containing 25 μL end repaired genomic DNA sample, 1 μL annealed adaptor, 1 μL T4 DNA ligase (Ion Torrent)

Table 5
Adaptor ligated into a bacteriophage termini

Tag Adaptor	Sequence
1	5′-AGTGTAGTAGT-3′
	3′-TCACAT CATCA-5′

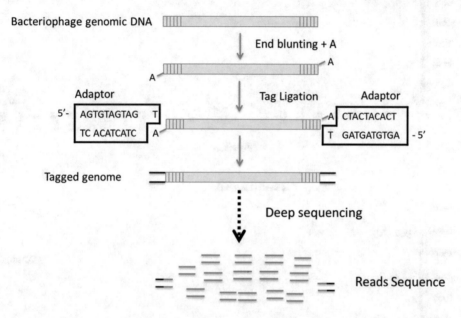

Fig. 5 Tag adaptor ligation to the genome termini of a bacteriophage

and 10 μL 10× ligase buffer (Ion Torrent) followed by incubation at 25 °C for 10 min., then the ligated DNA was purified using AMPure beads.

We choose a well-studied T3 bacteriophage as a model. First the termini of T3 genome sequences were identified using our theory. Then the termini of T3 genomic DNA were tagged using the synthetic dsDNA adaptor (Table 5). A sample of un-tagged T3 bacteriophage genome was also prepared as a control group (*see* **Note 2**). The HTS results of tagged T3 genome and untagged T3 genome were analyzed. As shown in Fig. 6, the respective two HFS of the tagged T3 genome and the untagged T3 genome are rightly located at their termini. As shown in Table 2, the occurrence frequency ratios of terminal reads to general reads are 659 (positive strand of tagged T3 genome), 525 (negative strand of tagged T3 genome), 306 (positive strand of un-tagged T3 genome), and 246 (negative strand of un-tagged T3 genome), respectively. Moreover, the terminal tagged T3 genome sequences were identical to the terminal un-tagged T3 genome sequences, and also consistent

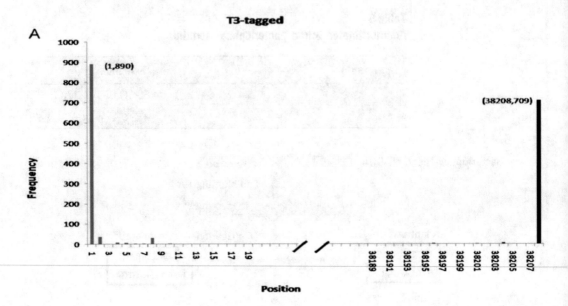

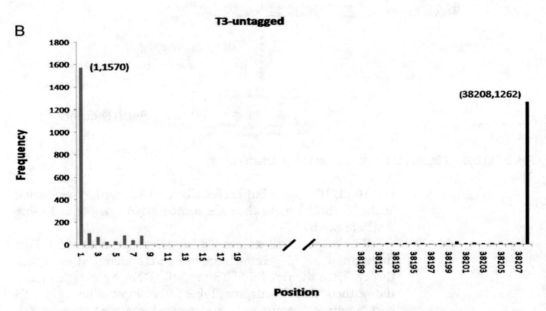

Fig. 6 Distribution of high occurrence frequency reads in the T3 genome. Reads with highest occurrence frequencies are the T3 genome termini (**a**) Adaptor tagged T3 genome. (**b**) Un-tagged T3 genome. Strand orientation and nucleotide numbering are adopted from the complete genome sequence record of T3 phage in GenBank (NC_003298)

with the previously reported T3 terminal sequences (NC_003298). This result confirms the validity of the High Occurrence Read Termini theory. After identifying the termini of T3 and assembling its complete genome using CLC genomics workbench, its replication is acquired and shown in Fig. 1, where the full length of T3 is 38,208 bp with 230 bp repeat.

3.4.2 Consensus Termini Determination of T4-Like Bacteriophage Using High Occurrence Read Termini Theory

T4-like bacteriophages is a model species contributing broadly to our understanding of molecular biology. T4-like DNA replication and packaging share various mechanisms with human double-stranded DNA viruses such as herpes virus. Our HTS data and experimental results illustrate that T4-like bacteriophage has consensus termini with sequence-preferred terminase cleavage. More details can be reached at [19].

Here, taking T4-like bacteriophage IME-08 as an example, we use High Occurrence Read Termini theory to analyze its genome termini.

HTS Read Occurrence Frequency Statistics

The IME-08 was high-throughput sequenced using the Solexa Genome Analyzer, which generated 5,011,480 pairs of reads. Each pair of read was separately stored in 1.fq and 2.fq files, with average lengths of 73 bp and 75 bp respectively [20]. The occurrence frequencies of HTS generated reads were calculated using High Occurrence Read Termini theory (*see* **Note 3**). As shown in Fig. 7, about 70% reads have 6–22 occurrence frequencies with the most occurrence frequency 13. The top 20 HFS are listed in Table 3. Compared with the average occurrence frequency of 13, all the occurrence frequencies of the top 20 HFS have more

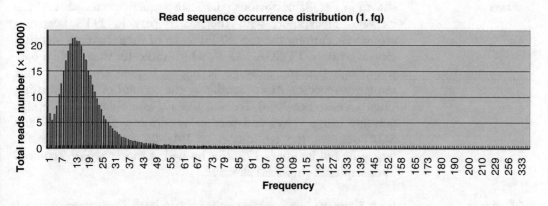

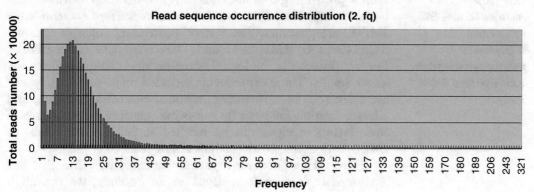

Fig. 7 Read sequence distribution in T4-like bacteriophage IME-08 genome

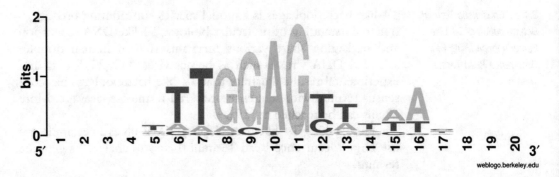

Fig. 8 Sequence logo of the upstream-contained read with top 20 highest occurrence frequencies (This figure was generated by using Weblogo [21])

than 400 times. For further start base statistics, the genome sequences containing each read upstream sequence is also listed in Table 3, where the bases are highlighted when they are both in read sequence and in upstream-contained sequence.

Consensus Termini Reveal Sequence-Preferred Cleavage by Bacteriophage Terminase

Weblogo [21] was used to generate the sequence logo of the top 20 HFS along with their upstream sequences, which were retrieved from the assembled IME-08 genome (full length: 172,253 bp). As shown in Fig. 8, an obvious consensus sequence exists around the cleavage breakpoint, with the major part in HTS upstream sequences. Among the top 20 HFS, 16 of them have an identical cleavage site 5'-TTGGA...G-3', which indicates that T4-like bacteriophage genome cleavage is highly sequence-preferred (not sequence-specific). After identifying the termini of IME-08 using High Occurrence Read Termini theory, assembling its complete genome using Velvet and further verification using ABYSS and SOAPdenovo, the replication of IME-08 is shown in Fig. 1, where its full length is 172,253 bp with a repeat region acquired after sequence-preferred cleavage.

3.4.3 Nine bp 3′ Protruding Cohesive Ends Determination of Enterococcus Bacteriophage Using High Occurrence Read Termini Theory

9 bp 3′ protruding cohesive ends exist in both *Enterococcus faecalis* bacteriophage IME-EF4 and *Enterococcus faecium* bacteriophage IME-EFm1 genomes. The 9 nt 3′ protruding cohesive ends are TCATCACCG (IME-EF4) and GGGTCAGCG (IME-EFm1). Further molecular biological experiments could confirm the above results. These experiments included mega-primer polymerase chain reaction sequencing, terminal run-off sequencing, and adaptor ligation followed by run-off sequencing. More details of their termini analysis can be reached at [22]. More details of IME-EFm1 characteristic analysis can be reached at [23].

Here, taking *Enterococcus faecalis* bacteriophage IME-EF4 and *Enterococcus faecium* IME-EFm1 as an example, we use High Occurrence Read Termini theory to identify the termini of *Enterococcus* bacteriophages.

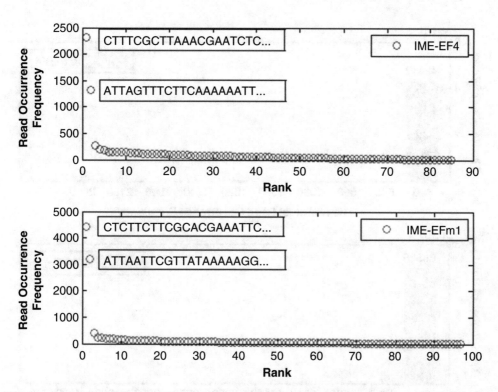

Fig. 9 Read occurrence frequency statistics. *Blue* and *red circles* represent the HTS reads of IME-EF4 and IME-EFm1 samples, respectively

HTS Read Occurrence
Frequency Statistics

As shown in Fig. 9, IME-EF4 and IME-EFm1 HTS read data have two significant high frequency read sequences (HFS), beginning with CTTTCGCTTAAACGAATCTC and ATTAGTTTCTT-CAAAAAATT, CTCTTCTTCGCACGAAATTC and ATTAATTCGTTATAAAAAGG, respectively. The HTS data of IME-EF4 and IME-EFm1 share similar sequence occurrence frequency curves. Figure 10 shows that more than 99% of reads have occurrence frequencies less than 237.2 (IME-EF4) and 446.6 (IME-EFm1). Using the High Occurrence Read Termini theory, we concluded that the two reads with the highest occurrence frequencies rightly represent the termini of bacteriophage IME-EF4. As shown in Table 4, the occurrence frequency ratios of terminal reads to general reads are 196 (positive strand of IME-EF4), 344 (negative strand of IME-EF4), 246 (positive strand of IME-EFm1), and 341 (negative strand of IME-EFm1), respectively.

Nine bp 3′ Protruding
Cohesive Ends of IME-EF4
and IME-EFm1

CLC genomics workbench was used to assemble the complete genomes of IME-EF4 and IME-EFm1. All related HTS reads were mapped to assembled IME-EF4 and IME-EFm1 genome sequences, separately. The mapping results are shown in Fig. 11, where the occurrence frequency of reads having the 9 nt protruding

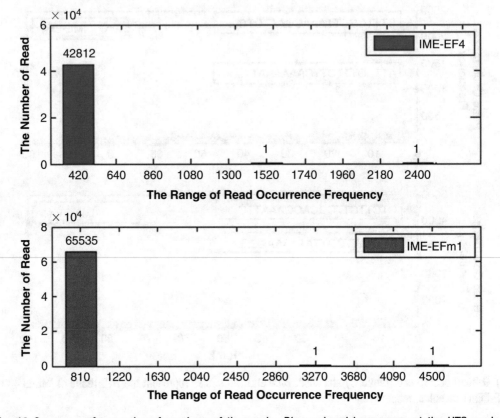

Fig. 10 Occurrence frequencies of numbers of the reads. *Blue* and *red bars* represent the HTS reads of IME-EF4 and IME-EFm1 samples, respectively. The number on the *x-axis* represents the range from the last number to the next number, i.e., "2318" means the read occurrence frequency during the range [2086.8,2318]

cohesive end is less than 20 times. Using the termini and complete sequence analysis results, their replications are depicted in Fig. 1, where IME-EF4 and IME-EFm1 both have 9 bp 3′ protruding cohesive ends.

Mega-Primer PCR Sequencing (Optional Verification Step)

We carried out PCR amplification for the IME-EF4 DNA genome. The PCR implements a mega-primer guided polymerization through the protruding cohesive end.

The genome sequence snapshot including the 9 base is shown in Fig. 12 (a), where the upstream sequence is "…GAGATTCGTTTAAGCGAAAG" and the downstream sequence is "ATTAGTTTCTTCAAAAAATT…". The PCR result is consistent with our HTS data statistical mapping result (*see* Fig. 11 (a)). It proves that both the IME-EF4 complete genome and the protruding cohesive ends acquired from the HTS data statistics are correct.

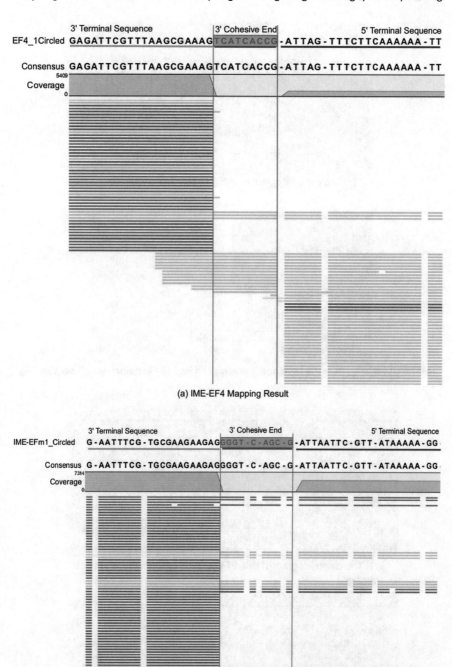

(a) IME-EF4 Mapping Result

(b) IME-EFm1 Mapping Result

Fig. 11 IME-EF4 and IME-EFm1 mapping results. The mapped reads are acquired from the original HTS data. The 3′ terminal sequences are underlined in orange, the 5′ terminal sequences are underlined in *dark red*, and the 3′ protruding cohesive ends are underlined in *blue*.

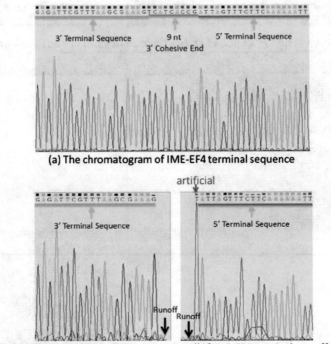

(a) The chromatogram of IME-EF4 terminal sequence

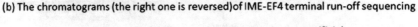

(b) The chromatograms (the right one is reversed) of IME-EF4 terminal run-off sequencing

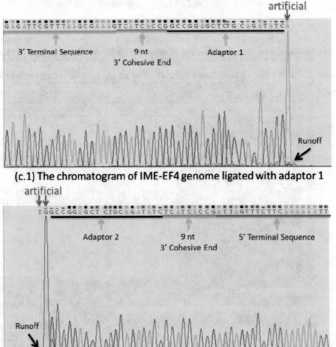

(c.1) The chromatogram of IME-EF4 genome ligated with adaptor 1

(c.2) The chromatogram (reversed) of IME-EF4 genome ligated with adaptor 2

Fig. 12 Chromatograms of the three molecular biology experiments, including IME-EF4 complete genome sequencing (**a**) IME-EF4 terminal run-off sequencing (**b**) and IME-EF4 adaptor ligation to the terminal sequences (**c**)

Terminal Run-off Sequencing (Optional Verification Step)

The IME-EF4 complete genome was used as the template for terminal run-off sequencing, the process of which is described in [11]. The 3′ end of the IME-EF4 genome was marked using the primer P1 (5′-CTCTAGTTTGTTGCGTGCGTAAATC-3′). The 5′ end of the IME-EF4 genome was marked using the primer P4 (5′-AGGTACGGACCGCAATGGGTTGGGA-3′).

As illustrated in Fig. 13, four different dsDNA protruding cohesive end cases hypothetically exist. Case 1 describes the situation of a negative strand protruding cohesive end, case 2 represents the positive strand protruding cohesive end, case 3 shows the 5′ protruding cohesive end situation, and case 4 describes our putative IME-EF4 situation of the 3′ protruding cohesive end.

If the IME-EF4 has the protruding cohesive end situation according to cases 1, 2, or 3, then we would find signals after either the terminal sequence "GAGATTCGTTTAAGCGAAAG", "ATTTTTTGAAGAAACTAATA", or both of them in the terminal run-off sequencing result. However, as shown in Fig. 12 (b), no signal was detected after the termini "GAGATTCGTTT AAGC-GAAAG" in the positive strand or after the termini "ATTTTTT-GAAGAAACTAATA" in the negative strand. This result further confirmed our conclusion that IME-EF4 has a linear, double-stranded DNA genome with a 9 nt 3′ protruding cohesive end, represented by case 4 of the 3′ protruding cohesive end situation.

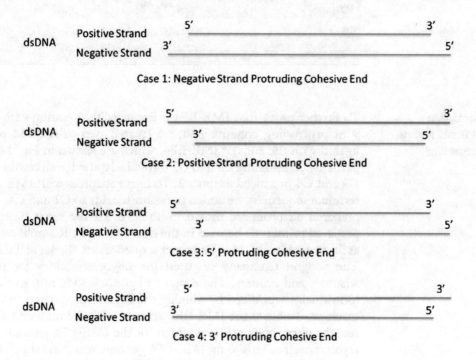

Fig. 13 Hypothetical genome dsDNA protruding cohesive end situations

Fig. 14 Illustration of the adaptors and primers used in the experiments. Adaptor 1 includes C1 and C2, and adaptor 2 includes C3 and C4. P2 is highlighted in green, and the P3 is highlighted in *blue*

Table 6
Prepared adaptor list

Adaptor		Sequence (From 5′ to 3′)
Adaptor	C1-P	P-GCCGGAGCTCTGCAGATATC
1	C2	GATATCTGCAGAGCTCCGGC-CGGTGATGA
Adaptor	C3	GCCGGAGCTCTGCAGATATC-TCATCACCG
2	C4-P	P-GATATCTGCAGAGCTCCGGC

Table 7
Prepared primer list

Primer	Sequence (From 5′ to 3′)
P1	CTCTAGTTTGTTGCGTGCGTAAATC
P2	GATATCTGCAGAGCTCCGGC
P3	GCCGGAGCTCTGCAGATATC
P4	AGGTACGGACCGCAATGGGTTGGGA

Adaptor Ligation to the Termini (Optional Verification Step)

To further prove that IME-EF4 is linear dsDNA ending with a 3′ 9 nt protruding cohesive end, we created two pairs of adaptors ligated with the ends of IME-EF4, which are shown in Fig. 14. In particular, combining C1 and C2 formed adaptor 1, and combining C3 and C4 produced adaptor 2. To ligate adaptors with IME-EF4 terminal sequences, we added phosphoric acids to C1 and C4. The prepared adaptors are shown in Table 6. At the same time, we prepared primers to be used in the next step of PCR amplification, as illustrated in Fig. 14. The primer sequences are shown in Table 7. The Sangon Company prepared the oligonucleotides for these adaptors and primers. The adaptor oligonucleotide mix was then individually hybridized by running the ligation program on a PCR machine. To ligate the IME-EF4 genome with the prepared adaptor, the phosphoric acid was added for the IME-EF4 genome end repair. Specifically, 800 ng IME-EF4 genome was diluted to 16 μL

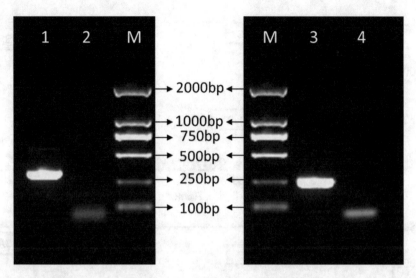

Fig. 15 Agarose gel electrophoresis to confirm adaptor ligation. "M" represents marker, "1" is the PCR result of the 3′ IME-EF4 genome sequence ligated with adaptor 1 (primer P1 and P2), "3" is the PCR result of the 5′ IME-EF4 genome sequence ligated with adaptor 2 (primer P3 and P4), and "2" and "4" are the negative control PCR results of the 3′ and 5′ IME-EF4 genome sequence without any adaptor

using nuclease-free water, and then the end repair mix was prepared. The phosphate groups were added also by running the program on a PCR machine. The adaptor ligated DNA was purified according to the instruction manual described in the NEBNext® Fast DNA Library Prep Set for Ion Torrent™ (version 3.1). The purified DNA samples were used as templates for PCR amplification and sequencing with the primer group of P1 and P2, and group of P3 and P4.

As shown in Fig. 15, the size of the 3′ IME-EF4 genome sequence ligated with adaptor 1 is about 280 bp, and the size of 5′ IME-EF4 genome sequence ligated with adaptor 2 is about 250 bp, which are consistent with our experimental design (please referring to the P1 and P2 illustration in Fig. 14). The sequencing results are shown in Fig. 12 (c). It further proves that the two adaptors (adaptor 1 and adaptor 2) have successfully ligated with the IME-EF4 genome, which means the IME-EF4 genome has a 9 bp 3′ protruding cohesive end.

3.4.4 Long Direct Termini of Twort-Like Bacteriophage

Here, taking Twort-like bacteriophage IME-SA1 and IME-SA2 as examples, we use High Occurrence Read Termini theory to identify the termini of Twort-like bacteriophage. About 8 kb direct terminal repeats (DTR) exist in both IME-SA1 (with 7592 kb DTR) and IME-SA2 (with 8118 kb DTR) genomes. More details can be reached at [24].

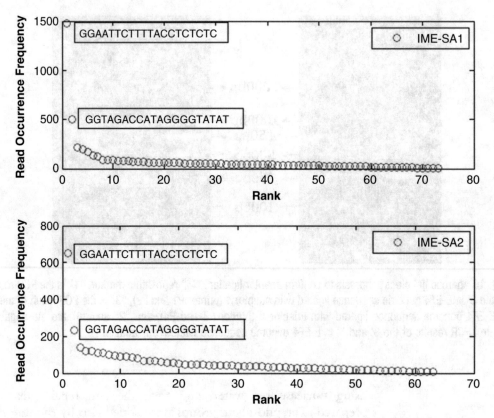

Fig. 16 Read occurrence frequency statistics. *Blue* and *red circles* represent the HTS reads of IME-SA1 and IME-SA2, respectively

HTS Read Occurrence
Frequency Statistics

For the IME-SA1 and IME-SA2 samples, we statistically analyzed all the HTS reads and then ranked all the read occurrence frequencies in descending order (*see* **Note 3**). As shown in Fig. 16, IME-SA1 and IME-SA2 HTS read data have two significant high-occurrence frequency reads, beginning with GGAATTCTTT-TACCTCTCTC and GGTAGACCATAGGGGTATAT, respectively. The HTS data of IME-SA1 and IME-SA2 share similar sequence occurrence frequency curves. Figure 17 shows that more than 99% of reads have occurrence frequencies less than 330 (IME-SA1) and 74 (IME-SA2). Using the High Occurrence Read Termini theory, we concluded that the two reads with the highest occurrence frequencies rightly represent the termini of bacteriophages IME-SA1 and IME-SA2. As shown in Table 8, the occurrence frequency ratios of terminal reads to general reads are 184 (positive strand of IME-SA1), 62 (negative strand of IME-SA1), 650 (positive strand of IME-SA2), and 234 (negative strand of IME-SA2), respectively.

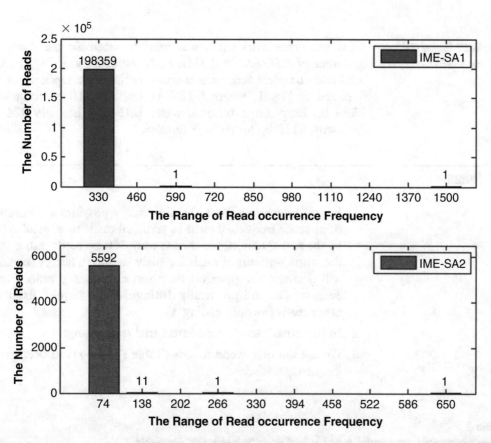

Fig. 17 Occurrence frequencies of numbers of the reads. *Blue* and *red bars* represent the HTS reads of IME-SA1 and IME-SA2, respectively. The number on the x-axis represents the range from the last number to the next number, i.e., "74" means the read occurrence frequency during the range [0,74]

Table 8
IME-SA1 and IME-SA2 terminal sequence occurrence frequency statistics

Bacteriophage	Strand	Ave. freq.	Ter. freq.	Occurrence frequency ratios	Terminal sequence
IME-SA1	Positive	8	1473	184	GGAATTCTTTTACCTCTCTC
	Negative		498	62	GGTAGACCATAGGGGTATAT
IME-SA2	Positive	1	650	650	GGAATTCTTTTACCTCTCTC
	Negative		234	234	GGTAGACCATAGGGGTATAT

| Long Direct Termini of IME-SA1 and IME-SA2 | CLC genomics workbench was used to assemble the complete genomes of IME-SA1 and IME-SA2, respectively. Using the termini and complete sequence analysis results, their replications are depicted in Fig. 1, where IME-SA1 has 147,810 bp long with 7592 bp long direct termini, while IME-SA2 has 149,024 bp long with 8118 bp long direct termini. |

4 Notes

1. High Occurrence Read Termini theory provides a theoretical occurrence frequency ratio of terminal reads to general reads. In the real situation, we choose only 300 bp reads rather than the whole amount of reads for analysis. Such a library selection will decrease the practical occurrence frequency ratio. However, we can still practically distinguish the termini from the other reads (*see* Subheading 3).

2. In the experiments, the control trial is necessary.

3. We use the below commands (Table 9) to do read occurrence frequency statistics.

Table 9
Linux commands for termini analysis and genome sequence assembly

Command	Usage
awk 'NR % 4 = = 2' input.fastq | sort | uniq -c | sort -g -r -o output.Freq	Do HTS read occurrence frequency statistics based on .fastq file
awk 'NR % 2 ==0' input.fasta | sort | uniq -c | sort -g -r -o output.freq	Do HTS read occurrence frequency statistics based on. fasta file
sed 'li temp' input.fastq | awk 'NR % 4 > 1' | sed 's/^@/\>/g' > output.fasta	Convert .fasta file to .fastq file
awk 'length ($1) > 50' input.fasta | cat -n | sed 's/\t/\n/g' | sed 's/\ //g' > output.fasta	Choose the reads of which the lengths are longer than 50 bases.
grep 'HFS in 5″ | sed 's/^.*HFS in 5′/HFS in 5″| grep 'HFS in 3″ | sed 's/HFS in 3′.*/HFS in 3′/g'	Choose out the high frequency read sequences (HFS)
echo sequences | rev | tr [ATCG] [TAGC] | tr [atcg] [tgac]	Convert a sequence into its reverse-complement counterpart
cat inputfile | awk '{print length, $0}' | sort -g | cut -d '' -f2	Order lines according to their lengths
grep 'Unmapped' ReadStatus.txt | cut -f 1 > unMappedReads	Extract the unmapped reads from HTS result

References

1. Hershey AD, Chase M (1952) Independent functions of viral protein and nucleic acid in growth of bacteriophage. J Gen Physiol 36 (1):39–56

2. McKenna M (2014) Drugs: gut response. Nature 508(7495):182–183

3. Li S, Fan H, An X, Fan H, Jiang H, Chen Y, Tong Y (2014) Scrutinizing virus genome termini by high-throughput sequencing. PLoS One 9(1):e85806

4. Adams MH (1959) Bacteriophages. Bacteriophages

5. Carlson K (2005) Appendix: working with bacteriophages: common techniques and methodological approaches. Bacteriophages:437–494

6. Green MR, Sambrook J (2012) Molecular cloning: a laboratory manual. Cold Spring Harbor Laboratory Press, New York

7. Loman NJ, Misra RV, Dallman TJ, Constantinidou C, Gharbia SE, Wain J, Pallen MJ (2012) Performance comparison of benchtop high-throughput sequencing platforms. Nat Biotechnol 30(5):434–439

8. Pennisi E (2010) Semiconductors inspire new sequencing technologies. Science 327 (5970):1190–1190

9. Zhang J, Chiodini R, Badr A, Zhang G (2011) The impact of next-generation sequencing on genomics. J Genet Genomics 38(3):95–109

10. Bentley DR, Balasubramanian S, Swerdlow HP, Smith GP, Milton J, Brown CG, Hall KP, Evers DJ, Barnes CL, Bignell HR (2008) Accurate whole human genome sequencing using reversible terminator chemistry. Nature 456 (7218):53–59

11. Lu S, Le S, Tan Y, Zhu J, Li M, Rao X, Zou L, Li S, Wang J, Jin X (2013) Genomic and proteomic analyses of the terminally redundant genome of the Pseudomonas aeruginosa phage PaP1: establishment of genus PaP1-like phages. PLoS One 8(5):e62933

12. Zerbino DR, Birney E (2008) Velvet: algorithms for de novo short read assembly using de Bruijn graphs. Genome Res 18(5):821–829

13. Simpson JT, Wong K, Jackman SD, Schein JE, Jones SJ, Birol I (2009) ABySS: a parallel assembler for short read sequence data. Genome Res 19(6):1117–1123

14. Li R, Li Y, Kristiansen K, Wang J (2008) SOAP: short oligonucleotide alignment program. Bioinformatics 24(5):713–714

15. Gordon A (2011) FASTX-Toolkit. Hannon Lab. http://hannonlab.cshl.edu/fastx_toolkit/index.html. Accessed 26 Nov 2014

16. NCBI (2014.) Nucleotide Blast http://blast. ncbi.nlm.nih.gov/Blast.cgi?PROGRAM=blastn&PAGE_TYPE=BlastSearch& LINK_LOC=blasthome. Accessed 26 Nov 2014

17. Aziz RK, Bartels D, Best AA, DeJongh M, Disz T, Edwards RA, Formsma K, Gerdes S, Glass EM, Kubal M (2008) The RAST server: rapid annotations using subsystems technology. BMC Genomics 9(1):75

18. Lowe TM, Eddy SR (1997) tRNAscan-SE: a program for improved detection of transfer RNA genes in genomic sequence. Nucleic Acids Res 25(5):0955–0964

19. Jiang X, Jiang H, Li C, Wang S, Mi Z, An X, Chen J, Tong Y (2011) Sequence characteristics of T4-like bacteriophage IME08 benome termini revealed by high throughput sequencing. Virol J 8:194

20. Sheng W, Huanhuan J, Jiankui C, Dabin L, Cun L, Bo P, Xiaoping A, Xin Z, Yusen Z, Yigang T (2010) Isolation and rapid genetic characterization of a novel T4-like bacteriophage. J Med Coll PLA 25(6):331–340

21. Crooks GE, Hon G, Chandonia J-M, Brenner SE (2004) WebLogo: a sequence logo generator. Genome Res 14(6):1188–1190

22. Zhang X, Wang Y, Li S et al. (2015) A novel termini analysis theory using HTS data alone for the identification of Enterococcus phage EF4-like genome termini. BMC Genomics 16 (414): DOI 10.1186/s12864-015-1612-3

23. Wang Y, Wang W, Lv Y, Zheng W, Mi Z, Pei G, An X, Xu X, Han C, Liu J (2014) Characterization and complete genome sequence analysis of novel bacteriophage IME-EFm1 infecting enterococcus faecium. J Gen Virol 95 (Pt 11):2565–2575

24. Zhang X, Kang H, Li Y et al. (2015) Conserved termini and adjacent variable region of Twortlikevirus Staphylococcus phages. Virologica Sinica 30(6):433–440

Chapter 12

Amplification for Whole Genome Sequencing of Bacteriophages from Single Isolated Plaques Using SISPA

Derick E. Fouts

Abstract

Genomics has greatly transformed our understanding of phage biology; however, traditional methods of DNA isolation for whole genome sequencing have required phages to be grown to high titers in large-scale preparations, potentially selecting for only those phages that can grow efficiently under laboratory conditions. This may also select for mutations or deletions that enable more efficient growth in culture. The ability to sequence a bacteriophage genome from a single isolated plaque reduces these risks while decreasing the time and complexity of bacteriophage genome sequencing. A method of amplification and library preparation is described, utilizing Sequence Independent Single Primer Amplification (SISPA), that can be used for whole genome shotgun sequencing of bacteriophages from a single isolated plaque.

Key words SISPA, Bacteriophage, Genomics, Single plaque, Sequencing

1 Introduction

Several methods for sequence-independent amplification of viral genomes for the purpose of whole genome sequencing have been developed [1–9]; however, none of them have been optimized for the purposes of whole genome sequencing from a single viral plaque until our method was first published [10]. Our method is based on the modifications that Djikeng et al. [6] made to the original SISPA method [11] to randomly amplify rather than ligate tagged primers and to clone via TOPO TA cloning instead of blunt-end ligation. One of the biggest advantages of the SISPA method is the simultaneous amplification and fragmentation of the genome. We further modified the Djikeng et al. protocol to increase sensitivity by 1000-fold, from 10 ng to as little as 10 pg of starting DNA. Specific modifications include: more robust removal of host nucleic acids, altered denaturation and annealing conditions, reduced reaction volumes, increased primer concentrations, additional primer and small fragment cleanup steps, and PCR conditions optimized for products between 300 and 850 bp. The continued popularity of

Martha R.J. Clokie et al. (eds.), *Bacteriophages: Methods and Protocols, Volume 3*, Methods in Molecular Biology, vol. 1681, https://doi.org/10.1007/978-1-4939-7343-9_12, © Springer Science+Business Media LLC 2018

the SISPA method in general is emphasized by recent applications to sequence positive-strand RNA viruses [12], and for rapid whole genome surveys of bacterial isolates [13].

The SISPA method described here incorporates high sensitivity optimizations for low yields of nucleic acids [10], but is geared more toward bacteriophages rather than eukaryotic viruses. It begins with the isolation of single, well-isolated plaques on a lawn of indicator host cells (Fig. 1). Phage particles are released from the agar plug through diffusion and filtered to remove host cells and

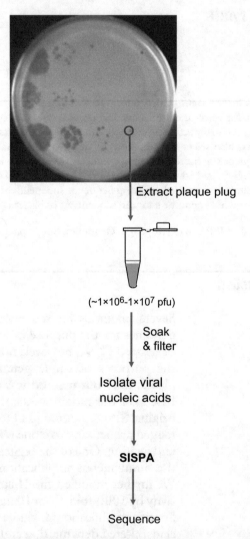

Extract plaque plug

(~1×10^6-1×10^7 pfu)

Soak
& filter

Isolate viral
nucleic acids

SISPA

Sequence

Fig. 1 Genome sequencing from a single plaque workflow. Single phage plaques are recovered using a sterile Pasteur pipette, soaked in SM buffer to release viral particles and filtered to remove bacterial and agar debris. Viral nucleic acids are extracted and libraries are created and amplified using a modified SISPA procedure. Gel-purified fragments are then sequenced following platform-specific methods

debris. Host and unpackaged phage nucleic acids are removed by cocktails of nucleases prior to extraction of the genomes using a proteinase K/SDS and phenol extraction method and precipitation aided by the addition of linear acrylamide [14]. The SISPA method begins with a mixture of template nucleic acids, primers and DMSO, denatured by heat and snap cooled on ice (Fig. 2). For DNA templates, two rounds of random-primed extension are

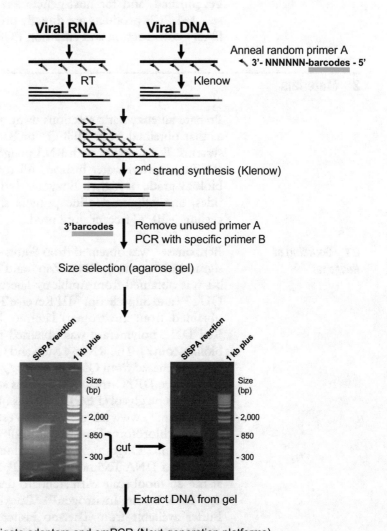

Fig. 2 Overview of the SISPA workflow. Oligonucleotide primers with a 3′ random hexamer and a 5′ bar-coded "tag", referred here as the "A" primer, are used to generate random single-stranded DNA products using either reverse transcriptase (for RNA genomes) or the large (Klenow) fragment of *E. coli* DNA polymerase I lacking exonuclease activity. A complementary second strand is generated using the Klenow fragment and amplified using primers specific to the 5′ tag sequence (the "B" primer). Fragments between 300 and 850 bp are extracted from agarose gels and ready for DNA sequencing

conducted using the large (Klenow) fragment of *Escherichia coli* DNA polymerase I. Unincorporated primers, small fragments (i.e., <200 bp), and single-stranded DNA are removed using PEG/MgCl$_2$ [15] and *E. coli* Exonuclease (Exo) I. The Klenow reaction product is then used as a template for PCR using a single primer representing the 20 bp barcode sequence portion of the SISPA random priming oligonucleotide. Amplicons of 500–850 bp are gel purified, and for next-generation sequencing platforms, size selected PCR products are directly processed for adaptor ligation, library construction and emulsion PCR.

2   Materials

Prepare all enzymatic reactions using sterile deionized water, such as that obtained from Milli-Q® or Barnstead™ water purification systems. For working with RNA phages, use diethylpyrocarbonate (DEPC)-treated water instead. All reagents should be molecular biology grade or higher. Enzymes, buffers, water aliquots, nucleotides, and oligonucleotide primers should be stored in a non-cycling −20 °C freezer until used.

2.1 Sourcing of Reagents

Benzonase® was obtained from Sigma-Aldrich Co. (https://www.sigmaaldrich.com/). RNaseZap® and a cocktail of RNases A and T1 was obtained from Ambion/Thermo Fisher Scientific, RNase-OUT™ and SuperScript® III Reverse Transcriptase (SSIIIRT) were obtained from Invitrogen/Thermo Fisher Scientific. BioMix™ Red DNA polymerase was obtained from Bioline (http://www.bioline.com/). The RNeasy Mini and QIAquick Gel Extraction kits were purchased from Qiagen (https://www.qiagen.com/). Sterile, RNase-free DEPC-treated water was sourced from Sigma-Aldrich. Polyethylene glycol (PEG) 8000 was obtained from USB Corporation (http://www.affymetrix.com/estore/browse/brand/usb/). Phenol/chloroform/isoamyl alcohol (PCI) was purchased from Invitrogen. Oligonucleotide primers were obtained from Integrated DNA Technologies (IDT; https://www.idtdna.com/) at the 25 nmol scale with standard desalting. The 6× gel loading dye used is the Invitrogen™ TrackIt™ Cyan/Yellow Loading Buffer available from Thermo Fisher Scientific (catalog number 10482035). Ethidium bromide (EtBr), and the 1 kb Plus DNA Ladder were purchased form Invitrogen.

2.2 Buffers and Solutions

1. SM buffer: 0.01% gelatin, 250 mM NaCl, 8.5 mM MgSO$_4$, 50 mM Tris–HCl, pH 7.5. Filter sterilize through a 0.22 μm filter and stored at 4 °C. For 1 l, mix together the following reagents:

   (a) 20 ml or 5 M NaCl.

(b) 8.5 ml of 1 M MgSO$_4$.

(c) 50 ml of 1 M Tris–HCl, pH 7.5.

(d) 10 ml of 1% gelatin (1 g/100 ml Milli-Q® water).

(e) Milli-Q® Water to 1 l in a graduated cylinder.

2. 10× Proteinase K digestion mix:

Reagent	10× Conc.	For 100 µl	For 200 µl	For 500 µl	For 1 ml
20 mg/ml Proteinase K	500 µg/ml	2.5 µl	5 µl	12.5 µl	25 µl
0.5 M EDTA	80 mM	16 µl	32 µl	80 µl	160 µl
10% SDS	5%	50 µl	100 µl	250 µl	500 µl
Sterile water[a]	–	31.5 µl	63 µl	157.5 µl	315 µl

[a]Use DEPC-treated water if isolating RNA

3. 10× TAE buffer.

Reagent	Total volume			Final Conc.
	500 ml	1 l	2 l	
Tris base	24.2 g	48.4 g	96.8 g	400 mM
Glacial acetic acid	5.72 ml	11.44 ml	22.88 ml	400 mM
0.5 M EDTA, pH 8.0	10 ml	20 ml	40 ml	10 mM

Procedure

1. Weigh out the Tris base, and add half the final volume of autoclaved Milli-Q water.

2. Stir (with stir bar) to dissolve while adding the next two reagents. Check the pH if needed, but recipe should produce a final solution at pH 8.0.

3. Bring up to the final volume with autoclaved Milli-Q water.

4. Do not autoclave, store at room temperature and dilute aliquots to 1× as needed with autoclaved Milli-Q water.

2.3 SISPA Primer Sequences

Primer name	Sequence	Usage
27F-YM	AGAGTTTGATY MTGGCTCAG	bacterial contamination check
1492R	TACCTTGTTACGACTT	bacterial contamination check
FR20RV-N	GCCGGAGCTCTGCAGATATCNNNNNN	SISPA random priming
FR20RV	GCCGGAGCTCTGCAGATATC	SISPA PCR step
BC004N	CGTAGTACACTCTAGAGCACTANNNNNN	SISPA random priming

BC004CG	CGTAGTACACTCTAGAGCACTA	SISPA PCR step
BC009N	CGAGCTCTATACGTGTAGTCTCNNNNNN	SISPA random priming
BC009CG	CGAGCTCTATACGTGTAGTCTC	SISPA PCR step
BC015N	CGTCGTACGCTGTCGTCGCGATNNNNNN	SISPA random priming
BC015CG	CGTCGTACGCTGTCGTCGCGAT	SISPA PCR step
BC019N	CGAGTATACGTACGTCTCAGTCNNNNNN	SISPA random priming
BC019CG	CGAGTATACGTACGTCTCAGTC	SISPA PCR step
BC024N	CGTAGTAGATAGTCACTCTACGNNNNNN	SISPA random priming
BC024CG	CGTAGTAGATAGTCACTCTACG	SISPA PCR step
BC025N	CGTCTATCATACGACTGTCTACNNNNNN	SISPA random priming
BC025CG	CGTCTATCATACGACTGTCTAC	SISPA PCR step
BC026N	CGCGTCTAGATACTCTGTAGAGNNNNNN	SISPA random priming
BC026CG	CGCGTCTAGATACTCTGTAGAG	SISPA PCR step
BC031N	CGTACATGTGTCGTATACACTCNNNNNN	SISPA random priming
BC031CG	CGTACATGTGTCGTATACACTC	SISPA PCR step
BC034N	CGAGACACTCATACGACTACTANNNNNN	SISPA random priming
BC034CG	CGAGACACTCATACGACTACTA	SISPA PCR step
BC035N	CGAGATGACGAGACGCACGACGNNNNNN	SISPA random priming
BC035CG	CGAGATGACGAGACGCACGACG	SISPA PCR step
BC044N	CGAGTAGACGATCGACGCGCTGNNNNNN	SISPA random priming
BC044CG	CGAGTAGACGATCGACGCGCTG	SISPA PCR step
BC045N	CGTGTCGTCTCGACGTGTGTGTNNNNNN	SISPA random priming
BC045CG	CGTGTCGTCTCGACGTGTGTGT	SISPA PCR step
BC081N	CGAGAGATACTGTACTAGAGCGNNNNNN	SISPA random priming
BC081CG	CGAGAGATACTGTACTAGAGCG	SISPA PCR step
BC0391N	CGTGACTATCTCGCGAGTACGANNNNNN	SISPA random priming
BC0391CG	CGTGACTATCTCGCGAGTACGA	SISPA PCR step

3 Methods

3.1 Phage Plaque Purification

1. Pick single, well-isolated plaques from agar plates, that contained a lawn of host cells using sterile 5 in. Pasteur pipettes (*see* **Note 1**).

2. Add a single agar plug to a sterile 1.5 ml Eppendorf-style tube containing 100 µl of SM buffer and soak overnight at 4 °C.

3. The following morning, increase the total volume to 400 μl with SM buffer before filtration through a 0.22 μm syringe filter to remove bacterial cells and debris.

3.2 Removal of Unpackaged and Host Nucleic Acids

1. For 500 μl volume of lysate, add 5 μl of 1 M $MgCl_2$ and 2 μl of 1 M $CaCl_2$ to bring the final concentration of $MgCl_2$ and $CaCl_2$ to 10 mM and 4 mM, respectively (*see* **Note 2**).

2. Add 0.5 μl (125 U) of Benzonase, 5 μl (10 U) of DNase I, 10 μl RNase cocktail (a mixture of 50 U of RNase A and 200 U of RNase T1) (*see* **Note 3**). Incubate at 37 °C for 1 h.

3. Remove 1 μl aliquot into a new sterile thin walled PCR tube for PCR-based bacterial DNA contamination check using 16S primers 27F-YM/1492R (*see* **Note 4**).

4. If plug suspension is free of contaminating DNA, add 50 μl of 0.5 M EDTA pH 8.0 and 50 μl of 0.5 M EGTA pH 8.0 to inactivate DNases (*see* **Note 5**).

3.3 PCR-Based Bacterial Contamination Check

1. *PCR Master Mix* (24 μl/rxn) plus 1 μl of the aliquot from Subheading 3.2, **step 3** or positive control

Reagent	Volume (μl)
Sterile water	19.5
10× PCR Buffer	2.5
50 mM $MgCl_2$	0.75
10 mM dNTPs	0.50
10 μM 27F-YM Primer	0.25
10 μM 1492R Primer	0.25
Platinum Taq Polymerase	0.25

Protocol

1. Reaction or master mix should be made by adding the reagents in the order listed in the table. Master mix is made by multiplying each reagent by the number of samples plus 1.

2. Include a negative control consisting of everything but template (add 1 μl of water instead of template) and include a positive control of bacterial genomic DNA.

3. Positive control template should have 25 ng for a 25 μl reaction. Dilute template with water.

4. Thermal Cycler Conditions:

94 °C for 2 min	
94 °C for 30 s 55 °C for 30 s 72 °C for 2 min	—35 cycles
72 °C for 7 min	
4 °C forever	

5. Gel Check Reaction: Run 5 μl of the reaction (plus 1 μl of 6×
gel loading dye) on a 1.2% agarose gel (90 V, 50 min) to
visualize products. The product of this PCR is a band ~1.5 kb
in length.

3.4 Extraction of Packaged Phage Nucleic Acids

1. Add 1/10 volume of 10× proteinase K digestion mix to phage
suspension from Subheading 3.2, **step 4** and incubate at 55 °C
for 1 h. Cool down at room temperature naturally in a tube
rack (*see* **Note 6**).

2. Add 1/10 volume of 5 M NaCl and 1 volume of PCI. Shake
vigorously and spin in a table top microcentrifuge at room
temperature for 10 min.

3. Transfer aqueous layer to a new 1.5 ml Eppendorf-style tube.

4. Add 1 volume of PCI to the aqueous phase and shake and
centrifuge as before.

5. Transfer aqueous layer to a new 1.5 ml Eppendorf-style tube.

6. Add 2 volumes of isopropanol and 2 μl of linear acrylamide (*see*
Note 7). Shake gently to mix and place on dry ice for 5 min.

7. Pellet nucleic acids in a microfuge at $16,100 \times g$ (max speed in
a microcentrifuge), 4 °C for 30 min.

8. Decant supernatant and wash the pellet with 1 volume of ice
cold 70% ethanol. Pellet nucleic acids again in a microcentrifuge at $16,100 \times g$, 4 °C for 20 min.

9. Decant supernatant using a sterile pipette and dry the pellet by
leaving the tubes open in a thoroughly cleaned PCR hood (*see*
Note 8). Resuspend pellet in 10 μl of EB (*see* **Note 9**) and let
set at room temperature for 10 min. The purified viral nucleic
acids are ready for SISPA.

3.5 SISPA Part 1: Denaturation and Annealing of the SISPA Bar-Coded Primers

3.5.1 For DNA Phages

1. Combine 2 μl of resuspended phage dsDNA from Subheading
3.4, **step 9** with 1 μl of 100 μM bar-coded SISPA "A" random
hexamer primer and 0.5 μl of 50% DMSO (7% final) in a sterile
PCR tube (*see* **Note 10**). Flick the side of the tube with your
finger and spin briefly in a microfuge to bring the liquid to the
bottom of the tube (*see* **Note 11**).

2. Heat in a thermocycler to 96 °C, 5 min. Snap cool on ice (*see*
Note 12).

3.5.2 For RNA Phages

1. Combine the following in duplicate reactions (10 μl total annealing reaction volume):

    (a) Purified RNA template from Subheading 3.4 in a total volume of 6.4 μl using DEPC-treated water.

    (b) Add 3.6 μl of a master mix containing:
    - 2 μl of 100 μM bar-coded SISPA "A" random hexamer primer
    - 0.6 μl of 10% DMSO
    - 1 μl of 10 mM dNTPs

2. Incubate 96 °C for 5 min.

3. Snap cool on ice

3.6 SISPA Part 2: Klenow Elongation Step

3.6.1 For DNA Phages

1. In a PCR hood, mix the following into a "Klenow Master Mix", multiplying volumes by $n + 1$ where n is the number of reactions that you want to amplify:

    (a) 0.6 μl sterile water

    (b) 0.5 μl 10× NEB Buffer 2

    (c) 0.1 μl 10 mM dNTPs

    (d) 0.3 μl (1.5 U) Klenow fragment (exo-) (5 U/μl)

2. Add 1.5 μl of this master mix per tube of annealed template from Subheading 3.5.1, **step 2**, resulting in a 5 μl final reaction volume.

3. Incubate in a heat block set at 37 °C for 1 h.

4. Add another 0.5 μl (2.5 U) of exo- Klenow fragment and continue incubation at 37 °C for 1 h (*see* **Note 13**).

5. Heat-inactivate the DNA polymerase at 75 °C for 15 min.

3.6.2 For RNA Phages

1. In a PCR hood, and with all surfaces and pipettes treated with RNaseZap®, mix the following to make a first strand master mix, multiplying each reagent by $n + 1$ as noted above:

    (a) 3.3 μl DEPC-treated water

    (b) 4 μl 5× First Strand Synthesis Buffer

    (c) 2 μl 0.1 M DTT

    (d) 0.2 μl RNaseOUT™ (40 U/μl)

    (e) 0.5 μl SSIIIRT (200 U/μl)

2. Add 10 μl of first strand master mix to 10 μl of primer-annealed RNA template from 3.5.2, **step 3**.

3. Incubate in a thermocycler using these conditions:

    (a) 25 °C for 10 min

    (b) 50 °C for 50 min

    (c) 85 °C for 10 min

(d) Snap cool on ice

4. Remove RNA template by adding 1 μl RNaseH (5 U/μl)

5. Incubate in a thermocycler 37 °C for 20 min, 85 °C for 10 min and hold at 4 °C forever.

6. Add 1 μl of Klenow (exo-) (5 U/μl) to generate second strand.

7. Incubate 37 °C for 1 h.

8. Heat-inactivate the DNA polymerase at 75 °C for 15 min and place on ice.

3.7 SISPA Part 3: Removal of Primers and Short Fragments (<200 Nucleotides)

1. If starting from DNA-containing genome, add 15 μl of sterile water to bring the volume up to 20 μl and transfer the entire volume to a new sterile 1.5 ml Eppendorf-style tube.

2. Add 10 μl of a 30% PEG 8000 supplemented with 10 mM $MgCl_2$ (*see* **Note 14**).

3. Mix well by finger flicking and let sit on ice for 15 min.

4. Pellet DNA at $16,100 \times g$ (max speed in a microcentrifuge) at 4 °C for 30 min.

5. Carefully decant the supernatant with a micropipette and resuspend the pellet in 20 μl of sterile water. Incubate at room temperature for 45 min (*see* **Note 15**).

6. Transfer entire volume into 0.2 ml PCR tubes.

7. Add 2.5 μl of 10× ExoI buffer and 1 μl of ExoI, finger flick the tube and quickspin.

8. Incubate at 37 °C for 30 min.

9. Heat-inactivate at 80 °C for 20 min (*see* **Note 16**).

10. Pool duplicate reactions into a single tube.

3.8 SISPA Part 4: Amplification of Random Fragments

1. Per reaction, mix together the following reagents in a 0.2 ml PCR tube in the following order on ice:

(a) 18 μl of sterile water

(b) 25 μl of 2× BioMix Red

(c) 2 μl of 10 μM of SISPA primer lacking the 3′ random hexamers

(d) 5 μl of pooled Klenow reactions from Subheading 3.7, **step 10**.

2. Place in a thermocycler and run using the following program:

98 °C for 30 s	
98 °C for 10 s 54 °C for 20 s 72 °C for 45 s	—35 cycles
72 °C for 5 min	
4 °C forever	

3. Run out a 1% or 1.2% agarose gel in $1 \times$ TAE, 90 V, 50 min to check the size of fragments.

3.9 SISPA Part 4: Gel Purification

1. If product has a smear between 300 and 1000 bp, run out the whole reaction on 1.2% agarose TAE gel containing EtBr (0.1 µg/ml final concentration) using a wide comb, 90 V, 2 h to get good separation. Add 2 µl of 10 mg/ml EtBr stock solution to the anode to sufficiently stain the lower third of the gel.

2. Cut out the band between 300 and 850 bp using an ethanol-sterilized razor blade and viewing the band with either blue light or a long wavelength UV light source (*see* **Note 17**).

3. Extract DNA from the gel fragment using a QIAquick gel extraction kit (catalog number 28704 or 28706). Follow the manufacturer's instructions and do all suggested extra steps.

4. Elute DNA with 25 µl of warm sterile water.

5. Quantitate DNA concentration using SYBR Gold or similar fluorescent method (*see* **Note 18**).

6. For Sanger sequencing, clone into pCR4-Topo or similar vector, pick colonies, template and sequencing following Sanger sequencing protocols.

7. For Next-generation sequencing methods, clean up fragments further using Agencourt AMPure XP beads (https://www.beckmancoulter.com/) and libraries are quantitated and QC'd using the Agilent High Sensitivity DNA Kit and by qPCR using a KAPA Biosystems (https://www.kapabiosystems.com/) Library Quantification Kit [10].

4 Notes

1. Can also use 1 ml sterile plastic pipettes.

2. If the buffer contains Mg^{2+} and Ca^{2+} in sufficient amounts, skip this step. If the volume recovered is <500 µl, adjust the amount of $MgCl_2$ and $CaCl_2$ to add accordingly.

3. Alternatively, one can use the Ambion® TURBO™ DNase instead of DNase I, which is more active than DNase I and recently used in a viral enrichment method [16].

4. Will only work if using just DNAse I since Benzonase (and the TURBO™DNase) can not be inactivated by heat alone. If lysate contains human/eukaryotic DNA (e.g., if purifying a human/eukaryotic virus), use other primer sets targeting the 18S rRNA gene or Alu repeat sequences.

5. At this point, the plug suspension can be frozen or stored in a refrigerator (overnight only).

6. For RNA genome purification, we recommend using the RNeasy Mini Kit from Qiagen rather than the proteinase K/SDS/EDTA method. Elute the RNA in 25 μl of RNase-free DEPC-treated water.

7. The linear acrylamide is a coprecipitant/carrier that aids in the precipitation of low amounts of nucleic acids [14] without risk of contamination from other carriers such as glycogen [17].

8. Can also decant by pouring and then quick spin the tubes and remove the remaining liquid using a micropipette. It takes ~1 h to evaporate the remaining ethanol.

9. If extracting RNA, resuspend in DEPC-treated water instead. EDTA in TE may interfere with the sensitivity of the downstream SISPA reactions.

10. This reaction volume was optimized to amplify as little as 10 pg of DNA (Fig. 3) [10]. The addition of DMSO has been shown

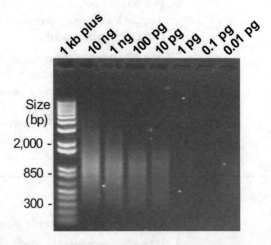

Fig. 3 Modified SISPA reaction can amplify as little as 10 pg of DNA. SISPA reactions were conducted from purified phage DNA serially diluted from 10 ng to 0.01 pg. Amplified products were run on a 1% agarose gel stained with EtBr. The DNA molecular marker is the Invitrogen™ 1 kb Plus DNA Ladder

to disrupt secondary structure of DNA to achieve higher yields in PCR [18], but may not be necessary in all cases.

11. Set up two reactions per phage to increase randomness.

12. We have also added the "A" primer after snap cooling. Alternatively, after snap cooling on ice, reactions can be incubated in a thermocycler with a 1 °C/min ramp from 4 to 37 °C as described previously [10]. This may increase sensitivity by facilitating the annealing of the random hexamer portion of the "A" primer.

13. Alternatively, add 0.5 µl of a Master mix: 0.1 µl 10× NEB buffer 2, 0.1 µl 10 mN dNTPs, 0.6 µl Klenow (exo-), and 0.2 µl water. Adjust volumes of master mix ingredients by multiplying by the number of reactions.

14. The final concentration of PEG is 8.7%, which will remove <200 bp [15].

15. Or store at −20 °C overnight.

16. Both Subheading 3.7, **steps 8** and **9** can be done in a thermocycler.

17. Do not use short wavelength UV—it will damage the DNA!

18. We've found that Nanodrop (http://www.nanodrop.com/) is not as accurate as fluorometric assays in quantitating small amounts of DNA. The QIAquick gel extraction kit will remove EtBr, according to the manufacturer.

Acknowledgments

The author wishes to thank Galina Koroleva, Jessica DePew, Janaki Purushe, Bin Zhou, and Manolito Torralba for their contributions to the development of these protocols. This work was supported with funds from the United States National Institutes of Health (R21-DE018063 and U54-AI84844) and in part with funds from the National Institute of Allergy and Infectious Diseases, National Institutes of Health, Department of Health and Human Services under contract number HHSN272200900007C.

References

1. Allander T, Emerson SU, Engle RE, Purcell RH, Bukh J (2001) A virus discovery method incorporating DNase treatment and its application to the identification of two bovine parvovirus species. Proc Natl Acad Sci U S A 98 (20):11609–11614

2. Breitbart M, Rohwer F (2005) Method for discovering novel DNA viruses in blood using viral particle selection and shotgun sequencing. Biotechniques 39(5):729–736

3. van der Hoek L, Pyrc K, Jebbink MF, Vermeulen-Oost W, Berkhout RJ, Wolthers KC et al (2004) Identification of a new human coronavirus. Nat Med 10(4):368–373

4. Jones MS, Kapoor A, Lukashov VV, Simmonds P, Hecht F, Delwart E (2005) New DNA viruses identified in patients with acute viral

infection syndrome. J Virol 79 (13):8230–8236

5. Palmenberg AC, Spiro D, Kuzmickas R, Wang S, Djikeng A, Rathe JA et al (2009) Sequencing and analyses of all known human rhinovirus genomes reveal structure and evolution. Science 324(5923):55–59

6. Djikeng A, Halpin R, Kuzmickas R, Depasse J, Feldblyum J, Sengamalay N et al (2008) Viral genome sequencing by random priming methods. BMC Genomics 9:5

7. Agindotan BO, Ahonsi MO, Domier LL, Gray ME, Bradley CA (2010) Application of sequence-independent amplification (SIA) for the identification of RNA viruses in bioenergy crops. J Virol Methods 169(1):119–128

8. Victoria JG, Kapoor A, Dupuis K, Schnurr DP, Delwart EL (2008) Rapid identification of known and new RNA viruses from animal tissues. PLoS Pathog 4(9):e1000163

9. Froussard P (1992) A random-PCR method (rPCR) to construct whole cDNA library from low amounts of RNA. Nucleic Acids Res 20(11):2900

10. Depew J, Zhou B, McCorrison JM, Wentworth DE, Purushe J, Koroleva G et al (2013) Sequencing viral genomes from a single isolated plaque. Virol J 10:181

11. Reyes GR, Kim JP (1991) Sequence-independent, single-primer amplification (SISPA) of complex DNA populations. Mol Cell Probes 5(6):473–481

12. Moser LA, Ramirez-Carvajal L, Puri V, Pauszek SJ, Matthews K, Dilley KA et al (2016) A universal next-generation sequencing protocol to generate noninfectious barcoded cdna libraries from high-containment rna viruses. mSystems 1(3):e00039

13. Wright MS, Stockwell TB, Beck E, Busam DA, Bajaksouzian S, Jacobs MR et al (2015) SISPA-Seq for rapid whole genome surveys of bacterial isolates. Infect Genet Evol 32:191–198

14. Gaillard C, Strauss F (1990) Ethanol precipitation of DNA with linear polyacrylamide as carrier. Nucleic Acids Res 18(2):378

15. Hartley JL, Bowen H (1996) PEG precipitation for selective removal of small DNA fragments. Focus 18(1):27

16. Hall RJ, Wang J, Todd AK, Bissielo AB, Yen S, Strydom H et al (2014) Evaluation of rapid and simple techniques for the enrichment of viruses prior to metagenomic virus discovery. J Virol Methods 195:194–204

17. Bartram A, Poon C, Neufeld J (2009) Nucleic acid contamination of glycogen used in nucleic acid precipitation and assessment of linear polyacrylamide as an alternative co-precipitant. Biotechniques 47(6):1019–1022

18. Frackman S, Kobs G, Simpson D, Storts D (1998) Betaine and DMSO: Enhancing Agents for PCR. Promega Notes 65:27–30

Chapter 13

Genome Sequencing of dsDNA-Containing Bacteriophages Directly from a Single Plaque

Witold Kot

Abstract

The sequencing of phage genomes has become a routine procedure for phage characterization. The protocol presented here allows rapid isolation of DNA from a single phage plaque followed by building ready-to-sequence Illumina-compatible library.

    Key words DNA sequencing, Single plaque, Genome, Illumina sequencing, *Caudovirales*

1 Introduction

Sequencing of bacteriophage DNA started in 1977 with the genome of φX174 [1], the first genome to be completely sequenced. The first double-stranded genome of DNA containing bacteriophage, λ (lambda), was determined 5 years later [2]. The vast majority of known bacteriophages (96%) belongs to *Caudovirales* order and contains dsDNA as genetic material [3]. Traditionally, in order to obtain high-quality DNA samples for sequencing, high-titer phage stocks were purified by CsCl gradient centrifugation followed by dialysis, proteinase K treatment and phenol--chloroform extraction of DNA [4]. This method provides a large amount of excellent quality DNA that is compatible with all sequencing platforms in use, including the third generation platforms [5] and allows additional genome analysis, e.g., restriction endonuclease analysis, analysis of the termini of the chromosomes etc.

    If there are difficulties in obtaining a sufficient quantity of DNA for library construction and sequencing, a range of DNA amplifying methods like multiple displacement amplification (MDA) [6] or sequence-independent single primer amplification (SISPA) [7] can be used. However, these methods slow down the process and

Martha R.J. Clokie et al. (eds.), *Bacteriophages: Methods and Protocols, Volume 3,* Methods in Molecular Biology, vol. 1681,
https://doi.org/10.1007/978-1-4939-7343-9_13, © Springer Science+Business Media LLC 2018

increase the overall cost of sequencing and might introduce additional biases.

With the introduction of transposon-based sequencing library preparation kits, such as the Nextera XT (Illumina, CA, USA), it is possible to sequence libraries from as little as 1 ng of DNA [8, 9].

The protocol presented below allows the isolation of DNA from a single phage plaque and the building of a ready-to-sequence Illumina-compatible library within just a few hours. This method was tested with variety of phages of for example *Escherichia coli*, *Bacillus subtilis*, *Pseudomonas aeruginosa*, and *Lactococcus lactis* [8, 10–12].

2 Materials

1. A double agar overlay plate with well-separated, single plaques of phages. Preferably, the overlay should be solidified with lower concentration of agarose of (5–6 g/l).

2. Variable volume pipettors for covering volumes in range 2 μl–1 ml.

3. Broad end 1 ml pipette tips or regular 1 ml pipette tips broadened to 1–2 mm of inside diameter with a sterile scalpel.

4. 1.5 ml capped microcentrifuge tubes, sterile.

5. Heating block for microcentrifuge tubes.

6. Laboratory vortex.

7. Ultrafiltration spin-columns (0.45 μm) which can be fitted in 1.5 ml microcentrifuge tube (Merck Millipore 20-218).

8. Centrifuge for microcentrifuge tubes.

9. DNase I and buffer for DNase I (1 U/μl, e.g., Thermo Fisher EN0525).

10. 50 mM EDTA solution.

11. 1% SDS solution.

12. Proteinase K solution (e.g., Thermo Fisher EO0491).

13. DNA Clean & Concentrator-5 kit (Zymo Research cat # D4013, contains: DNA binding buffer, Zymo-Spin columns, DNA Wash Buffer, and DNA elution buffer).

14. Nextera XT DNA Library Prep Kit (Illumina, cat # FC-131-1024 or -1096, contains: Tagment DNA buffer, Amplicon Tagment Mix, Neutralize Tagment buffer, Nextera PCR mix and Resuspension buffer).

15. Nextera XT Index Kit (Illumina, e.g., FC-131-1001, contains: index 1 and 2 primers).

16. 0.2 ml PCR-tubes.

17. Thermocyler.

18. Magnetic stand for 1.5 ml microcentrifuge tubes (e.g., Dyna-Mag™-2 Magnet or similar).

19. Agencourt AMPure XP beads (Beckman Coulter, cat # A63880).

20. 80% ethanol.

21. Qubit fluorometer with dsDNA HS kit (Thermo Fisher, cat # Q32851).

3 Methods

3.1 Isolation of DNA from a Single Plaque

1. Carefully pick up a single plaque using 1 ml pipettor equipped with broad-end tip. Pick up only the overlay layer; avoid bottom agar and bacterial lawn. Avoid transferring too much of material as it might affect subsequent DNase I activity.

2. Suspend the material in 100 µl of $1\times$ DNase I buffer (without DNase I enzyme) in a 1.5 ml microcentrifuge tube.

3. Allow phages to diffuse for minimum 30 min at 37 °C.

4. Transfer the solution into a 0.45 µm ultrafiltration spin-column fitted in a new 1.5 ml microcentrifuge tube and centrifuge at $2500 \times g$ for 1 min. This step will remove host cells and thus reduce amount of host DNA.

5. Add 5 U of DNase I (5 µl) and incubate for 30 min at 37 °C. This step will reduce amount of host DNA that is not protected by protein capsid.

6. Add 10 µl of 50 mM EDTA and 10 µl of 1% SDS to inactivate DNase I and enhance the proteinase K activity.

7. Add 5 µl of proteinase K (approx. 3 U) and incubate for 45 min at 55 °C. This step will digest phages capsids and release phage DNA.

8. Purify the DNA using DNA Clean & Concentrator-5 kit. Add 2 volumes of DNA binding buffer (200 µl) and load the mixture to a Zymo-Spin Column fitted in a collection tube

9. Centrifuge at full speed ($>10,000 \times g$) for 30 s.

10. Add 200 µl of DNA Wash Buffer and centrifuge for 30 s at full speed. Repeat wash step.

11. Place the Zymo-Spin column into a new 1.5 ml microcentrifuge tube. Add 6 µl of DNA elution buffer directly to the column matrix. Centrifuge for 30 s at full speed.

12. Use eluted DNA directly for the Nextera XT DNA Library Prep Kit.

3.2 Sequencing Library Building

1. Label a new 0.2 ml PCR tube with sample name.

2. Add 10 µl of Tagment DNA buffer.

3. Add 5 µl of Amplicon Tagment Mix.

4. Add 5 µl of eluted phage DNA from a previous step without adjusting the DNA concentration.

5. Vortex briefly and spin down for 30 s.

6. Place the tube in a thermocycler and run the following program:

    (a) 55 °C for 5 min

    (b) Hold at 10 °C

7. When sample reaches 10 °C add immediately 5 µl of Neutralize Tagment buffer.

8. Vortex briefly and spin down for 30 s and incubate for 5 min at room temperature.

9. Add 15 µl of Nextera PCR master mix to the PCR tube.

10. Add 5 µl of index 1 primer to the PCR tube.

11. Add 5 µl of index 2 primer to the PCR tube.

12. Vortex briefly and spin down for 30 s.

13. Place the tube in a thermocycler and run the following program:
    (a) 72 °C for 3 min
    (b) 95 °C for 30 s
    (c) 16 cycles of:
        • 95 °C for 10 s
        • 55 °C for 30 s
        • 72 °C for 30 s
    (d) 72 °C for 5 min
    (e) Hold at 10 °C

The higher number of cycles in comparison to the manufacturer's protocol is to compensate for <1 ng of DNA input.

14. Proceed to PCR cleanup. Label a new 1.5 ml tube with your sample name and transfer 50 µl of Nextera XT library.

15. Add 25 µl of AMPure XP beads.

16. Vortex briefly and incubate at room temperature for 5 min.

17. Place the tube on a magnetic stand. Leave for 2 min.

18. Carefully remove and discard the supernatant.

19. Add 300 µl of 80% ethanol. Incubate at room temperature for 30 s. Carefully remove and discard the supernatant and repeat the wash.

20. Carefully remove and discard the supernatant.

21. With the tube still on the magnetic stand, allow the beads to air-dry for 10 min.

22. Remove the tube from the magnetic stand. Add 52 μl of Resuspension buffer. Vortex briefly and incubate in room temperature for 2 min.

23. Place the tube on a magnetic stand and leave for 2 min.

24. Transfer 50 μl of ready library into a new, labeled 1.5 ml tube.

25. Keep library on ice until use or store at −20 °C.

4 Notes

1. Use filtered tips to avoid cross-contamination.

2. Try to lower the concentration of agarose in double layer plates. This will result in larger and easier to pick plaques.

3. Use fresh plates as they produce better results.

4. Method can be adjusted to specific phage-host pair by use of ultrafiltration spin-column with different cutoff value, e.g., 0.2 μm.

5. This method might not provide good read coverage at the regions proximal to the genomic termini. Nextera XT technology has an expected drop in sequence coverage of about 50 bp from each distal end. This seems not to be a problem for phages that have terminally redundant genome.

6. The DNA Clean & Concentrator-5 kit is designed to purify DNA fragments up to 23 kb; however, it works well with larger phage genome sizes (e.g., T4 bacteriophage with a 168 kb genome).

Acknowledgment

The author gratefully acknowledges support from the Danish Council for Independent Research (grant number 4093-00198 awarded to Witold Kot).

References

1. Sanger F, Coulson AR, Friedmann T, Air GM, Barrell BG, Brown NL et al (1978) The nucleotide sequence of bacteriophage φX174. J Mol Biol 125:225–246

2. Sanger F, Coulson AR, Hong GF, Hill C, Petersen GB (1982) Nucleotide sequence of bacteriophage lambda DNA. J Mol Biol 162:729–773

3. Ackermann HW (2011) Bacteriophage taxonomy. Microbiol Aust 32:90–94

4. Russell DW, Sambrook J (1989) Molecular cloning: a laboratory manual. Book 1, 2nd

edn. Cold Spring Harbour Laboratory Press, Cold Spring Harbour, NY

5. Schmuki MM, Erne D, Loessner MJ, Klumpp J (2012) Bacteriophage P70: Unique Morphology and unrelatedness to other *Listeria* bacteriophages. J Virol 86:13099–13102. doi:10.1128/JVI.02350-12

6. Dean FB, Nelson JR, Giesler TL, Lasken RS, Dean FB, Nelson JR et al (2001) Rapid amplification of plasmid and phage DNA using phi29 DNA polymerase and multiply-primed rolling circle amplification. Genome Res 11:1095–1099. doi:10.1101/gr.180501

7. DePew J, Zhou B, McCorrison JM, Wentworth DE, Purushe J, Koroleva G et al (2013) Sequencing viral genomes from a single isolated plaque. Virol J 10:181. doi:10.1186/1743-422X-10-181

8. Kot W, Vogensen FK, Sørensen SJ, Hansen LH (2014) DPS - a rapid method for genome sequencing of DNA-containing bacteriophages directly from a single plaque. J Virol Methods 196:152–156

9. Marine R, Polson SW, Ravel J, Hatfull G, Russell D, Sullivan M et al (2011) Evaluation of a transposase protocol for rapid generation of shotgun high-throughput sequencing libraries from nanogram quantities of DNA. Appl Environ Microbiol 77:8071–8079

10. Alves DR, Perez-Esteban P, Kot W, Bean JE, Arnot T, Hansen LH et al (2015) A novel bacteriophage cocktail reduces and disperses *Pseudomonas aeruginosa* biofilms under static and flow conditions. J Microbial Biotechnol 9:61–74. doi:10.1111/1751-7915.12316

11. Kot W, Neve H, Vogensen FK, Heller KJ, Sørensen SJ, Hansen LH (2014) Complete genome sequences of four novel *Lactococcus lactis* phages distantly related to the rare 1706 phage species. Genome Announc 2:4. doi:10.1128/genomeA.00265-14

12. Carstens AB, Kot W, Hansen LH (2015) Complete genome sequences of four novel *Escherichia coli* bacteriophages belonging to new phage groups. Genome Announc 3(4): e00741–e00715. doi:10.1128/genomeA.00741-15

Chapter 14

Preparing cDNA Libraries from Lytic Phage-Infected Cells for Whole Transcriptome Analysis by RNA-Seq

Bob Blasdel, Pieter-Jan Ceyssens, and Rob Lavigne

Abstract

Whole genome wide analysis of transcription using RNA-Seq methods is a powerful way to elucidate differential expression of gene features in bacteria across different conditions as well as for discovering previously exotic RNA species. Indeed, RNA sequencing has revolutionized the study of bacterial transcription with the diversity and quantity of small noncoding RNA elements that have been found and its ability to clearly define operons, promoters, and terminators. We discuss our experience with applying RNA sequencing technology to analyzing the lytic cycle, including extraction, processing, and a guide to the customized statistical analysis necessary for analyzing differential host and phage transcription.

Key words Bacteriophage, RNA-Seq, Library preparation, Transcriptome, RNA, Gene expression

1 Introduction

RNA sequencing (RNA-Seq), also known as Whole Transcriptome Shotgun Sequencing, is the use of second-generation platforms to sequence cDNA libraries that have been reverse transcribed from RNA populations present in target cells. When applied to phage-infected cells it allows for the identification of both phage and host encoded mRNAs, tRNAs, and sRNAs while quantifying them in relation to each other in a single experiment. RNA-Seq presents a number of significant advantages over microarray-based techniques, as it is not biased by hybridization efficiencies between oligonucleotides and allows the precise definition of RNA species to the single nucleotide level for both host and phage. It is also able to capture a faithful sample of the target population of RNAs across a much wider range of expression levels as it does not rely on direct detection methods like radioactivity or light, which can become oversaturated when enough material is used to detect low abundance transcripts (1).

As the number of both phage and bacterial genomes published in public databases continues to increase exponentially, our ability

Martha R.J. Clokie et al. (eds.), *Bacteriophages: Methods and Protocols, Volume 3*, Methods in Molecular Biology, vol. 1681, https://doi.org/10.1007/978-1-4939-7343-9_14, © Springer Science+Business Media LLC 2018

to understand and annotate the gene features that give those genomes useful function has not kept pace. Published gene features have almost entirely been predicted in silico, based on the presence of open reading frames and often distant orthology to other often hypothetical features. By experimentally defining the shape and location of transcripts in both phage and host, directional RNA-Seq has the ability to discover novel coding sequences, particularly for small phage peptides falling below gene prediction thresholds (2), and refine annotations of existing coding sequences. Additionally, directional RNA-Seq allows detection of a plethora of noncoding RNA species. For example, it can define *cis* antisense encoded RNA, which has been described in N4-like phage (3), that are not possible to predict in silico and exist on conditionally bidirectionally transcribed regions and block translation or other functions of sense transcripts (4).

It is important to consider that, excluding the smallest types, phage typically progressively express multiple transcriptional schemes—changing expression over time to fit the temporally distinct needs of the phage. Where, classically according to the T4 model, phage will first transcribe genes involved in shutting down the host's self-defense capability while converting its metabolism toward viral production in an "early phase" of expression. Next, genes involved in genome replication and the production of structural proteins are transcribed in a "middle phase" before genes related to assembly, packaging, and lysis in a "late phase." When RNA-Seq is performed on a synchronously infected population of cells, each phase can be captured individually in separate samples and compared quantitatively.

With the biological replicates necessary to demonstrate statistical significance, RNA-Seq can also qualitatively evaluate differences in gene expression imposed on the host relative to phage-negative controls. Even as phage transcripts rapidly replace host RNA species, RNA-Seq will detect both host operons specifically targeted by the phage for modulation as well as the host mediated response to phage infection. Whether differential expression is mediated by the host or phage can be distinguished by performing RNA-Seq on multiple phage infecting the same host.

When assessing whether RNA-Seq of the phage lytic cycle is adaptable to a given phage-host model system, it is important to consider that it requires accurately sequenced genomes for both phage and host to align RNA-Seq reads to. Additionally, producing synchronously infected cultures requires the ability to generate high titers of phage that adsorb quickly relative to the timespan of infection. Moreover, developing an educated guess for when to take samples requires controlled infection parameters such as when the latent phase ends and when lysis occurs in the system being sampled from.

1.1 Design

Performing RNA Seq can be divided into three distinct parts, collecting nucleic acid samples from various phases of a synchronous infection (**Part A**, *see* Fig. 1), processing those samples into a collection of sequencing reads that are representative of the RNA population in the infected culture (**Part B**, *see* Fig. 1), and aligning those reads to both the host and phage genomes (**Part C**, *see* Fig. 1).

A. To collect data that is specific to the various phases of phage transcription, a synchronous infection must first be prepared. To do this, a culture of ~1×10^8 cells growing in the early exponential phase is infected at a high MOI under conditions that allow fewer than 5% of bacterial survivors to be remaining within 5 min (*see* **Note 1**). Then, at time points selected to represent early, middle, and late transcription, one third of the infected culture is removed and halted by rapid cooling in diluted phenol, which also temporarily stabilizes the RNA population. Generating statistically significant differential expression data requires that this be performed in triplicate to create biological replicates.

B. To process collected samples into cDNA libraries for sequencing, cells are first lysed and RNases present in the cells and media are inactivated to produce a stable suspension of nucleic acids (**step 1**). Then all genomic DNA from both the phage and host must be enzymatically removed from the suspension (**step 2**). Optionally, rRNA may then be removed from the sample with commercially available kits to better economize on available sequencing depth (**step 3**). The RNA population is then reverse transcribed using commercially available kits into a cDNA library that can be shotgun sequenced (**step 4**).

C. The obtained sequencing reads for each sample must then be processed to remove adaptors and low quality reads before they can be aligned to the genomes of both phage and host using either open source programs or commercially available pipeline software. Once aligned to phage genomes, the distribution of reads can be used to correct gene annotations, define operons and upstream untranslated regions, as well as discover new gene features such as sRNA and small peptides that fall below ordinary gene prediction thresholds in size. Additionally, with replicates, the read counts that align to annotated gene features can be compared between samples to statistically test for differential expression.

This chapter is primarily focused on **Part B**, but discusses aspects of **Part C**. The methods required for **Part A** are described in detail by Kropinski in Chapter 2.

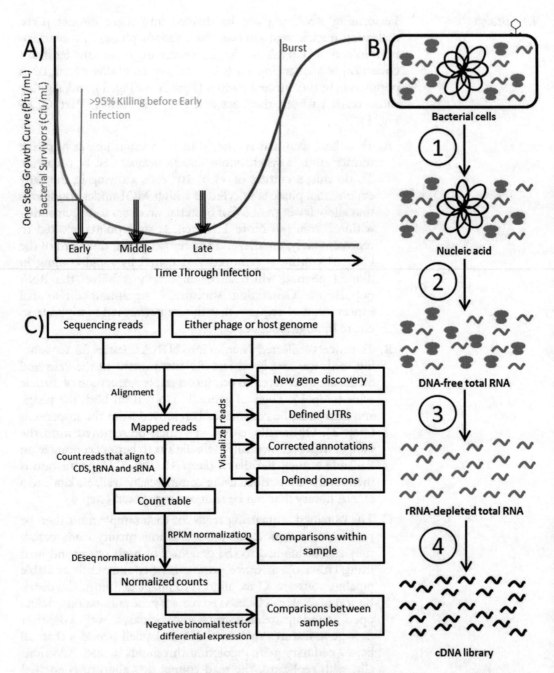

Fig. 1 Workflow for RNA-Seq analysis of cells infected by lytic phage. Biologically relevant samples are first collected in triplicate from a phage-negative control as well as time points in a synchronous infection (**Part A**). The samples must then be processed independently to liberate nucleic acids, remove both phage and host genomic DNA, deplete rRNA, and convert the remaining RNA into cDNA libraries for sequencing (**Part B**). Once sequencing is complete, the resulting reads must be aligned to their relevant reference genomes where they can be visualized to show the shape of the transcriptome. Additionally, they can be counted to make statistical comparisons between the abundance of reads that align to different gene features within a sample or between samples (**Part C**)

2 Materials

TRIzol® (Life Technologies)

Chloroform.RNase-free ethanol.RNase-free water.RNase-free 3 M NaOAc pH 5.2.RNase-free DNase.RNase-free disposables such as pipette tips and microcentrifuge tubes.Titers of phage in excess of 1×10^{11}/ml.Stop Solution: One part RNA buffered phenol to nine parts absolute ethanol by volume, kept ice cold. Lysis Buffer: Solution of lysozyme prepared according to manufacturer's instructions to 4 mg/ml.

3 Methods (Part B)

3.1 Organic Extraction of RNA (Step 1)

1. Before infection, prepare one centrifuge tube large enough to contain 1/3 of the infection per time point to be taken, each with one part Stop Solution for every nine parts of cell suspension that RNA is being extracted from and place on ice.

2. Over the course of infection, remove samples of ~2.5×10^7 cells at the desired time points and pipette them into one tenth volume of prepared Stop Solution before immediately shaking vigorously and placing back on ice.

3. After infection, centrifuge at $5000 \times g$ for 15 min, to securely pellet the stopped cells and remove the supernatant.

4. Resuspend pellet in 400 µl of Lysis Buffer before transferring to a 1.5 ml microcentrifuge tube.

5. Incubate for 10 min, but not longer, at room temperature before freezing with liquid nitrogen and thawing in a water bath at 45 °C. Repeat the freeze–thaw cycle three times and look under microscope to confirm cell lysis (*see* **Note 2**).

6. Add 500 µl TRIzol® to pellet, thoroughly pipette mix, and incubate for 10 min at room temperature (*see* **Note 3**).

7. Add 200 µl chloroform and mix before incubating for 10 min at room temperature.

8. Centrifuge ~$16,000 \times g$ for 15 min 4 °C.

9. Carefully transfer aqueous upper phase into new tube without disturbing the organic phase or the protein layer found at the interphase.

10. Add 1/10 volume RNase-free 3 M NaOAc and 2 volumes 96% EtOH, splitting the sample into multiple tubes.

11. Store at −20 °C overnight to ensure precipitation of small RNA species and centrifuge ~$16,000 \times g$ for 1 h at 4 °C.

12. Remove supernatant and wash pellet with ice cold 70% EtOH before centrifuging again ~$16,000 \times g$ for 15 min at 4 °C.

13. Remove supernatant, centrifuge again for 1 min to, remove remaining supernatant and air-dry pellet for 5 min.

14. Resuspend the pellets for each sample in 200 μl total RNase-free water and combine into a single tube.

15. Analyze sample on NanoDrop spectrophotometer (Thermo Scientific, Wilmington, DE) to ensure adequate concentration and purity: $OD_{260/280 \& 260/230} > 1.8$ (*see* **Note 4**).

16. Store at −20 °C.

3.2 Removal of Genomic DNA (Step 2)

Complete removal of both phage and host gDNA is essential to obtain accurate sequence information, as cDNA and gDNA reads will be indistinguishable. Eliminating genomic DNA contamination can be challenging, as the high concentrations of RNA present will act as a competitive inhibitor to commercially available DNase enzymes, impeding their function. It is also important to consider that the DNase used may be sensitive to noncanonical nucleotides commonly present in phage DNA. Using standard DNase according to manufacturers' instructions may work, though below is an expanded protocol that optimizes enzyme function, and even this may need to be repeated several times.

1. Add RNase-free DNase buffer to 1× concentration. Incubate for 5 min at 65 °C to ensure remnant DNA is fully in solution (*see* **Note 5**).

2. Return to room temperature before adding the recommended amount of RNase-free DNase and incubating for 1 h at 37 °C.

3. Add the same amount of DNase a second time and incubate for another 1 h at 37 °C.

4. Analyze each sample for residual phage and host DNA by performing PCR using primers that amplify small products and have been verified to be sensitive to low concentrations.

5. Store at −20 °C.

3.3 rRNA Depletion (Step 3—Optional)

Depending on the sequencing resources available, it may be desirable to use commercially available rRNA depletion kits for bacterial total RNA to increase the depth of coverage for desired RNA species (Table 1). We have had variable success using the

Ribo-Zero kit available from Illumina (San Diego, California), which captures rRNA using oligo hybridization to beads that are then removed with a strong magnet. Although commercially available kits are typically regarded as less reliable than advertised, even a reduction of the rRNA fraction from ~95% to ~50% of the sample can result in enrichment of the output for other RNA species by an order of magnitude.

Table 1
Commercially available rRNA removal methods appropriate for bacteria

Name	Supplier	Catalog Number
Ribo-Zero™ rRNA Removal Kit (Bacteria)	Illumina	#MRZMB126
MICROBExpress™ Bacterial mRNA Enrichment Kit	Life Technologies (Thermo Fisher Scientific)	#AM1905
Terminator™ Exonuclease (*see* **Note 6**)	Epicenter Biotechnologies	#TER51020

3.4 cDNA Library Preparation and Sequencing (Step 4)

DNA and rRNA depleted RNA is typically transformed into double stranded cDNA libraries through a process that uses random hexamer primed reverse transcription, followed by synthesis of a second strand. Through this process, both strands of cDNA are then sequenced identically in a way that scrambles the natural strand specificity inherent to transcription. However, particularly with the extraordinary coding density of phage genomes the various strand specific methods that have been devised for cDNA library preparation have special value for understanding the transcriptomes of phage (5). Indeed, there have been significant amounts of antisense RNA that have been characterized (3) that would be impossible to distinguish with un-stranded RNA-Seq, and transcript features often overlap in ways that strand specificity can aid in defining appropriately.

While there are many established techniques for accomplishing strand specificity in RNA-Seq a comprehensive comparative analysis of strand-specific RNA sequencing methods has convincingly argued that the Illumina RNA ligation methods (6) and the dUTP second strand marking methods (7) provide better results for the effort expended (8), *see* (5) for additional discussion. We have had success with Illumina's TruSeq® Stranded Total RNA Sample Prep Kit, which uses a method similar to the dUTP second strand marking method (Catalog #: RS-122-2201). However, once a cDNA library is generated it can be sequenced using standard high throughput platforms to generate the list of millions of short reads that will be used in the next section.

4 Experimental Analysis (Part C)

4.1 Mapping Reads

The first step in making sense of the millions of short reads generated by RNA-Seq is to turn those reads into a quantification of localized transcript abundance by aligning them to either the phage or host genomes. This involves attempting to match each read to a corresponding sequence in each potential reference genome, a process that is complicated by short reads aligning to multiple locations, RNA-Seq sequencing errors, reference genome sequencing mistakes, and RNA editing events. Current protocols use either the Burrows Wheeler transform or hash table based methods to assemble a list of candidate matches available in the reference for each read and then pick between them. Alignment can be performed using various free and open-source software packages such as the Burrows Wheeler Aligner (9) or TopHat (10).

4.2 Generating Transcription Maps

Once aligned to both phage and host genomes, the reads form a map revealing the abundance of RNA transcribed from any given locus in the infected cell accurate to the single nucleotide level. These maps can be used to precisely determine transcription start and end sites allowing promoters and terminators to be predicted and their operons to be characterized. With defined operons, 5′ upstream untranslated regions can be annotated and their effects on translation hypothesized. Additionally, unannotated yet transcribed regions can be scrutinized for peptides that are too small to definitively predict from sequence alone, indeed using RNA-Seq Ceyssens et al. (2) updated the ΦKZ genome with 63 (20.5%) additional coding sequences. Noncoding sRNAs will also be highlighted as transcribed features without plausible open reading frames while both *cis* and *trans*-encoded antisense RNAs will map to the antisense strand of coding sequences. These maps can also point out faulty annotations, when previously defined open reading frames lack sense transcripts or the start of transcription indicates a different start codon.

4.3 Differential Expression Analysis

Differential analysis of the number of sequencing reads that align to specifically annotated host gene features between an uninfected control, sampled immediately before infection, and various time points after infection also provides a valuable window into how phage infection affects host transcript abundance. This is accomplished by summarizing expression data into a table of the number of reads that map to each host CDS, ncRNA, and tRNA with three biological replicates before infection and comparing them statistically to three biological replicates after infection. To perform this differential analysis we recommend using DESeq as a R/Bioconductor package to normalize read counts between samples and then to test for differential expression and thus infer signal within the noise inherent to RNA-Seq. DESeq uses a method based on the negative binomial distribution to model the differences that would be expected due to natural variation and thus determine if an observed difference in read counts is statistically significant. This is more appropriate than other methods based on the Poisson distribution for modeling the variance inherent to phage infection (11).

While the alignment of sequencing reads to either host or phage has remained clear and distinct in our experience, determining whether the host or the phage is causing observed changes in the host transcript abundance during phage infection can become muddied. Differential analysis highlights changes in the abundance of specific transcripts imposed on the host by the phage such as the promotion and repression of particular transcripts or targeted degradation. However, depending the presence or success of phage mechanisms for shutting down host systems, it will also highlight host responses to phage infection for defense or as a reaction to various stresses that are inherent to phage infection. The difference between the two can be potentially distinguished by context and other sources of data, but can also be highlighted by performing RNA-Seq on infections by several diverse phages in the same host. As taxonomically divergent phages are unlikely to affect even a common host in the same way, a similar transcriptomic response to many phages will indicate that it is performed as a host response.

When interpreting your results it is important to consider that, aside from dramatic examples such as those produced by prophages in the host sensing infection and attempting to escape (2), most host transcripts will downshift in abundance relative to the total RNA in the cell during a successful infection due to the rapid synthesis of phage transcripts. Specific modulation of host transcripts needs to be tested for independently of this global depletion, which is done when normalizing only host read counts in one sample to host read counts in another sample while excluding phage reads. This will faithfully highlight how the distribution of reads transcribed from the host genome changes, but will not on its own show changes in abundance relative to the total transcript population in the cell as it hides the natural relative decrease in host reads.

5 Notes

1. If phage binding efficiency proves inadequate, the addition of 1–20 mM $CaCl_2$ and/or $MgCl_2$ to the medium may be needed to assist the phage (12).

2. If this proves inadequate to lyse the host, additional preferred methods can supplement or replace incubation with lysozyme such as beating with microbeads. It is important to ensure that the time that the sample spends at room temperature before **step 6** is kept to an absolute minimum.

3. The RNA samples suspended in TRIzol® at **step 6** can be safely frozen at −20 °C for up to 3 months. It is important to note that this stage is intended to inactivate the RNases made by the host and present in the media. From this point on, all materials that interact with the sample after this point must be RNase-

free, gloves must be worn, and the samples should be kept on ice when worked with on the bench.

4. A poor $OD_{260/280}$ ratio indicates an unacceptable level of either phenol contamination, from the phenol contained in the TRIzol®, or protein contamination, from the white interphase layer in **step 9**, relative to the RNA concentration. It can be addressed by starting again from **step 6**. A poor 260/230 concentration indicates an unacceptable level of salt contamination, from the NaOAc used in **step 10** as well as the media, relative to the RNA concentration. It can be addressed by starting again from **step 10**.

5. The 2′-OH group of RNA is capable of catalyzing autocleavage of RNA strands at high temperatures and high pH.

6. Epicenter Bioscience's Terminator™ 5′-Phosphate-Dependent Exonuclease is an inexpensive and especially effective way to deplete rRNA, which are posttranscriptionally modified with a 5′-monophosphate, but will also remove any other RNA species that could have been similarly modified.

References

1. Oshlack A, Robinson MD, Young MD (2010) From RNA-Seq reads to differential expression results. Genome Biol 11:220. doi:10.1186/gb-2010-11-12-220

2. Ceyssens P, Minakhin L, Van den Bossche A, Yakunina M, Klimuk E, Blasdel B, De Smet J, Noben J, Bläsi U, Severinov K, Lavigne R (2014) Development of giant bacteriophage ΦKZ is independent of the host transcription apparatus. J Virol 88(18):10501–10510

3. Wagemans J, Blasdel B, Van den Bossche A, Uytterhoeven B, De Smet J, Paeshuyse J, Cenens W, Aertsen A, Uetz P, Delattre A, Ceyssens P, Lavigne R (2014) Functional elucidation of antibacterial phage ORFans targeting Pseudomonas aeruginosa. Cell Microbiol 16(12):1822–1835

4. Georg J, Hess WR (2011) cis-antisense RNA, another level of gene regulation in bacteria. Microbiol Mol Biol Rev 75(2):286–300

5. Mills JD, Kawahara Y, Janitz M (2013) Strand-specific RNA-Seq provides greater resolution of transcriptome profiling. Curr Genomics 14(3):173–181

6. Croucher NJ, Fookes MC, Perkins TT, Turner DJ, Marguerat SB, Keane T, Quail MA, He M, Assefa S, Bähler J, Kingsley RA, Parkhill J, Bentley SD, Dougan G, Thomson NR (2009) A simple method for directional transcriptome

sequencing using Illumina technology. Nucleic Acids Res 37(22):e148

7. Zhang Z, Theurkauf WE, Weng Z, Zamore PD (2012) Strand-specific libraries for high throughput RNA sequencing (RNA-Seq) prepared without poly(A) selection. Silence 3:9. doi:10.1186/1758-907X-3-9

8. Levin JZ, Yassour M, Adiconis X, Nusbaum C, Thompson DA, Friedman N, Gnirke A, Regev A (2010) Comprehensive comparative analysis of strand-specific RNA sequencing methods. Nat Methods 7(9):709–715

9. Li H, Durbin R (2010) Fast and accurate long-read alignment with Burrows-Wheeler Transform. Bioinformatics 1(5):589–595

10. Trapnell C, Roberts A, Goff L, Pertea G, Kim D, Kelley DR, Pimentel H, Salzberg SL, Rinn JL, Pachter L (2012) Differential gene and transcript expression analysis of RNA-Seq experiments with TopHat and Cufflinks. Nat Protoc 1(3):562–578

11. Anders S, Huber W (2010) Differential expression analysis for sequence count data. Genome Biol 11:10. doi:10.1186/gb-2010-11-10-r106

12. Rountree PM (1951) The role of certain electrolytes in the adsorption of staphylococcal bacteriophages. Microbiology 5(4):673–680

Part III

Phage-Related Bioinformatics Tools

Essential Steps in Characterizing Bacteriophages: Biology, Taxonomy, and Genome Analysis

Ramy Karam Aziz, Hans-Wolfgang Ackermann, Nicola K. Petty, and Andrew M. Kropinski

Abstract

Because of the rise in antimicrobial resistance there has been a significant increase in interest in phages for therapeutic use. Furthermore, the cost of sequencing phage genomes has decreased to the point where it is being used as a teaching tool for genomics. Unfortunately, the quality of the descriptions of the phage and its annotation frequently are substandard. The following chapter is designed to help people working on phages, particularly those new to the field, to accurately describe their newly isolated viruses.

Key words Annotation, CDS, Electron microscopy, Genomes, Locus tag, ORF, Phage, Promoter, Software, Taxonomy, Terminator

1 Introduction

The phage community, journals, public databases, and the International Committee on Taxonomy of Viruses (ICTV) are seeing a marked increase in poor quality descriptions of newly isolated phages. To fully characterize a phage, one should accurately describe the plaque and particle morphologies; characterize the phage adsorption kinetics and host range, including, ideally, the identification of the surface receptor. Many of these features have been dealt with in previous chapters in this book.

The problems that we increasingly see in manuscripts include: (1) incomplete description of the phage life cycle and poor electron micrographs, (2) taxonomy which is not borne out by the sequence data, (3) incomplete phage genomes passed off as being complete, (4) incorrectly assembled and chimeric genomes, (5) genomes for which the annotation is incomplete or wrong, and, last but not least, (6) poor, inaccurate, or inexistent metadata associated with submitted phage genome sequences.

Martha R.J. Clokie et al. (eds.), *Bacteriophages: Methods and Protocols, Volume 3*, Methods in Molecular Biology, vol. 1681, https://doi.org/10.1007/978-1-4939-7343-9_15, © Springer Science+Business Media LLC 2018

This chapter is written to offer authors some hints on how to fully and accurately describe their new phages and to recommend that they validate their data prior to submission of the genome data to one of the primary databases (GenBank, EMBL or DDBJ), to phage-specific databases (e.g., PhagesDB (http://phagesdb.org/), ACLAME (http://aclame.ulb.ac.be/) [1], PhAnToMe (http://www.phantome.org) [2]), and for publication.

2   Naming

Scientists working on phages have the right to name them as they see fit. In the phage community, in particular, naming traditions have ranged from repetitive (e.g., P1, N4, S2 [3]) to too creative (e.g., SheldonCooper and Jabbawokkie). Unfortunately, this has resulted, though the general use of a combination of Greco-Roman characters and Arabic numerals in names such as β, λ, φX174, K, T4, and P22, which are meaningless to the general reader. In addition, because of lack of awareness, different names have been applied to the same virus, and the same name has been applied to different viruses. Authors are encouraged to peruse Bacteriophage Names 2000 (http://www.phage.org/names/2000/), GenBank, and PubMed when naming their new virus. The use of Greek letters in phage names should be strongly discouraged since, in databases, φ and Φ will be represented as phi and Phi. In addition, some caution should be used when "1" (one) and "l" (small letter L), or O (letter) and 0 (zero) simultaneously occur in the name of the virus, such as SIO1. The *Mycobacterium* phage community opted for the use of more fanciful names such as Rosebush, Corndog, Seabiscuit, and Jabbawokkie (http://phagesdb.org/phages/)—a system which worked well until it was also applied to phages infecting *Bacillus* and *Streptomyces* strains. The major problem with all of these systems is that the names, in isolation, provide no information about what the named entity is, nor the host or the taxonomic position of the virus. To address these problems, Kropinski et al. [4] proposed a formal four-part naming system for newly characterized viruses. A phage recognition signal "vB" (*v*irus of *b*acteria), analogous to the small "p" which precedes many plasmid names, precedes all names. This is followed by a three letter abbreviation for the host genus and species, usually derived from REBASE [5]; and, a single letter for the phage morphotype, for example EcoP indicates that the virus is a podovirus infecting *Escherichia coli*. Lastly, the name in common usage in the laboratory is appended. Therefore, coliphage λ and *Salmonella enterica* serovar Typhimurium phage P22 would, if newly isolated, receive the formal names vB_EcoS_Lambda and vB_SenP_P22. After using the formal name once in the paper's title and/or abstract, the phage would be referred to as Lambda or P22, respectively in the remainder of the manuscript.

3 Morphology

Over 6400 bacterial viruses have been examined in the electron microscope [6]. Interestingly we don't have even 1/6th of the genomes of those microscopically described phages. This has generated valuable insights as well as a lot of worthless data [7]. Briefly, morphology is essential for viral family and, often, genus identification. While some investigators seem to consider that morphology can be replaced by genomics, this is definitely untrue. On the contrary, electron microscopy is a short-cut for phage identifications by sequencing, while purely genomic investigations are often beset by poor or absent electron microscopic data and abound in questionable identifications.

Some manuscripts describing novel phages show no micrographs at all, but instead offer vague family assignments such as "myovirus" or "podovirus," or give poor quality (unsharp, contrastless) pictures with no scale markers, dimensions, or structural details (collars, base plates, or tail fibers). Many lack information on the types of electron microscope and stains employed, or how the virus was purified. Some descriptions affirm that isometric phage heads are "icosahedral" without a shadow of proof (N.B., they could be dodecahedral or tetrahedral). If dimensions are given, the type of magnification control is not indicated and this makes the whole description of limited usefulness. Finally, it is not uncommon that the virus is shown on a stamp-sized micrograph, where it resembles more a nail than a phage. Such publications, at least with respect to electron microscopy, are misleading to genomics studies and worthless to virology.

Cryo-electron microscopy is of no help here as it produces image reconstructions and not original phage pictures. It cannot replace conventional transmission phage images obtained after negative staining. A way out of the present dismal situation can be summarized in a few guidelines [8]:

1. Purify your phage, best by repeated washing.
2. Always indicate the type of electron microscope used, purification method, final magnification, calibration method, and stain.
3. Show a scale marker.
4. If uranyl acetate is not satisfactory, try phosphotungstate (or vice versa).
5. Describe your phage in detail with complete dimensions

6. Submit to journals micrographs that are highly contrasted (not grey-on-grey) of a sufficiently high magnification for reproduction (at least $150,000-300,000\times$).

7. Do your electron microscopy yourself. Do not "farm it out" to, possibly, unexperienced technicians.

4   Preliminary Determination of Genome Relationships

With the rapid increase in fully sequenced bacterial and viral genomes being deposited in public databases, there is a stronger possibility that a new virus will show homology at the DNA or protein levels with the genomes or proteomes of existing phages. This can readily be assessed by similarity search algorithms, such as BLASTN and BLASTX, against the nr (nonredundant) database or "Organism, optional" Viruses (taxid:10239). We would most definitely recommend the latter when dealing with temperate phages.

5   Accurate and Complete Metadata

When investigators submit one or a few virus genomes, they are usually focused on the accuracy and completion of the genomic data, which is very important; however, metadata associated with the virus itself are equally important, especially for future comparative genomic studies, or other studies related to viral/bacterial genomics or metagenomics. The simplest definition of metadata is that they represent a data set describing the submitted data. For a viral genome, metadata may include anything describing the virus or the genome, especially those data that cannot be deduced or computed from the genomic information. For example, a genome's length (in base pairs), %G+C, nucleotide bias, and codon usage are all useful metadata; however, they are easily computed from the submitted DNA sequence. On the other hand, information such as morphology, naming, taxonomy, source of isolation, and host range is invaluable for several types of analyses. Comparative genomics and metagenomic analyses often stop short of reaching important conclusions because the metadata associated with the genomes analyzed are incomplete, spurious, or jumbled. Metadata problems include spelling inconsistencies (e.g., enterobacteria vs. Enterobacteriaceae; temperate vs. lysogenic in describing phage lifestyle), absence of controlled vocabulary (e.g., naming issues described above), and irrelevant data (e.g., providing geographic coordinates for a virus isolated from bacteria causing bovine mastitis to a cow in North France, while not providing the name of the bacterium or the site of bacterial infection within the animal). We recommend submitting all possible metadata about a viral genome, provided that these data are supported by strong evidence. As with other genomic information, usually no data at all is better than entering inaccurate data.

6   Genome Organization and Sequence Checking Prior to Annotation

It is imperative that the sequence deposited with the databases minimally represent the nonredundant sequence of the phage and be error free. It should be oriented in the same manner as the type virus (*see* Fig. 1) to which it is related and be accompanied by an accurate taxonomy. Lastly, the submission should include enough metadata, i.e., information on who isolated it, when, from what source and on what strain (see above). Commonly, phage genomes have short 3′ or 5′-cohesive termini, or terminal redundancies, ranging from hundreds to thousands of base pairs. The latter may be accompanied by circular permutation. It must be rigorously understood that circular permutation does not mean that the genome is circular. Circular genomes do not exist in the *Caudovirales*, being excluded by the packaging mechanisms. Information on the nature of the ends can be derived from restriction analysis [9], sequence data [10], pulsed-field gel electrophoresis [11, 12], the nature of the large subunit terminase [9], or direct sequencing off the ends of the phage genomic DNA [13–16].

If the new phage is similar to an existing virus in one of the databases, possible errors in the assembled sequence can be indicated by BLASTX analysis against the related 'reference' phage,

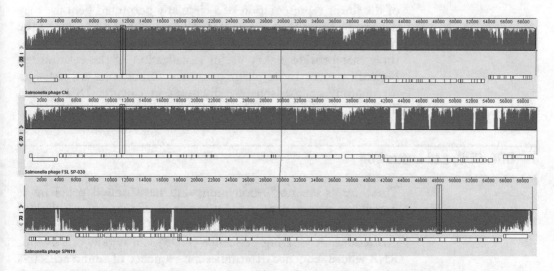

Fig. 1 progressiveMauve [17] alignment of three *Salmonella* phage genomes taken from NCBI GenBank. The genomes show strong overall sequence similarity except in certain (*white*) regions—a very typical phenomenon in phage genomics. The *top* represents the type species; phage genome A (*middle*) is collinear, while the genome of phage B is the inverse complement (*bottom*)

entered in the "Organism, optional" window of NCBI's BLAST interface. If the genome is >50 kb we recommend that the investigator divide the genome into 25 kb segments before analysis. Dividing or fragmenting a genome can readily be accomplished with Segmenter at http://lfz.corefacility.ca/segmenter/. Alternatively, Artemis Comparison Tool (ACT) [18, 22] is excellent for comparing two or more entire genomes via BLASTN or TBLASTX. Using this tool, the new phage can be compared to related phages at the whole genome level, or zoomed in for comparison at the amino acid or nucleotide level. If the genome of the new phage did not assemble into a single contiguous sequence (contig), ACT is useful for reordering the contigs based on the reference genome. To check for sequencing errors and confirm the genome assembly is correct, the sequencing reads should be mapped to the new genome assembly and carefully scrutinized. ACT can also be used to visualize the mapped reads and the genome should be carefully scrolled through to screen for potential errors in the genome assembly. Particular attention should be paid to coding sequences which have frameshifts compared to their homologs in the reference genome. These are often caused by homopolymeric tract sequencing errors, which are a common problem with (454) pyrosequencing [19]. If there is no evidence in the sequencing reads to support the "correction" of a frameshift, the sequence should be confirmed by PCR to determine if it is indeed an intact homolog or if it is a *bona fide* frameshift (which could be read through by ribosomal slippage or result in a pseudogene).

Lastly, again based upon the closest reference genome, the start of the linear representation of a circularly permuted genome may need to be rearranged so that it is collinear with the genome of the type species. Easyfig [20] and progressiveMauve [17] are other tools that provide a very useful visualization of the relationship between two or more phage genomes. Other software packages for genome comparisons are discussed in Chapter 18 by Dann Turner.

7   ORF, CDS, and Locus Tag

These terms are not synonymous and have caused problems is describing phage genes. "Most protein sequences are derived from translations of CoDing Sequence (CDS) derived from gene predictions. A CoDing Sequence (CDS) is a region of DNA or RNA whose sequence determines the sequence of amino acids in a protein. It should not be mixed up with an Open Reading Frame (ORF), which is a series of DNA codons that does not contain any STOP codons. All CDS are ORFs, but not all ORFs are CDS ..." (http://www.uniprot.org/help/cds_protein_definition). A CDS displays three essential features, an initiation codon, ribosome-binding site and a stop codon. Bacterial and bacteriophage initiation codons are commonly ATG (Methionine) or GTG (Valine), occasionally TTG (Leucine) and rarely CTG (Leucine), ATA, ATC, or ATT (Isoleucine). The most likely initiation codon will be immediately downstream from a sequence showing similarity to AGGAGGT which functions as the ribosome-binding site (Shine–Dalgarno sequence/box). Please note that whatever the initiation codon, methionine is the first amino acid incorporated in the nascent protein.

"The locus_tag is a systematic gene identifier that is assigned to each gene. Each genome project have the same unique locus_tag prefix to ensure that a locus_tag is specific for a particular genome project, which is why we require that the locus_tag prefix be registered. The locus_tag prefix must be 3–12 alphanumeric characters and the first character may not be a digit. Additionally locus_tag prefixes are case-sensitive. The locus_tag prefix is followed by an underscore and then an alphanumeric identification number that is unique within the given genome. Other than the single underscore used to separate the prefix from the identification number, no other special characters can be used in the locus_tag." (http://www.ncbi.nlm.nih.gov/genbank/genomesubmit/#locus_tag). We recommend that you submit locus tags based upon the name of your virus; otherwise GenBank will issue its own, which can lead to confusion in future comparative genomics studies.

With rare exceptions the genes of phages show short overlaps with the upstream gene or short intergenic regions (Figs. 2 and 3). The shortest CDSs are for λ Ral (28 AA) and Sf6 gp45 (27 AA) while the longest in the current databases, at 20,798 bp, is for

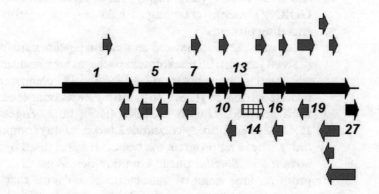

Fig. 2 An example from GenBank of a portion of a poorly annotated phage genome in which the real coding sequences are illustrated in *black*; the initiation codon for gene 14 is incorrect, leading to significant overlap in coding sequences. The genes illustrated as *stippled arrows* do not exist. The identity of this sequence has been concealed to protect the guilty

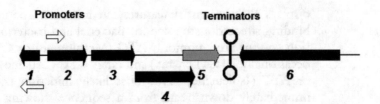

Fig. 3 The normal relationship of genes to one another with *1* and *2* being divergently transcribed; and *2–5* in an "operon" with *3* and *4* showing minimal overlap of coding sequence while *4* and *5* show the far more rarely observed significant overlap. Since transcription of *5* and *6* would lead to dampened expression of the downstream gene, these elements are frequently separated by bidirectional rho-independent terminators

cyanophage S-SSM4 locus_tag CYXG_00059 (N.B. please note the discrepancy between the name of the virus and the locus tag designator).

8   Primary Gene Annotation

Bacteriophage genomes typically have a high coding density, and many of their coding regions are small. By way of a simple illustration, T4 phage encodes 278 proteins with an average size of 197 amino acids, while its host, *E. coli* K12 W3110, encodes 4213 proteins with an average size of 317 amino acids. Since most of the gene calling algorithms are trained on larger genes, there is a problem in distinguishing between small coding sequences and artifacts [21]. Most phage genes are arranged with short overlaps or with small intergenic gaps, though occasionally gene wastelands occur [22]. Please note that many phages also encode tRNAs for which two online resources are recommended tRNAscan-SE ([23]; http://lowelab.ucsc.edu/tRNAscan-SE/) and ARAGORN ([24]; http://mbio-serv2.mbioekol.lu.se/ARAGORN/). Genes encoding tRNAs do not overlap with genes encoding proteins.

A number of automated annotation pipelines are freely available (Table 1) along with free software packages for visualizing and manually curating the annotation on Mac and PC computers. The latter include Artemis ([25, 26]; http://www.sanger.ac.uk/science/tools/artemis), Unipro UGENE ([27]; http://ugene.net/), and DNA Master (http://cobamide2.bio.pitt.edu/computer.htm; PC only). The latter program has been extensively used by the *Mycobacterium* and *Bacillus* phage communities. What is essential at the proof-reading stage of annotation is software that will present the protein sequence on the DNA sequence so that the user can easily verify the start codon and position of the ribosome-binding site.

In the case of auto-annotated genomes, the results should be meticulously scanned for missed genes, incorrectly called genes,

Table 1
Automated annotation pipelines for phage genomes

Name	URL	Comment
RAST	http://rast.nmpdr.org	Allows online annotation only [28]
RASTtk	To download: https://github.com/TheSEED/RASTtk-Distribution/releases/ For tutorials: http://tutorial.theseed.org	Allows batch genome annotation [29]
MyRAST	http://blog.theseed.org/servers/installation/distribution-of-the-seed-server-packages.html Download: http://blog.theseed.org/downloads/myRAST-Intel.dmg	Is becoming dated (not updated) [30]
Prokka	http://www.vicbioinformatics.com/software.prokka.shtml	Rapid annotation of prokaryote genomes. Command-line only [31]
phAST	http://www.phantome.org/PhageSeed/Phage.cgi?page=phast	Offers alternative genome-calling algorithms
BASys	http://basys.ca/	Uses more than 30 programs to determine nearly 60 annotation subfields for each gene [32]
GenSAS v3.0	https://www.gensas.org	Requires registration; uses Glimmer3 for prokaryote gene identification
IGS Prokaryotic Annotation Pipeline	http://ae.igs.umaryland.edu	Fasta-formatted genome submitted by email. Provides access to Manatee [33]
MAKER Web Annotation Service (MWAS)	http://www.yandell-lab.org/software/mwas.html	Web-accessible genome annotation pipeline [34]

genes with incorrect initiation codons, and misannotations before submission to one of the databases. The availability of multiple genomes from closely related phages often helps detecting errors in start codon assignments by comparative genomics, provided one or two of those closely related phages have been annotated and published by a reliable research group. The opposite is also true: with the spread of automated annotation pipelines and the difficulty of reviewing every CDS in every submitted sequence, propagation of annotation errors is quite common. Thus, when checking a start codon or a gene call for accuracy, it is important not to give weight to the majority of sequences but rather to the reliability of annotations. More reliable sequences are usually those of type species/reference sequences; those with experimental evidence (e.g., proteomic evidence or mRNA sequences; and those with well described annotation evidence) .

9   Naming of Gene Products

In several religious traditions major and minor sins are recognized. The same applies to annotation: describing a product based upon limited or no evidence as a "DNA polymerase" is far worse than referring to it as a "hypothetical protein". The use of "gp#" (*gene product*) to describe a gene or protein product should be discouraged since it leads to confusion specifically with the T4 phage community. More importantly, such names or symbols lead to disastrous annotations with the application of automated pipelines because computers know how to perfectly match patterns; however, they cannot tell that a gp3 of some phage family is not the same as a gp3 of another family. The designation gp43, as used by the T4 community refers to DNA polymerases. This same gp43 designation has been applied to radically different protein products in *Myoviridae* active on *Bacillus*, *Brochothrix*, *Burkholderia*, *Erwinia*, *Listeria*, *Mycobacterium*, and *Sphingomonas*. In addition, the same product name has been used for numerous *Podoviridae* including those active on *Burkholderia*, *Escherichia*, *Salmonella*, and *Xylella*; and 12 different bacterial genera have siphoviral phage proteins described as gp43. Other problematic product designations include: "UboA", "NrdA", "hypothetical protein SA5_0153/152", "ORF184" (as bad as gp184), "RNAP1", and "32 kDa protein" since they do not mean anything to the casual (or even informed) reader. Full functional descriptions are always superior to symbols (e.g., DNA-dependent RNA polymerase type 1 is much more meaningful and specific than RNAP1). Finally, for computational annotation pipelines to work, consistency is a must. Humans can easily tell that "DNA polymerase" is the same as "DNA Polymerase" (uppercase P); however, computers will classify those two identical names as two different enzymes. Asking computers to ignore case will only introduce more problems.

We would recommend that you include in the "note" section of the database submission file (Sequin in the case of GenBank) a statement to the effect of "similar to NP_049662 gp43 DNA polymerase [Enterobacteria phage T4]."

10   Defining Function of Encoded Proteins

Once the preliminary annotation of the genome has been completed one will want to define the function of the numerous "hypothetical proteins" which abound in phage genomes. The protein sequences of the genes defined in the GenBank flatfile (*.gbk) can be extracted by GenBank to Fasta converters (e.g., http://rocaplab.ocean.washington.edu/tools/genbank_to_fasta or gbk2faa). In Windows-based computers, you can examine the results using

Notepad or Wordpad. On Mac computers you can use a text editor such as TextWrangler. Homology searches can then be run, on a one-by-one or batch basis with BLASTP (protein-protein BLAST), PSI-BLAST (Position-Specific Iterated BLAST), or the more recent DeltaBLAST at NCBI. If you have a temperate phage it is recommended that you limit the search to "Viruses (taxid:10239)" since prophage genomes are notoriously poorly annotated in host genome sequences. Similar algorithms are available with the FASTA family of similarity searches [35]. The batch feature is not available at EMBL-EBI or GenomeNet. There are various options for running both BLAST and FASTA on all or selected CDSs directly from within Artemis. For a long time people solely relied on BLAST analysis against the nonredundant (nr) protein databases to assign function to their proteins. While useful, the sole reliance on the results of these searches has several problems. First, poor annotation, particularly automated annotation of prophage proteins can lead one astray. Second, the relationship between experimental data and in silico analysis is getting more distant. We offer two suggestions (a) display considerable caution in accepting sequence relatedness—Are the proteins of similar size? Are the percentage identities high enough? Is there sufficient sequence similarity over the full length of the proteins? Are the E-value scores stringent enough ($>10^{-5}$)? Do the results make biological sense? And; (b) back up the proposed nomenclature with motif analysis. For the latter we recommend Pfam (http://pfam.xfam.org/search; [36]), InterProScan 5 ([37]; http://www.ebi.ac.uk/interpro/), the Conserved Domain Database (CDD; http://www.ncbi.nlm.nih.gov/Structure/cdd/wrpsb.cgi; [38]) or HHpred [39] at https://toolkit.tuebingen.mpg.de/#/tools/hhpred. Please note that Pfam and CDD can be run in batch mode.

11 Caution with Interpreting Protein Motifs

Again, the results of searching protein motif databases need to be interpreted with caution. Two examples will suffice. *Cronobacter* phage GAP32 gp335 is a 43 kDa protein containing pfam05816 (*E*-value 1.03e−45) which is defined as "Toxic anion resistance protein (TelA)." Both coliphage PBECO4 and *Klebsiella* phage RaK2 contain homologs, while, outside these phage, homologs are only to be found in bacteria such as the *Staphylococcus aureus* tellurite resistance protein TelA (WP_000138402; BLASTP *E*-value 4e−34). In *Staphylococcus aureus* tellurite resistance is mediated by its (TeO_3^{2-}; $Te(IV)$) reduction to $Te(0)$, which has been used in the selective and differential medium, Baird-Parker agar [40]. In *S. aureus*, catalase [41] and cysteine synthase appear to be responsible for tellurite resistance, but these proteins display no homology to gp335 or to proteins designated as TelA. Pfam family

TelA (PF05816) motif is based upon a protein (TelA/KlaB), which is found in a "tellurite resistance" operon in plasmids and *Rhodobacter* [42–44], yet this motif is found in ORFans [45] in *S. aureus*. Unless you are a bioinformatician with expertise in protein domain analysis, we would suggest that you ignore protein motifs with scores $<10^{-4}$, and even then be conservative reporting results.

Salmonella phage vB_SnwM_CGG4-1 gp100 encodes a Hoc homolog, which HHpred suggests is highly related to titin (Protein Data Bank accession number 3b43_A)—a protein that contributes to myofibril elasticity. Again we would recommend that you only report HHpred hits with probability values >90% where they make sense.

It is important to take all available evidence into account when annotating a gene, including (where possible) data from more than one protein domain database, similarity with annotated genes in related genomes, position within the genome and percentage identity to an experimentally characterized protein.

12 Toward a Consistent Nomenclature for Phage Gene Products

Unfortunately, at present there is no consistent method for describing the proteins encoded by genes. An example of the diversity of names, both logical and illogical, derived from a BLASTP search using coliphage T4 rIIA protein is shown in Table 2. As indicated

Table 2
Some names used to describe the product of T4 *rIIA* gene

rIIA protector from prophage-induced early lysis
protector from prophage-induced early lysis
protector from prophage-induced early lysis rIIA
membrane-associated affects host membrane ATPase
rIIA membrane-associated affects host membrane ATPase
phage rIIA lysis inhibitor
rIIA protector
rIIA
RIIA
rIIA protein
putative rIIa-like protein
putative rIIA
membrane integrity protector
orf001 gene product
1 gene product
hypothetical protein
unnamed protein product
protein of unknown function

above, the use of controlled vocabulary and spelling consistency are imperative for the proper functioning of computational genome annotation pipelines.

13 Analysis of Promoters and Terminators

Promoters are located in the 3'-ends of upstream genes or in the intergenic regions. Two types of promoters can be found in phage genomes, those recognized by the host RNA polymerase (RNP) and those by phage-specified polymerases. The former promoters fall into two classes, those recognized by unmodified host RNP which are similar to the hosts' kitchen variety promoter—often a variant of TTGACA(N15-18)TATAAT and those recognized by phage-modified host RNPs. In the absence of experimental data we would recommend very conservative reporting on in silico identified promoters and only allowing a 2 bp mismatch to the consensus.

The best examples of phage RNP-recognized promoters are to be found in the genomes of T7-like viruses. These are best discovered using the program extractupstreamDNA (https://github.com/ajvilleg/extractUpStreamDNA) coupled with MEME [46, 47] at http://meme-suite.org or; the Windows-based program PHIRE (PHage In silico Regulatory Elements) [48, 49] on the whole genome. Please note that the latter program, which is written in Visual Basic, is very slow. The former program is excellent for discovering host promoters plus phage-modified RNP promoters, such as the middle promoters of T4-like phages.

Like promoters, transcriptional terminators are found at the 3'-end of genes and in intergenic spaces. The structure of a typical rho-independent terminator is illustrated in Fig. 4, which shows a GC-rich stem, a small loop and a polythymidylate tail. The length of the tail has been correlated with the efficiency of termination [50].

Online resources to identify terminators include those that select for terminators immediately downstream of genes, which include WebGeSTer [51] at http://pallab.serc.iisc.ernet.in/gester/; ARNold ([52]; http://rna.igmors.u-psud.fr/toolbox/arnold/) and FindTerm ([53]; http://linux1.softberry.com/berry.phtml?topic=findterm&group=programs&subgroup=gfindb). With the latter program, choose to display "All putative terminators" with "Energy threshold value" of >-10. With both these resources you will have to check on the location of the terminator, and exclude any which occur in the middle of genes.

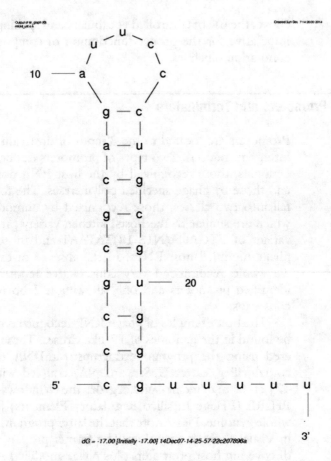

Fig. 4 Rho-independent terminator from *Pseudomonas* phage φKMV graphically represented using MFOLD [54]

14   Comparative Genomics and Proteomics

A variety of tools are available for comparing the genomes and proteomes of phages. Linear genome comparisons require that the genomes are collinear, while proteome comparison requires that the genes have been correctly annotated. At the DNA level EMBOSS Stretcher [55] will provide you with quantitative data on the sequence identity of the two genomes. As an alternatives, you might consider NCBI BLASTN (multiplying the % coverage by the % identity to give an overall % identity value), ANI (Average Nucleotide Identity; http://enve-omics.ce.gatech.edu/ani/), GGDC 2.0 (Genome to Genome Distance Calculator; http://ggdc. dsmz.de/distcalc2.php) or jSpeciesWS (also an ANI program; http://jspecies.ribohost.com/jspeciesws/ [56–62]).

For qualitative comparisons we recommend Easyfig ([20]; http://mjsull.github.io/Easyfig/) for gene maps and linear genome comparisons and BLAST Ring Image Generator [63] (BRIG; http://sourceforge.net/projects/brig) for circular comparisons, both of which are available for Windows, Mac and Unix; progressiveMauve ([17]; http://darlinglab.org/mauve/mauve.html), which is available for Mac and PC computers; CGView using the BLAST features ([64]; http://stothard.afns.ualberta.ca/cgview_server/) or WebACT ([65]; http://www.webact.org/WebACT/generate). Easy comparisons at the protein level can be made using CoreGenes ([47]; http://www.binf.gmu.edu/genometools.html). With the latter tool you can calculate the percentage of homologous proteins, plus how the homologs are arranged on the genomes. As mentioned above, other software packages for genome and proteome comparisons are discussed in Chapter 18 by Dann Turner.

15 Molecular Taxonomy

The Bacterial and Archaeal Virus Subcommittee of ICTV often discusses how viral genera should be defined. In the past, phages that possessed a total proteome 40% in common (as determined using CoreGenes) with another phage were considered to be members of the same genus [66, 67]. While this clearly shows close relationship, it is not without problems, the first of which is that both type and comparator phages must be equally well annotated (their genes should be called by similar algorithms using similar training sets; their start and stop codons should be defined by the same standards). The other problem is that the two phages may possess limited or no DNA sequence relatedness, which is the first line of molecular taxonomy of their hosts. Three examples of such problems include the relationships between *Pseudomonas* phage gh-1 [68] and coliphage T7 (20.6% sequence identity); coliphages T1 and vB_EcoS_Rogue1 [69] (14.2% sequence identity); and *Escherichia* phage N4 and *Sulfitobacter* phage EE36φ1 [70] (4.0% sequence identity). These pairs of phages are commonly recognized as being members of the *T7likevirus*, *Tunalikevirus*, and *N4likevirus* genera, respectively. That DNA similarity is a great method for grouping phages is recognized by the *Mycobacterium* phage community [71–73], and more recently to apply to all enterobacterial phages [74]. These results strongly suggest that ICTV has been lumping divergent phages into the same genera, and that overall DNA sequence identity can be effectively used to, at least, demonstrate relationships at the genus and subfamily level. Implementing those directives is currently being addressed.

Reference

1. Leplae R, Hebrant A, Wodak SJ, Toussaint A (2004) ACLAME: a CLAssification of Mobile genetic Elements. Nucleic Acids Res 32: D45–D49

2. McNair K, Bailey BA, Edwards RA (2012) PHACTS, a computational approach to classifying the lifestyle of phages. Bioinformatics 28:614–618

3. Abedon ST, Ackermann H-W (2001) Bacteriophage names 2000. The Bacteriophage Ecology Group (BEG). http://www.phage.org/names.htm

4. Kropinski AM, Prangishvili D, Lavigne R (2009) Position paper: the creation of a rational scheme for the nomenclature of viruses of Bacteria and Archaea. Environ Microbiol 11:2775–2777

5. Roberts RJ, Vincze T, Posfai J, Macelis D (2003) REBASE: restriction enzymes and methyltransferases. Nucleic Acids Res 31:418–420

6. Ackermann HW, Prangishvili D (2012) Prokaryote viruses studied by electron microscopy. Arch Virol 157:1843–1849

7. Ackermann H-W (2014) Sad state of phage electron microscopy. Please shoot the messenger. Microorganisms 2:1–10

8. Ackermann HW, Tiekotter KL (2012) Murphy's law-if anything can go wrong, it will: Problems in phage electron microscopy. Bacteriophage 2:122–129

9. Casjens SR, Gilcrease EB (2009) Determining DNA packaging strategy by analysis of the termini of the chromosomes in tailed-bacteriophage virions. Methods Mol Biol 502:91–111

10. Li SS, Fan H, An XP, Fan HH, Jiang HH, Mi ZQ, Tong YG (2013) Utility of high throughput sequencing technology in analyzing the terminal sequence of caudovirales bacteriophage genome. Bing Du Xue Bao 29:39–43

11. Lingohr E, Frost S, Johnson RP (2009) Determination of bacteriophage genome size by pulsed-field gel electrophoresis. Methods Mol Biol 502:19–25

12. Tamakoshi M, Murakami A, Sugisawa M, Tsuneizumi K, Takeda S, Saheki T, Izumi T, Akiba T, Mitsuoka K, Toh H, Yamashita A, Arisaka F, Hattori M, Oshima T, Yamagishi A (2011) Genomic and proteomic characterization of the large Myoviridae bacteriophage φTMA of the extreme thermophile Thermus thermophilus. Bacteriophage 1:152–164

13. Sharp R, Jansons IS, Gertman E, Kropinski AM (1996) Genetic and sequence analysis of the cos region of the temperate Pseudomonas aeruginosa bacteriophage, D3. Gene 177:47–53

14. Juhala RJ, Ford ME, Duda RL, Youlton A, Hatfull GF, Hendrix RW (2000) Genetic sequences of bacteriophages HK97 and HK022: Pervasive genetic mosaicism in the lambdoid bacteriophages. J Mol Biol 299:27–51

15. Ceyssens PJ, Lavigne R, Mattheus W, Chibeu A, Hertveldt K, Mast J, Robben J, Volckaert G (2006) Genomic analysis of Pseudomonas aeruginosa phages LKD16 and LKA1: establishment of the φKMV subgroup within the T7 supergroup. J Bacteriol 188:6924–6931

16. Glukhov AS, Krutilina AI, Shlyapnikov MG, Severinov K, Lavysh D, Kochetkov VV, McGrath JW, de LC SOV, Krylov VN, Akulenko NV, Kulakov LA (2012) Genomic analysis of Pseudomonas putida phage tf with localized single-strand DNA interruptions. PLoS One 7:e51163

17. Darling AE, Mau B (2010) Perna NT: progressiveMauve: multiple genome alignment with gene gain, loss and rearrangement. PLoS One 5:e11147

18. Carver TJ, Rutherford KM, Berriman M, Rajandream M-A, Barrell BG, Parkhill J (2005) ACT: the Artemis comparison tool. Bioinformatics 21:3422–3423

19. Becker EA, Burns CM, Leon EJ, Rajabojan S, Friedman R, Friedrich TC, O'Connor SL, Hughes AL (2012) Experimental analysis of sources of error in evolutionary studies based on Roche/454 pyrosequencing of viral genomes. Genome Biol Evol 4:457–465

20. Sullivan MJ, Petty NK, Beatson SA (2011) Easyfig: a genome comparison visualizer. Bioinformatics 27:1009–1010

21. Basrai MA, Hieter P, Boeke JD (1997) Small open reading frames: beautiful needles in the haystack. Genome Res 7:768–771

22. Kropinski AM, Waddell T, Meng J, Franklin K, Ackermann HW, Ahmed R, Mazzocco A, Yates J, Lingohr EJ, Johnson RP (2013) The host-range, genomics and proteomics of Escherichia coli O157:H7 bacteriophage rV5. Virol J 10:76

23. Lowe TM, Eddy SR (1997) tRNAscan-SE: a program for improved detection of transfer RNA genes in genomic sequence. Nucleic Acids Res 25:955–964

24. Laslett D, Canback B (2004) ARAGORN, a program to detect tRNA genes and tmRNA genes in nucleotide sequences. Nucleic Acids Res 32:11–16

25. Carver T, Berriman M, Tivey A, Patel C, Bohme U, Barrell BG, Parkhill J, Rajandream MA (2008) Artemis and ACT: viewing, annotating and comparing sequences stored in a relational database. Bioinformatics 24:2672–2676

26. Kropinski AM, Borodovsky M, Carver TJ, Cerdeno-Tarraga AM, Darling A, Lomsadze A, Mahadevan P, Stothard P, Seto D, Van DG, Wishart DS (2009) In silico identification of genes in bacteriophage DNA. Methods Mol Biol 502:57–89

27. Okonechnikov K, Golosova O, Fursov M (2012) Unipro UGENE: a unified bioinformatics toolkit. Bioinformatics 28:1166–1167

28. Aziz RK, Bartels D, Best AA, DeJongh M, Disz T, Edwards RA, Formsma K, Gerdes S, Glass EM, Kubal M, Meyer F, Olsen GJ, Olson R, Osterman AL, Overbeek RA, McNeil LK, Paarmann D, Paczian T, Parrello B, Pusch GD, Reich C, Stevens R, Vassieva O, Vonstein V, Wilke A, Zagnitko O (2008) The RAST Server: rapid annotations using subsystems technology. BMC Genomics 9:75

29. Brettin T, Davis JJ, Disz T, Edwards RA, Gerdes S, Olsen GJ, Olson R, Overbeek R, Parrello B, Pusch GD, Shukla M, Thomason JA III, Stevens R, Vonstein V, Wattam AR, Xia F (2015) RASTtk: a modular and extensible implementation of the RAST algorithm for building custom annotation pipelines and annotating batches of genomes. Sci Rep 5:8365

30. Aziz RK, Devoid S, Disz T, Edwards RA, Henry CS, Olsen GJ, Olson R, Overbeek R, Parrello B, Pusch GD, Stevens RL, Vonstein V, Xia F (2012) SEED servers: high-performance access to the SEED genomes, annotations, and metabolic models. PLoS One 7:e48053

31. Seemann T (2014) Prokka: rapid prokaryotic genome annotation. Bioinformatics 30:2068–2069

32. Van Domselaar GH, Stothard P, Shrivastava S, Cruz JA, Guo A, Dong X, Lu P, Szafron D, Greiner R, Wishart DS (2005) BASys: a web server for automated bacterial genome annotation. Nucleic Acids Res 33:W455–W459

33. Galens K, Orvis J, Daugherty S, Creasy HH, Angiuoli S, White O, Wortman J, Mahurkar A, Giglio MG (2011) The IGS standard operating procedure for automated prokaryotic annotation. Stand Genomic Sci 4:244–251

34. Campbell MS, Holt C, Moore B, Yandell M (2014) Genome annotation and curation using MAKER and MAKER-P. Curr Protoc Bioinformatics 48:4.11.1–4.11.39. doi:10.1002/0471250953.bi0411s48.:4

35. Pearson WR (2013) An introduction to sequence similarity ("homology") searching. Curr Protoc Bioinformatics, Chapter 3: Unit3.1.:Unit3

36. Finn RD, Mistry J, Tate J, Coggill P, Heger A, Pollington JE, Gavin OL, Gunasekaran P, Ceric G, Forslund K, Holm L, Sonnhammer EL, Eddy SR, Bateman A (2010) The Pfam protein families database. Nucleic Acids Res 38:D211–D222

37. Jones P, Binns D, Chang HY, Fraser M, Li W, McAnulla C, McWilliam H, Maslen J, Mitchell A, Nuka G, Pesseat S, Quinn AF, Sangrador-Vegas A, Scheremetjew M, Yong SY, Lopez R, Hunter S (2014) InterProScan 5: genome-scale protein function classification. Bioinformatics 30:1236–1240

38. Marchler-Bauer A, Lu S, Anderson JB, Chitsaz F, Derbyshire MK, DeWeese-Scott C, Fong JH, Geer LY, Geer RC, Gonzales NR, Gwadz M, Hurwitz DI, Jackson JD, Ke Z, Lanczycki CJ, Lu F, Marchler GH, Mullokandov M, Omelchenko MV, Robertson CL, Song JS, Thanki N, Yamashita RA, Zhang D, Zhang N, Zheng C, Bryant SH (2011) CDD: a Conserved Domain Database for the functional annotation of proteins. Nucleic Acids Res 39: D225–D229

39. Soding J, Biegert A, Lupas AN (2005) The HHpred interactive server for protein homology detection and structure prediction. Nucleic Acids Res 33:W244–W248

40. Holbrook R, Anderson JM, Baird-Parker AC (1969) The performance of a stable version of Baird-Parker's medium for isolating Staphylococcus aureus. J Appl Bacteriol 32:187–192

41. Calderon IL, Arenas FA, Perez JM, Fuentes DE, Araya MA, Saavedra CP, Tantalean JC, Pichuantes SE, Youderian PA, Vasquez CC (2006) Catalases are NAD(P)H-dependent tellurite reductases. PLoS One 1:e70

42. Walter EG, Thomas CM, Ibbotson JP, Taylor DE (1991) Transcriptional analysis, translational analysis, and sequence of the kilA-tellurite resistance region of plasmid RK2Ter. J Bacteriol 173:1111–1119

43. Whelan KF, Colleran E, Taylor DE (1995) Phage inhibition, colicin resistance, and tellurite resistance are encoded by a single cluster of genes on the IncHI2 plasmid R478. J Bacteriol 177:5016–5027

44. O'Gara JP, Gomelsky M, Kaplan S (1997) Identification and molecular genetic analysis of multiple loci contributing to high-level tellurite resistance in Rhodobacter sphaeroides 2.4.1. Appl Environ Microbiol 63:4713–4720

45. Fischer D, Eisenberg D (1999) Finding families for genomic ORFans. Bioinformatics 15:759–762

46. Bailey TL, Elkan C (1994) Fitting a mixture model by expectation maximization to discover motifs in biopolymers. AAAI Press, Menlo Park, CA, pp 28–36

47. Bailey TL, Boden M, Buske FA, Frith M, Grant CE, Clementi L, Ren J, Li WW, Noble WS (2009) MEME SUITE: tools for motif discovery and searching. Nucleic Acids Res 37: W202–W208

48. Lavigne R, Sun WD, Volckaert G (2004) PHIRE, a deterministic approach to reveal regulatory elements in bacteriophage genomes. Bioinformatics 20:629–6135

49. Lavigne R, Villegas A, Kropinski AM (2009) In silico characterization of DNA motifs with particular reference to promoters and terminators. Methods Mol Biol 502:113–129. doi:10.1007/978-1-60327-565-1_8

50. Jeng ST, Lay SH, Lai HM (1997) Transcription termination by bacteriophage T3 and SP6 RNA polymerases at Rho-independent terminators. Can J Microbiol 43:1147–1156

51. Mitra A, Kesarwani AK, Pal D, Nagaraja V (2011) WebGeSTer DB–a transcription terminator database. Nucleic Acids Res 39: D129–D135

52. Naville M, Ghuillot-Gaudeffroy A, Marchais A, Gautheret D (2011) ARNold: a web tool for the prediction of Rho-independent transcription terminators. RNA Biol 8:11–13

53. Solovyev V, Salamov A (2011) Automatic annotation of microbial genomes and metagenomic sequences. In: Li RW (ed) Metagenomics and its applications in agriculture, biomedicine and environmental studies. Nova Science Publishers, Hauppauge, NY, pp 61–78

54. Zuker M (2003) Mfold web server for nucleic acid folding and hybridization prediction. Nucleic Acids Res 31:3406–3415

55. Rice P, Longden I, Bleasby A, Rice P, Longden I, Bleasby A (2000) EMBOSS: the European Molecular Biology Open Software Suite. Trends Genet 16:276–277

56. Figueras MJ, Beaz-Hidalgo R, Hossain MJ, Liles MR (2014) Taxonomic affiliation of new genomes should be verified using average nucleotide identity and multilocus phylogenetic analysis. Genome Announc 2: e00927–e00914

57. Goris J, Konstantinidis KT, Klappenbach JA, Coenye T, Vandamme P, Tiedje JM (2007) DNA-DNA hybridization values and their relationship to whole-genome sequence similarities. Int J Syst Evol Microbiol 57:81–91

58. Kim M, Oh HS, Park SC, Chun J (2014) Towards a taxonomic coherence between average nucleotide identity and 16S rRNA gene sequence similarity for species demarcation of prokaryotes. Int J Syst Evol Microbiol 64:346–351

59. Konstantinidis KT, Ramette A, Tiedje JM (2006) Toward a more robust assessment of intraspecies diversity, using fewer genetic markers. Appl Environ Microbiol 72:7286–7293

60. Konstantinidis KT, Tiedje JM (2005) Genomic insights that advance the species definition for prokaryotes. Proc Natl Acad Sci U S A 102:2567–2572

61. Thompson CC, Chimetto L, Edwards RA, Swings J, Stackebrandt E, Thompson FL (2013) Microbial genomic taxonomy. BMC Genomics 14:913. doi:10.1186/1471-2164-14-913.:913-914

62. Richter M, Rossello-Mora R (2009) Shifting the genomic gold standard for the prokaryotic species definition. Proc Natl Acad Sci U S A 106:19126–19131

63. Alikhan NF, Petty NK, Ben Zakour NL, Beatson SA (2011) BLAST Ring Image Generator (BRIG): simple prokaryote genome comparisons. BMC Genomics 12:402. doi:10.1186/1471-2164-12-402.:402-412

64. Stothard P, Wishart DS (2005) Circular genome visualization and exploration using CGView. Bioinformatics 21:537–539

65. Abbott JC, Aanensen DM, Rutherford K, Butcher S, Spratt BG (2005) WebACT–an online companion for the Artemis Comparison Tool. Bioinformatics 21:3665–3666

66. Lavigne R, Seto D, Mahadevan P, Ackermann H-W, Kropinski AM (2008) Unifying classical and molecular taxonomic classification: analysis of the Podoviridae using BLASTP-based tools. Res Microbiol 159:406–414

67. Lavigne R, Darius P, Summer EJ, Seto D, Mahadevan P, Nilsson AS, Ackermann H-W, Kropinski AM (2009) Classification of Myoviridae bacteriophages using protein sequence similarity. BMC Microbiol 9:224

68. Kovalyova IV, Kropinski AM (2003) The complete genomic sequence of lytic bacteriophage gh-1 infecting Pseudomonas putida-evidence for close relationship to the T7 group. Virology 311:305–315

69. Kropinski AM, Lingohr EJ, Moyles DM, Ojha S, Mazzocco A, She YM, Bach SJ, Rozema EA, Stanford K, McAllister TA, Johnson RP (2012) Endemic bacteriophages: a cautionary tale for evaluation of bacteriophage therapy and other interventions for infection control in animals. J Virol 9:207

70. Zhao Y, Wang K, Jiao N, Chen F (2009) Genome sequences of two novel phages infecting marine roseobacters. Environ Microbiol 11:2055–2064

71. Hatfull GF (2012) The secret lives of mycobacteriophages. Adv Virus Res 82:179–288

72. Hatfull GF (2012) Complete genome sequences of 138 mycobacteriophages. J Virol 86:2382–2384

73. Hatfull GF (2014) Molecular genetics of Mycobacteriophages. Microbiol Spect 2:1–36

74. Grose JH, Casjens SR (2014) Understanding the enormous diversity of bacteriophages: The tailed phages that infect the bacterial family Enterobacteriaceae. Virology 468-470:421–443

Chapter 16

Annotation of Bacteriophage Genome Sequences Using DNA Master: An Overview

Welkin H. Pope and Deborah Jacobs-Sera

Abstract

Current sequencing technologies allow for the rapid and inexpensive sequencing of complete bacteriophage genomes, using small quantities of nucleic acid as starting material. Determination of the location and function of the gene features within the genome sequence, or annotation, is a necessary next step prior to submission to a public database, publication in a scientific journal, or advanced comparative genomic and proteomic studies. Gene prediction can be largely accomplished through the use of several freely available programs. However, manual inspection and refinement is essential to the production of the most accurate genome annotations. Here, we describe an overview of the annotation of a bacteriophage genome sequence using the freely available program DNA Master.

Key words Bacteriophage, Genome, Annotation, DNA master, Sequence analysis

1 Introduction

The advent of next-generation sequencing technology has made it possible to obtain high quality complete genome sequences of bacteriophages from less than a few micrograms of DNA in a matter of days for a few hundred dollars or less (Sequencing of Bacteriophage Genomes, D. Russell, Chapter 9). Although availability of phage isolates remains the limiting factor in phage genomics, annotation and bioinformatic analyses are often more time consuming that the actual DNA sequence determination. The creation of a well-supported bacteriophage genome annotation is a three-step process beginning with rapid automated gene identification ("auto-annotation"), manual inspection and revision, followed by functional predictions. The auto-annotation is quick and effective, but ~5–10% of the genes are misannotated or have incorrect translational start site predictions. Manual inspection is important in the identification and correction of these errors, and for evaluation of short open reading frames that are often missed by automated applications.

Martha R.J. Clokie et al. (eds.), *Bacteriophages: Methods and Protocols, Volume 3*, Methods in Molecular Biology, vol. 1681,
https://doi.org/10.1007/978-1-4939-7343-9_16, © Springer Science+Business Media LLC 2018

The program DNA Master, written by Dr. Jeffrey Lawrence at the University of Pittsburgh, alleviates some of the impediments to the production of high quality annotations, as it is a single platform with an easy-to-use interface that integrates multiple annotation tools, interfaces with GenBank and NCBI, and has the ability directly to generate GenBank submission files. DNA Master is well-suited to phage genome annotation because of its integration of multiple individual gene prediction programs such as GeneMark [1] GLIMMER 3[2] Aragorn [3] and BLAST, simple editing of the gene assignments, and prediction of common prokaryotic elements including ribosome binding sites, promoters, conserved motifs, programmed translational frameshifts, and self-splicing introns. The overall quality of DNA Master-generated annotations is supported by proteomic analyses of phage-infected cells [4, 5]. Detailed descriptions of the gene prediction programs GeneMark [1] and Glimmer 3[2], the sequence alignment tool BLAST[6], and the tRNA-finding program Aragorn [3] are described elsewhere and we will not review them here. For more information on these programs, please see their websites:

https://ccb.jhu.edu/software/glimmer/

http://exon.gatech.edu/GeneMark/

http://blast.ncbi.nlm.nih.gov/Blast.cgi

http://mbio-serv2.mbioekol.lu.se/ARAGORN/

Beyond the basic annotation tools, DNA Master also contains many integrated analysis tools for the more advanced bioinformatician, including the ability to compare multiple genomes, analyze biases in codon usage and nucleotide composition, scan for specific DNA sequences, find origins of replication and more; these are also out of the scope of this chapter. Many of these tools are explained in the DNA Master Help files found within the program.

Although DNA Master was designed as a powerful research tool, it is sufficiently accessible to be used by annotation novices, and we have deployed it broadly for genome annotation within the Science Education Alliance-Phage Hunters Advancing Genomics and Evolutionary Science (SEA-PHAGES) program. With nearly 100 participating institutions and over 2600 students per year, DNA Master has found widespread utility as a facile tool for genome annotation, in an educational context.

This chapter discusses an overview of bacteriophage genome annotation using DNA Master; space does not permit a detailed description of program mechanics or annotation rationales. Our full manual on bacteriophage genome annotation with DNA Master that includes content for novice annotators [7] is available for download at http://phagesdb.org.

2 Installation

DNA Master runs only on Windows Operating Systems, versions Windows XP or higher; and is available to download at: http://cobamide2.bio.pitt.edu. In the menu on the left on the Lawrence Lab website, click "Software", and then select "DNA Master" on the Software page.

To install the program, download and run the DNA Master installer. This will launch a Set-Up Wizard. Follow the prompts in the Wizard to install the program.

For further help with DNA Master installation, as well as some useful tricks for troubleshooting see the DNA Master Installation Guide, available at http://phagesdb.org/Documents.

By default, all necessary files and helper programs will be downloaded to the directory:

C://Program Files (x86)/DNA Master

After installation, there are three critical components that are essential for the proper functioning of DNA Master. The first is that the files and folders within the /DNA Master directory must not be moved or rearranged. Within this directory, the file "DNAMas.exe" is the executable file for the program. If it is moved from this directory it cannot locate the various helper programs and databases present in the /DNA Master directory that it needs to access. Since the file structure makes launching the program time-consuming, it is advisable to create a "Shortcut" to DNAMas.exe, and place the "Shortcut" on the Windows Desktop. It is then possible to launch the program via clicking on the Shortcut without disrupting the file structure.

The second critical component is that the program must be run with full Administrator privileges. In Windows 7 or later, this is accomplished by logging into Windows with an account that has Administrator rights and right-clicking on the DNAMas.exe file icon, and then selecting "Run as Administrator" from the pop-up menu. Failure to give the program Administrator privileges results in errors in updating the program, saving files, auto-annotations, and a number of errors in various processes (*see* **Notes 1** and **2**).

The third critical component for the correct running of DNA Master is to immediately update the program after installation. At the time of this writing, the version of the program installed directly via download is from 2012; however, the most current version of the program is from Jan. 2015. It is possible to determine the version and release date of the program currently running on your machine by choosing Help → About at the top of the program window. Program updates are released online and can only be accessed and installed through the update function in within the program. Updating is accomplished by launching DNA Master with Administrator privileges, clicking on the Help menu, and selecting "Update DNA Master". After Updating, close and restart the program.

3 Preferences

Prior to annotation of any genomes, the preferences of the program should be set to accommodate the user of the program and the type of genome s/he is annotating. The majority of the default preferences are recommended to the novice user, with the possible exceptions of the Default File saving location and the Default colors of the program. The Preferences menu is found under File, at the top of the program window. It is possible to change the colors within the maps, six-frame translation, frames window, and other figures using the Local Settings tab, Colors sub-tab, to set the start codons for auto-annotations to include TTG and GTG using the Local Settings, tab, Translation sub-tab, and to change the default location for file saving on the Directories tab, using the "Archive to. . ." field.

4 Using a fasta-Formatted Sequence File to Generate a DNA Master *.dnam5 File

In order to annotate a bacteriophage genome sequence, the sequence must be correctly imported into DNA Master and a *.dnam5 file must be generated. This is a one-time per sequence process; once the *.dnam5 file is made, all annotations and data will be saved in the *.dnam5 file. While DNA Master can handle several types of sequence file formats, we will only discuss .fasta here, as this format is widely used in both public databases and sequencing cores.

Fasta-formatted files consist of a header line that begins with the character >, followed by some information about the sequence. Below the header line is the nucleotide sequence of the genome, without line breaks or periods, etc. Fasta files are available from GenBank and from many sequencing cores.

To import your fasta sequence file into DNA Master choose: File → Open → Multiple Sequence Fasta file. Browse to find your genome sequence file, select it, and press "Open".

This will bring up a window labeled "FastA sequences from 'yourfilename.here'"; where 'yourfilename.here' is the name of the file that was opened. The header line from the .fasta file should be visible in the central information pane in the window, as should the full length of the phage genome in the file. It is important to check that the known genome sequence length matches the length of the sequence in the imported file; this will help avoid annotation errors down the road due to incorrect sequence data. If the header information and the sequence length are correct, press the button at the lower left corner of the window marked "Export", and choose "Create Sequence from this entry only". This will generate a *.dnam5 file with the genome sequence correctly imported into the program (*see* **Note 3**).

5 *.dnam5 Sequence Files and the *.dnam5 File Structure

The primary file DNA Master generates and interacts with is a *.dnam5 archive file. DNA Master *.dnam5 files are database files comprised of a genome sequence and multiple tables that contain data about that sequence, including gene start and stop coordinates, functional assignments, BLAST data, etc. The term "archive" in most instances within the program refers to the structure within each *.dnam5 file: each file is a collection of many pieces of data about the sequence(s) within; and "archive" does not mean that this file has been hidden away because the user is not planning on frequently accessing it. Thus, as noted above, the location for saving current files is labeled "Archive to …" and the default location is c:/Program Files (x86)/DNA Master/Archives. For more information on file structure, see the DNA Master Help files.

The basic window of a *.dnam5 file shows five tabs on the left-hand side: Overview, Features, References, Sequence, and Documentation. With a newly imported genome, the bulk of the information in these tabs has not been determined yet, and so they will be blank or empty. If the file was generated through importing a .fasta formatted sequence file, the "Sequence" tab will contain the sequence from the .fasta formatted file.

Once the file is auto-annotated (*see* Subheading 6), the "Features" tab will become populated with genes and/or tRNAs. The central column of the Features tab is a display of all the information about the Features currently stored in the database file. It is not possible to alter the information in the central column by clicking on it directly; changes must be made through the "Description" tab for that Feature that appears immediately to the right of the central column.

The "Documentation" tab shows a static text document that reflects the information found in the Features database. As Features are changed or added, the Documentation should be periodically recreated to match the contents of the Features database; this improves the overall *.dnam5 file stability, make it less likely that the file will suffer a major corruption, and provides a mechanism for rebuilding the file in the case of major corruption. The Documentation does not change unless the user prompts it to, by pressing the "Recreate" button on the tab, or by setting the program to automatically adjust the Documentation during processes like Auto-annotation, or gene insertion. We recommend recreating the Documentation immediately prior to saving the file (*see* **Note 4**).

The "Parse" button found at the top of the Documentation tab uses the contents of the Documentation tab to populate the contents of the Features database, overwriting any features or information currently stored in the database. The "Parse" button should only be used if a complete overwrite of the Features table in the *.dnam5 file is the goal of the annotator. The "Parse" function is the opposite of the "Recreate" function; "Recreate" uses the information from the "Features" Table to write the Documentation; and "Parse" uses the notes in the Documentation to populate the Features database.

6   Auto-annotation

Much of the gene identification work in DNA Master is done using the gene prediction programs GeneMark, GLIMMER 3, and Aragorn. These programs are accessed through the auto-annotation function of DNA Master. During auto-annotation, the raw sequence is submitted by DNA Master to the servers at the National Center for Biotechnology Information (NCBI) that host the heuristic versions of GLIMMER and GeneMark (*see* **Note 5**). The coordinates of the protein-encoding genes that these two programs find are retrieved from the NCBI servers and written into the Features database of the *.dnam5 file. The two programs together are highly accurate, and the bulk of the gene predictions in the auto-annotation will be correct. However, in every genome, some genes will be missed, sometimes the two programs disagree on the inclusion of a particular gene or where a particular gene starts; and some genes are added erroneously. This is especially apparent in the case of mobile or parasitic elements such as homing-endonucleases and transposons; or in the case of genes recently acquired within the genome. Therefore each auto-annotation must be hand-curated to make these corrections; and to add functional assignments where appropriate.

Auto-annotation is accomplished through opening a saved or generating a new *.dnam5 sequence file, and then selecting "Auto-annotate" through the Genome menu. In the window that appears, it is possible to choose which of the two gene prediction programs should "win" if the two programs don't agree. It doesn't matter which one is selected, as the output from both will be recorded in the Notes for each gene Feature during Auto-annotation (*see* **Note 6**). On a separate tab, it is also possible to generate numerous analyses that may aid in the manual inspection of the auto-annotation, including an ORF map. Finally, it is possible to have all of the protein sequences of the Features found automatically compared to the GenBank protein sequence database via BLASTp, and to have those results retrieved from NCBI and permanently stored in the *.dnam5 file (*see* **Note 7**).

The program Aragorn is used to find tRNA genes within the sequence during auto-annotation. The stand-alone version of Aragorn embedded within DNA Master is an older version than the current online release, and so we recommend analyzing any genome sequence with the most current version of Aragorn available on the Aragorn website, and then adjusting the tRNA coordinates manually.

7   Manual Inspection and Refining the Annotation

Once an auto-annotation has been completed, it is necessary to examine each of the new features to assess its validity as a gene, to adjust its start coordinate if necessary, and to add functions where appropriate. During this examination, it is important to keep in mind the guiding principles of bacteriophage annotation [7] (some are listed below):

1. The majority of the sequence space in a bacteriophage genome is occupied by protein encoding genes (90% or greater).

2. Each piece of DNA is generally limited to one protein-encoding gene; that is, there are not usually two genes expressed from the same piece of DNA in different translational frames.

3. Genes are grouped in operons, and therefore change transcriptional direction on a gene-by-gene basis infrequently.

4. While some genes have canonical Shine–Dalgarno sequences, others do not; and therefore a good ribosome binding site is not essential for selecting a specific start (see Note 8).

Manual inspection of the genome can be most easily accomplished through the use of the Frames window in DNA Master. After the genome has been auto-annotated, open the *.dnam5 (if it isn't already), and select "Frames" from the "DNA" menu (see Note 9). This will bring up a window with six horizontal tracks, one for each translational frame. Vertical bars within the tracks represent start and stop codons; with stop codons running the full vertical height of the track, and start codons appearing half the height of the stop codons (Fig. 1). Buttons on the lower left-corner of the window control the navigation of the map; it is possible to scroll from left to right down the length of the genome, or to zoom in. It is possible to show all of the Features in the database of a *.dnam5 file on the Frames window by pressing the "ORFs" button in the lower-right hand corner of the window.

Shine–Dalgarno scores for each possible start codon can be obtained by first clicking on the ORF of interest in the frames window, and then on the "RBS" button in the lower right-hand corner. This will bring up a new window that lists all the possible

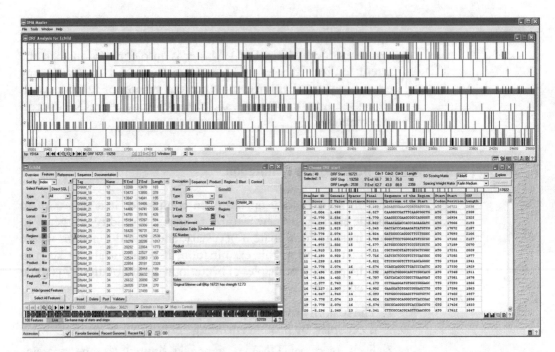

Fig. 1 Typical configuration of DNA Master windows for gene evaluation. *Top window* is the Frames window, showing six reading frames. Stops are *vertical lines* that bisect each row, starts are designated as lines half that length. The *thick horizontal green bars* are the annotated genes in the window, labeled by a number that matches the feature table feature name. The *bottom left window* is the *.dnam5 sequence file currently being annotated, with Feature tab selected and the Feature table displayed; the Feature table lists each feature and associated descriptors that are currently stored in the sequence database. The Features database can be edited using the fields in the Description sub-tab, shown to the *right* of the Feature table. The *thin green line* in the Frames window was generated by clicking on that Feature in the Frames window; subsequent pressing of the "ORF" button in the *lower-right corner* of the Frames window generated the "Choose Start" window that appears on the *lower right side*. Each row in the Choose Start window represents the data for each possible start in the open reading frame selected in the Frames window

start codons for the ORF, along with the upstream sequence. It is possible to select a scoring matrix to evaluate the strength of the upstream sequence match to the canonical Shine–Dalgarno sequence, in conjunction with the spacer distance between the location of that sequence and the putative start codon. The matrices available in DNA Master are Kibler 5, 6, 7, 9, 10A, or 10B [8]; and the spacer matrices available are Broad, Relaxed, Karlin Medium, Karlin High, or Karlin Ribosomal [9]. The default choices (Kibler 7, Medium work well to evaluate the majority of bacteriophage genes. In the evaluation of the final score, the least negative number represents the best match, and a Z-value above 2 indicated that the match is two standard deviations away from a match to a random sequence.

All changes to the auto-annotated Features in the *.dnam5 file must be made through the Features tab. Addition or deletion of gene features is accomplished through the use of the "Insert" or

"Delete" button at the bottom of the Features database table. To Insert a Feature, press the Insert button, and then enter the coordinates and type of Feature (CDS for "coding sequence" or "tRNA") in the pop-up window. To delete a Feature, click on it in the table, and then press "Delete". Modification of the start coordinate of any Feature is accomplished through the Description sub-tab associated with that Feature. To modify a start coordinate, click on the Feature to be altered in the Features table, and click on the Description tab on the right-side of the window (if it isn't already selected). The gene boundaries are listed in the fields marked "5′ end" and "3′ end". These labels refer to the 5′ and 3′ end of the genome as a whole, and not the 5′ and 3′ ends of the specific gene. For genes transcribed in the rightwards (forwards) direction, the start coordinate is shown in the 5′end box; simply click in the box and type the corrected start coordinate. For leftwards (reverse) transcribed genes, click in the 3′ box and type the new start coordinate. After changing the start, the gene length must be manually adjusted; this is accomplished by clicking the button next to the Length field that looks like a calculator. Clicking this button will also post the change to the database table automatically.

In addition to adjusting the gene content and start coordinates, functional assignments must be entered. Functional assignments should be listed in the field marked "Product". Functional assignments can be found using BLASTp and the GenBank database, the conserved domain database, or by using HHPred and the Protein Data Bank. All functional matches should be carefully considered in the context of quality and length of their alignments as well as with respect to the biology of the organism prior to their inclusion in an annotation file for GenBank submission. We prefer to err on the side of caution, and do not assign functions without significant supporting evidence. If there is no known function for a particular gene, this field should be filled in with "Hypothetical protein". This can be done automatically for all genes of no known function at the end of the annotation through the "Validation" tab.

8 Validation and Formatting for GenBank

At any point during the annotation, it is possible to validate the annotation by pressing the "Validate" button at the bottom of the Features table. During validation, the program will assess the annotation and report if any Features begin without a start codon, contain a stop codon within them, or are not the correct length in the Validation tab. All Features should be valid prior to submission to GenBank. Within the Validation tab (which automatically appears once the "Validate" button is pressed), it is also possible to renumber all of the Features in the Features table, to add a custom locus tag for GenBank submissions, and to assign "Hypothetical protein" to all of the genes without a known function. This is accomplished through the sub-tabs "Control" and "Numbering", which appear at the bottom of the Validation tab.

9 Generation of GenBank Files and Submission to GenBank

After the annotation has been completed, validated, and saved a final time, it is possible to generate GenBank flat file and submission files through the Submit to GenBank tool found in the Tools menu. The new file should be added as a New Project using the "Add" button on the lower left side of the window (Fig. 2). The other tabs contain various fields that correspond to headers in GenBank files. Critical information that is required by GenBank for submission includes the selection of the correct genetic code "Bacterial and Plant Plastid", if the genome sequence is complete, if the genome is linear or circular, the host strain, the lineage of the bacteriophage, the sequencing platform and fold coverage, the sequence assembly software and release date, and the annotation authors. DNA Master has a tab for almost all of this information, the exception is that of the sequencing platform, coverage, and assembly software; and this

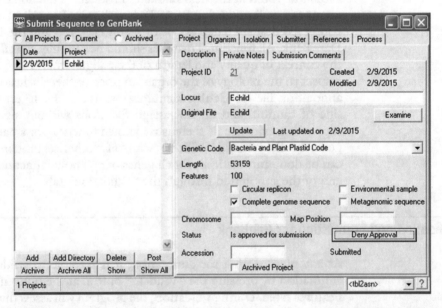

Fig. 2 Submit to GenBank window. This is the first window used to prepare a file to submit to GenBank. Note the tabs—Project, Organism, Isolation, Submitter, References, and Process—on the *right side*, each of which should be filled out prior to generation of the GenBank file for submission. Displayed is the "Project" tab, with the sub-tab "Description" correctly filled out. Other sub-tabs for Project are Private Notes, the contents of which will not be included in the final submission file, and Submission Comments, the contents of which will be included in the final files

can all be entered manually on the tab marked "Submission comments".

After all of the appropriate information has been entered, the final flat files and submission files can be generated from the Process tab. The Flat file is a text file that appears the way a final GenBank entry would. This should be carefully inspected for typographical or other errors prior to the submission of the *.asn1 final file to GenBank. At the time of this writing, the ability of DNA Master to automatically submit annotations has been disabled; however, the submission file can be saved as a simple .txt file, and then emailed to gb-sub@ncbi.nlm.nih.gov

10   Notes

1. Merely installing the program DNA Master using an account that has Administrator privileges and launching the program is NOT sufficient. (In Windows XP, it is sufficient to run the program merely when logged in with an account that has Administrator privileges.)

2. In Windows 7 or above, it is possible to permanently launch the program with Administrator privileges by right-clicking on the DNAMas.exe file, and selecting "Properties" from the pop-up menu. In the Compatibility tab, under "Run this program in compatibility mode" select "Windows XP Service Pack 3" from the drop-down menu; and under "Privilege Level" check the box next to "Run this program as an Administrator". It is also possible to then click "Change settings for all users", which allows all users of the Windows machine to correctly run DNA Master, even if they are logged into an account that does not have Administrator rights. This is particularly useful in a student computer lab setting, as Universities IT staff are generally unwilling to allow students to log into University computers with Administrator accounts.

3. It might be tempting to start a new annotation by creating a new *.dnam5 file via the menu "File → New", and pasting the genome sequence from a .fasta file into the sequence tab of the *.dnam5 file. Do not do this. Copying and pasting the sequence data can result in copying line breaks or other types of invisible-to-the-user marks in the file into the sequence pane that are then read by the program as part of the sequence. This leads to a corrupted sequence and may cause some parts of the program to not work correctly.

4. If the *.dnam5 file becomes corrupted during annotation, it is possible to recover the annotation features and notes that have been saved in the Documentation. This is accomplished by creating a fresh *.dnam5 file through importing the original.

Fasta formatted sequence as above; and then pasting the contents of the Documentation tab from the old file into the new one and pressing the "Parse" button. This only works if the Documentation has been recreated with regularity.

5. Auto-annotation will fail if the NCBI servers that host GLIMMER and GeneMark are offline; likewise, it will fail if the computer does not have an internet connection. As this functionality of DNA Master is reliant on a third-party's website, it is possible to lose this ability of the program without warning when changes are made to NCBI's servers. Once appropriate maintenance of DNA Master is completed, this function is usually restored to the user with a program update.

6. It is important to know that if both programs agree on a gene call, only the output from the "preferred" program, selected during auto-annotation, will be displayed. If the programs disagree, both outputs will be displayed.

7. Retrieving the BLAST alignment data from GenBank for an entire genome can take many hours depending on the time of day, during which time DNA Master cannot be used for anything else. If you decide to batch BLAST the genome, you may want to start at 9 pm EST, which is the beginning of Off-Peak hours for the NCBI servers.

8. The most common arrangement of genes within the phages of the Order *Actinomycetales* that we have observed is a 4 bp overlap between the stop codon of the upstream gene and the start codon of the downstream gene.

9. The Genome Menu and the DNA Menu can only be accessed when a *.dnam5 file has been opened. All of the selections in these menus are alphabetized.

Acknowledgments

We would like to acknowledge and sincerely thank Jeffrey Lawrence, for the creation of the program DNA Master and for his cheerful responses to our constant requests for new features; Graham Hatfull and Roger Hendrix for their leadership and expertise in the field of bacteriophage; Dan Russell, Charlie Bowman, and Steve Cresawn for sequencing and bioinformatics insights; the members of the Hatfull Lab for helpful discussions; David Asai and the leadership of the HHMI SEA-PHAGES program for support; and the students and faculty of the 80+ SEA-PHAGES institutions for the multitude of new phages isolated over the past 8 years.

References

1. Besemer J, Borodovsky M (2005) GeneMark: web software for gene finding in prokaryotes, eukaryotes and viruses. Nucleic Acids Res 33 (Web Server issue):W451–W454. doi:10.1093/nar/gki487

2. Delcher AL, Bratke KA, Powers EC, Salzberg SL (2007) Identifying bacterial genes and endosymbiont DNA with Glimmer. Bioinformatics 23(6):673–679. doi:10.1093/bioinformatics/btm009

3. Laslett D, Canback B (2004) ARAGORN, a program to detect tRNA genes and tmRNA genes in nucleotide sequences. Nucleic Acids Res 32(1):11–16. doi:10.1093/nar/gkh152

4. Pope WH, Jacobs-Sera D, Russell DA, Rubin DH, Kajee A, Msibi ZN, Larsen MH, Jacobs WR Jr, Lawrence JG, Hendrix RW, Hatfull GF (2014) Genomics and proteomics of mycobacteriophage patience, an accidental tourist in the *Mycobacterium* neighborhood. mBio 5(6): e02145. doi:10.1128/mBio.02145-14

5. Cresawn SG, Pope WH, Jacobs-Sera D, Bowman CA, Russell DA, Dedrick RM, Adair T, Anders KR, Ball S, Bollivar D, Breitenberger C, Burnett SH, Butela K, Byrnes D, Carzo S, Cornely KA, Cross T, Daniels RL, Dunbar D, Findley AM, Gissendanner CR, Golebiewska UP, Hartzog GA, Hatherill JR, Hughes LE, Jalloh CS, DeLos Santos C, Ekanam K, Khambule SL, King RA, King-Smith C, Klyczek K, Krukonis GP, Laing C, Lapin JS, Lopez AJ, Mkhwanazi SM, Molloy SD, Moran D, Munsamy V, Pacey E, Plymale R, Poxleitner M, Reyna N, Schildbach JF, Stukey J, Taylor S, Ware VC, Wellmann AL, Westholm D, Wodarski D, Zajko M, Zikalala TS, Hendrix RW, Hatfull GF (2015) Comparative genomics of cluster o mycobacteriophages. PLoS One 10:e0118725

6. Altschul SF, Gish W, Miller W, Myers EW, Lipman DJ (1990) Basic local alignment search tool. J Mol Biol 215(3):403–410. doi:10.1016/S0022-2836(05)80360-2

7. Jacobs-Sera D, Pope WH, Russell DA, Bowman CA, Cresawn SG, Hatfull GF (2014) Annotation and bioinformatic analysis of bacteriophage genomes: a user guide to DNA master. Software Guides on phagesdb.org. Last revision December 2014

8. Kibler DaH, S. Characterizing the E. coli Shine-Dalgarno motif: probability matrices and meight matrices. In: International Conference on Mathematical and Engineering Techniques in Medicine and Biological Science (METMBS-2002), Las Vegas, NV, 2002, METMBS-2002, pp 358–364

9. Ma J, Campbell A, Karlin S (2002) Correlations between Shine-Dalgarno sequences and gene features such as predicted expression levels and operon structures. J Bacteriol 184 (20):5733–5745

Chapter 17

Phage Genome Annotation Using the RAST Pipeline

Katelyn McNair, Ramy Karam Aziz, Gordon D. Pusch, Ross Overbeek, Bas E. Dutilh, and Robert Edwards

Abstract

Phages are complex biomolecular machineries that have to survive in a bacterial world. Phage genomes show many adaptations to their lifestyle such as shorter genes, reduced capacity for redundant DNA sequences, and the inclusion of tRNAs in their genomes. In addition, phages are not free-living, they require a host for replication and survival. These unique adaptations provide challenges for the bioinformatics analysis of phage genomes. In particular, ORF calling, genome annotation, noncoding RNA (ncRNA) identification, and the identification of transposons and insertions are all complicated in phage genome analysis. We provide a road map through the phage genome annotation pipeline, and discuss the challenges and solutions for phage genome annotation as we have implemented in the rapid annotation using subsystems (RAST) pipeline.

Key words Phage, Genome annotation, RAST, Functional annotation, Gene predictions

1 The Steps of Phage Genome Annotation

The essential steps in annotating any genome, whether phage, bacterial, or eukaryotic, consist of identifying the features in the genome and assigning terms describing roles or functions to those features. Typical features that can be found in a phage genome include protein-encoding genes, noncoding RNA genes, insertion elements and transposons, direct and indirect repeats, origins of replication, and attachment or integration sites. Annotations are routinely only added to protein and RNA-encoding genes, labels are often provided for insertion elements or transposons. Specific for phages, they are fundamentally dependent on a cellular host to replicate, and the functions on its genome can only be completely understood in the context of the genome of the host. Thus, identification of prediction of the bacterial or archaeal host is an important part of phage annotation. Together, these features provide the core annotation of phages and this annotation provides the first steps to understanding the function of the phage as it interacts with

Martha R.J. Clokie et al. (eds.), *Bacteriophages: Methods and Protocols, Volume 3*, Methods in Molecular Biology, vol. 1681, https://doi.org/10.1007/978-1-4939-7343-9_17, © Springer Science+Business Media LLC 2018

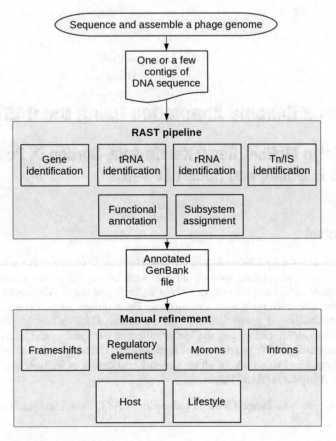

Fig. 1 Pipeline of phage genome annotation starting with DNA sequences and ending with an annotated genome

its host (Fig. 1). We discuss the approaches to identify and annotate each of these features below, and discuss how these annotations are performed in the Rapid Annotation Using Subsystems Technology approach (RAST) [1, 2].

Protein-encoding genes are the focus of most automated annotation systems, and more algorithms have been designed to handle these features than other features. Generally a protein-encoding gene can be identified as a long stretch of sequence in one reading frame that can be translated into protein sequence without including one of the three stop codons; these long stretches are called Open Reading Frames (ORFs). In gene calling, the stop codons are obvious because there is a choice of three codons to choose from and they are all stop codons (unless the phage encodes a suppressor tRNA which we do not discuss here). Most algorithms attempt to identify the longest nonoverlapping ORFs in a genome, based on the theory that the longer the open reading frame the less likely it is to occur by chance. There are many alternative gene-finding algorithms that have been developed over the last two decades, including CRITICA [3], GeneMark [4, 5], GISMO [6], Glimmer [7, 8], MetaGeneAnnotator [9], and Prodigal [10]. Most of the gene-finding algorithms find the same large genes because these are obvious and have high confidence. The algorithms may differ in the particular start sites that they identify; there may be multiple methionine (ATG) or valine (GTG) codons that could all be used as the start codon, and predicting exactly which start codon is the correct one for a given gene is difficult without a priori knowledge of the translation boundaries of the gene. In addition, the gene callers also differ in their ability to identify small protein-encoding genes. Short genes are statistically difficult to separate from the background noise of stretches of nucleotides that do not encode a stop codon, and often gene calling algorithms use an artificial cut off of (for example) 75 amino acids. It remains to be determined how many small proteins are encoded in phage genomes, and this is unlikely to be approached from a pure bioinformatics standpoint, as it will require biological validation of bioinformatics predictions or large-scale proteomic studies.

Most bacterial genomes are not thought to contain overlapping open reading frames, and these *shadow ORFs* are removed during the annotation step [10]. In viruses, including phages, however, there are several well-known examples of two different genes from the same stretch of DNA, such as the Rz/Rz1 system [11]. One study even suggests that new genes may be born via this process, providing evidence from the comparative genomics of *Rhabdoviridae* genomes [12]. These overlapping regions are generally not predicted using most bioinformatics approaches, as adding overlapping ORFs to gene prediction algorithms would include an enormous number of false positives to compensate for only a few false negatives. Therefore, most phage protein prediction schemes ignore overlapping proteins.

Following ORF identification, most bioinformatic gene prediction tools assign a confidence score to the ORFs using a model of what a gene is expected to look like, based on its nucleotide usage statistics. These statistics are specific for a species, and depend on properties like the codon usage and GC content of the genome. In bacterial genomes, the RAST pipeline starts by identifying highly conserved genes that are present in nearly every genome. The statistics from those genes are then used to build a genome-specific model for open reading frame identification that is applied to the rest of the genome. In phage genes, there are typically very few, if any, highly conserved genes, and never enough to build a reliable gene model. Therefore, most gene calling is performed by a generic model that is not trained on the specific genome being annotated but on the genomes of all phages. By default, the RAST pipeline uses Glimmer to identify the open reading frames, but options are available to use MetaGeneAnnotator [9], GeneMark [4], or Prodigal [10].

The functional annotation of protein-encoding phage genes is usually based on homology searches against existing phages. Historically, phage genes were named with a single letter starting at gpA, and either proceeding along the genome or assigning names based on the order in which the genes or their products were found. This resulted in several unrelated proteins from different phages all having the same names. For example both terminases and DNA replication initiation proteins have been annotated as gpA in different phage genomes available from GenBank. This confusion, amplified by the explosion of genome sequences in recent years, led to efforts to categorize phage proteins into either phage orthologous groups (POGs) [13] or subsystems [14] that have unified the annotation of many phage proteins. These common, descriptive, names provide a framework for comparing annotations among different phage genomes. The RAST system uses a combination of homology, chromosomal clustering, and subsystems to assign functions to proteins. First, proteins are annotated on the basis of homology to known proteins. If this initial search yields matches to proteins that are a component of a subsystem, RAST then tries to find other members of the subsystem that should be present in the same genome based on information from the previously annotated genomes. The advantage of this approach is that the RAST system can strengthen otherwise weak assertions of homology, based on predictions from subsystem annotations. Of note, the RAST tools allow the analysis of proteins in their chromosomal context, which sometimes helps determine the roles of proteins with unknown functions based on the functions of their chromosomal neighbors (e.g., protein subunits encoded by different genes, members of operons, or transporters of metabolites whose metabolizing enzymes are encoded on the same cluster). Phage genomes, like bacterial genomes, also order some of their genes, and this information can be leveraged to identify clusters of genes. For example, the small and large terminase (TerS and TerL) are frequently adjacent on the genome, and the identification of one leads to the identification of the other.

A major difficulty in the functional annotation of protein-encoding genes on phage genomes by homology searches is the fact that most proteins have no close homologs in the reference databases. Especially for novel phages, this results in the majority of encoded ORFs having no annotated function, or a hypothetical function at best. A possible solution includes homology-independent annotation, based on amino acid usage profiles of the proteins. One such approach, iVIREONS (https://vdm.sdsu.edu/ivireons/) uses machine learning to "learn" the characteristics of manually annotated phage proteins and then tests unknown proteins to see if they have similar characteristics [15].

Noncoding RNA (ncRNA) genes. Although Ribosomal RNAs have not yet been found in phage genomes, most pipelines, including RAST, look for them anyway as the pipelines have been developed for bacterial genome annotation and the computational cost of looking for rRNA genes is a minimal addition to the pipeline. Ribosomal RNA genes are highly conserved and are identified by extrinsic gene calling—using a database of known RNA genes to compare against. In contrast to rRNA genes that are recognized by homology, tRNA genes are recognized by intrinsic gene calling—using only features of the sequence. They are typically identified by computational tools built specifically to recognize the secondary structure of the tRNA molecule [16]. As with tRNAs, the function of other non-protein coding RNA genes also depends on the structure of the folded RNA molecule rather than the nucleotide sequence. Therefore, other noncoding RNA genes are also recognized by their conserved secondary structure rather than homology to existing sequences [17]. The RAST pipeline uses a manually curated database of ribosomal RNA genes to find them in a genome, and uses tRNAScan-SE [16] to identify tRNA genes. Many phages encode tRNA genes, and it has been proposed that these may supplement host-encoded tRNAs in translating phage proteins for anticodons that are insufficiently covered by the bacterial tRNAs [18]. These tRNA genes are also often used as phage integration sites in the host's genome (*attP*). Integration of the phage disrupts the host gene, and thus carrying complete, or near complete, tRNA genes allows the phage to reconstitute a tRNA into which it can integrate [19]. There has been little exploration of the role of ncRNA in phage lifestyle. Recent work with CRISPR/Cas systems have identified the presence of these systems in phage genomes [20] and metagenomes [21], and it is thought that they are being used to attack other phages that may be infecting the same host.

Insertion elements and transposons are currently identified by annotations of protein-encoding genes. Transposases (Tn) are readily identified as protein-encoding genes, and the similarity between members of the transposase family, and with other recombinases, is high enough that they usually receive accurate annotation. However, the repeats flanking the insertion sequence or transposon are not typically automatically annotated. There are boutique databases of these problematic mobile elements [22, 23], but often the classification of insertion (IS) elements is dependent on one or a few residues. Typically automatic annotation systems identify the Tn or IS elements but cannot identify the fine details responsible for the accurate categorization of these elements. More work is required to accurately denote the ends of these mobile elements in automatic phage annotation systems. Direct and indirect repeats are usually used to identify the ends of insertion elements and transposons [22], and to predict the ends of prophages that have been found in bacterial genomes [13]. Standard informatics approaches can easily identify repeats longer than approximately 14 nucleotides in a phage genome. Below that length, repeats are found too frequently to ascertain whether they are indeed the correct flanking repeats, or randomly occurring repeated sequence elements. A few websites can be used to identify repeats in DNA sequences (e.g., [24, 25]).

Phage attachment sites are impossible to detect de novo if only the phage is known, but if the phage and the host genome sequences are known, they are trivial to find. The phage carries the attachment site P (*attP*) that has sequence homology to the bacterial attachment site B (*attB*). Integration is initiated by recombination between *attP* and *attB*, resulting in *attL* and *attR* sites that flank the nascent prophage.

Accurately Annotating Phage Metadata. Annotating genomic metadata is a general challenge to genomics and metagenomics. With bacteriophages, this issue is even more problematic, given the lack of systematic nomenclature for viruses (as opposed to the binomial system used for cellular organisms, *see* Chapter 15 of this book). Some attempts were made to suggest systematic nomenclature for viruses similar to those used for plasmids [26], but they are not widely applied or enforced. In addition to accurate taxonomic descriptions of viruses, including metadata associated with the virus (e.g., its morphology, actual host, host range, and lifestyle) is equally important. These make comparative genomics studies possible, enable predictive tools such as those that identify the host of unknown phages [27], or predict the lifestyle of new phages [14] and improve metagenomic/microbiomic annotations. Other important types of metadata can be computed from the genomic information, e.g., a genome's length, %G+C, and codon usage [28]. These too have quite powerful applications in comparative genomics, prophage finding, and metagenomics. For example, information content of phage genomes has improved prophage finding [29] and is proposed to improve metagenomic analysis [30]. As with gene annotation, metadata annotation needs to use a controlled vocabulary (which has to be consistent but not necessarily rigid or hierarchical). Spelling inconsistencies (e.g., firmicutes vs. Firmicutes vs. gram-positive bacteria) or terminology inconsistencies (e.g., temperate vs. lysogenic lifestyles) are all obstacles against computational analysis and data propagation.

To summarize, phage annotation involves the identification and functional description of several types of features, including protein-encoding genes, RNA genes, insertion elements and transposons, repeats, and attachment sites. Moreover, phage–host associations are an important part of understanding phage biology that can be predicted using a range of computational tools [27]. The RAST pipeline provides an automated approach to phage genome annotation. The pipeline currently uses bacterial ORF-finding algorithms to identify the proteins in the genome, and a combination of homology-based and subsystems-based approaches to decorate those proteins with their functional annotation. RNA genes are detected by a combination of extrinsic and intrinsic gene calling methods. There remain several hurdles to accurate phage genome annotation, especially the assignment of functions to unknown proteins, the identification of small proteins in the genome, and the correct and unambiguous identification of insertion elements and transposons. The combinations of bioinformatics advances and a better understanding of phage biology will help to improve phage genome annotation, making this field a fertile area for further exploration.

Acknowledgments

This work was supported by grants from the National Science Foundation MCB-1330800 and DUE-1323809 to RAE. BED was supported by the Netherlands Organization for Scientific Research (NWO) Vidi grant 864.14.004.

References

1. Aziz RK, Bartels D, Best AA, DeJongh M, Disz T, Edwards RA, Formsma K, Gerdes S, Glass EM, Kubal M, Meyer F, Olsen GJ, Olson R, Osterman AL, Overbeek RA, McNeil LK, Paarmann D, Paczian T, Parrello B, Pusch GD, Reich C, Stevens R, Vassieva O, Vonstein V, Wilke A, Zagnitko O (2008) The RAST Server: rapid annotations using subsystems technology. BMC Genomics 9:75

2. Brettin T, Davis JJ, Disz T, Edwards RA, Gerdes S, Olsen GJ, Olson R, Overbeek R, Parrello B, Pusch GD, Shukla M, Thomason Iii JA, Stevens R, Vonstein V, Wattam AR, Xia F (2015) RASTtk: A modular and extensible implementation of the RAST algorithm for building custom annotation pipelines and annotating batches of genomes. Sci Rep 5:8365

3. Badger JH, Olsen GJ (1999) CRITICA: coding region identification tool invoking comparative analysis. Mol Biol Evol 16:512–524

4. Borodovsky M, McIninch JD, Koonin EV, Rudd KE, Médigue C, Danchin A (1995) Detection of new genes in a bacterial genome using Markov models for three gene classes. Nucleic Acids Res 23:3554–3562

5. Lukashin AV, Borodovsky M (1998) GeneMark.hmm: new solutions for gene finding. Nucleic Acids Res 26:1107–1115

6. Krause L, McHardy AC, Pühler A, Stoye J, Meyer F (2007) GISMO - Gene identification using a support vector machine for ORF classification. Nucleic Acids Res 35:540–549

7. Delcher AL, Harmon D, Kasif S, White O, Salzberg SL (1999) Improved microbial gene identification with GLIMMER. Nucleic Acids Res 27:4636–4641

8. Kelley DR, Liu B, Delcher AL, Pop M, Salzberg SL (2012) Gene prediction with Glimmer for metagenomic sequences augmented by classification and clustering. Nucleic Acids Res 40:e9–e9

9. Noguchi H, Taniguchi T, Itoh T (2008) MetaGeneAnnotator: Detecting species-specific patterns of ribosomal binding site for precise gene prediction in anonymous prokaryotic and phage genomes. DNA Res 15:387–396

10. Hyatt D, Chen G-L, LoCascio PF, Land ML, Larimer FW, Hauser LJ (2010) Prodigal: prokaryotic gene recognition and translation initiation site identification. BMC Bioinformatics 11:119

11. Summer EJ, Berry J, Tran TAT, Niu L, Struck DK, Young R (2007) Rz/Rz1 lysis gene equivalents in phages of Gram-negative hosts. J Mol Biol 373:1098–1112

12. Walker PJ, Firth C, Widen SG, Blasdell KR, Guzman H, Wood TG, Paradkar PN, Holmes EC, Tesh RB, Vasilakis N (2015) Evolution of genome size and complexity in the *Rhabdoviridae*. PLoS Pathog 11:e1004664

13. Kristensen DM, Waller AS, Yamada T, Bork P, Mushegian AR, Koonin EV (2013) Orthologous gene clusters and taxon signature genes for viruses of prokaryotes. J Bacteriol 195:941–950

14. McNair K, Bailey BA, Edwards RA (2012) PHACTS, a computational approach to classifying the lifestyle of phages. Bioinformatics 28:614–618

15. Seguritan V, Alves N, Arnoult M, Raymond A, Lorimer D, Burgin AB, Salamon P, Segall AM (2012) Artificial neural networks trained to detect viral and phage structural proteins. PLoS Comput Biol 8:e1002657

16. Lowe TM, Eddy SR (1997) tRNAscan-SE: a program for improved detection of transfer RNA genes in genomic sequence. Nucleic Acids Res 25:955–964

17. Nawrocki EP (2014) Annotating functional RNAs in genomes using Infernal. Methods Mol Biol 1097:163–197

18. Bailly-Bechet M, Vergassola M, Rocha E (2007) Causes for the intriguing presence of tRNAs in phages. Genome Res 17:1486–1495

19. Williams KP (2002) Integration sites for genetic elements in prokaryotic tRNA and tmRNA genes: sublocation preference of integrase subfamilies. Nucleic Acids Res 30:866–875

20. Seed KD, Lazinski DW, Calderwood SB, Camilli A (2013) A bacteriophage encodes its own CRISPR/Cas adaptive response to evade host innate immunity. Nature 494:489–491

21. Cassman N, Prieto-Davó A, Walsh K, Silva GGZ, Angly F, Akhter S, Barott K, Busch J, McDole T, Haggerty JM, Willner D, Alarcón G, Ulloa O, DeLong EF, Dutilh BE, Rohwer F, Dinsdale EA (2012) Oxygen minimum zones harbour novel viral communities with low diversity. Environ Microbiol 14:3043–3065

22. Aziz RK, Breitbart M, Edwards RA (2010) Transposases are the most abundant, most ubiquitous genes in nature. Nucleic Acids Res 38:4207–4217

23. Riadi G, Medina-Moenne C, Holmes DS (2012) TnpPred: a web service for the robust prediction of prokaryotic transposases. Comp Funct Genomics 2012:678761

24. Benson G (1999) Tandem repeats finder: a program to analyze DNA sequences. Nucleic Acids Res 27:573–580

25. Volfovsky N, Haas BJ, Salzberg SL (2001) A clustering method for repeat analysis in DNA sequences. Genome Biol 2:RESEARCH0027

26. Kropinski AM, Prangishvili D, Lavigne R (2009) Position paper: the creation of a rational scheme for the nomenclature of viruses of Bacteria and Archaea. Environ Microbiol 11:2775–2777

27. Edwards RA, McNair K, Faust K, Raes J, Dutilh BE (2016) Computational approaches to predict bacteriophage–host relationships. FEMS Microbiol Rev 40:58–72

28. Aziz RK, Dwivedi B, Akhter S, Breitbart M, Edwards RA (2015) Multidimensional metrics for estimating phage abundance, distribution, gene density, and sequence coverage in metagenomes. Front Microbiol 6:381

29. Akhter S, Aziz RK, Edwards RA (2012) PhiSpy: a novel algorithm for finding prophages in bacterial genomes that combines similarity- and composition-based strategies. Nucleic Acids Res 40:e126–e126

30. Akhter S, Bailey BA, Salamon P, Aziz RK, Edwards RA (2013) Applying Shannon's information theory to bacterial and phage genomes and metagenomes. Sci Rep 3:1033

Chapter 18

Visualization of Phage Genomic Data: Comparative Genomics and Publication-Quality Diagrams

Dann Turner, J. Mark Sutton, Darren M. Reynolds, Eby M. Sim, and Nicola K. Petty

Abstract

The presentation of bacteriophage genomes as diagrams allows the location and organization of features to be communicated in a clear and effective manner. A wide range of software applications are available for the clear and accurate visualization of genomic data. Several of these applications incorporate comparative analysis tools, allowing for insertions, deletions, rearrangements and variations in syntenic regions to be visualized. In this chapter, freely available software and resources for the generation of high-quality graphical maps of bacteriophage genomes are listed and discussed.

Key words Phage, Genomes, Comparative genomics, Visualization, Software

1 Introduction

The visualization of data is a key element of communicating research and ideas; to quote Tufte "graphics reveal data" [1]. Diagrams should seek to be intuitive, instructive, coherent, displaying the data without ambiguity.

Circular and linear DNA diagrams provide a powerful tool for illustrating the structure, organisation and comparisons of bacterial genomes. Compared to bacteria, phage genomes are relatively small, ranging between 2.4 kb of the inovirus *Leuconostoc* phage L5 [GenBank: L06183] to 497.5 kb for the giant myovirus *Bacillus* phage G [GenBank: JN638751]. This smaller size allows the genome and annotated features to be clearly presented in their entirety as diagrams.

In addition to illustrating the structural organization of annotated features in a genome, visualization serves as a powerful tool to

Electronic supplementary material: The online version of this chapter (doi:10.1007/978-1-4939-7343-9_18) contains supplementary material, which is available to authorized users.

Martha R.J. Clokie et al. (eds.), *Bacteriophages: Methods and Protocols, Volume 3*, Methods in Molecular Biology, vol. 1681, https://doi.org/10.1007/978-1-4939-7343-9_18, © Springer Science+Business Media LLC 2018

exemplify similarities and differences between closely and distantly related phages uncovered by comparative genomic analysis. These relationships can include the syntenic conservation of gene order, grouping of genes into functional modules, positional relationships such as insertions, deletions, rearrangements, localized areas of alignment and the identification of orthologous gene pairs. Represented as diagrams, these relationships can be grasped quickly and intuitively from complex data, which might otherwise be presented as dense tabular information.

This chapter provides an overview of applications available for creating genome diagrams. All applications listed in this chapter come with a manual and tutorials for their usage, distributed either with the application or on their website. Most of the applications have an easily accessible graphical user interface for use on a personal computer (**Note 1**), or can be used via a web-service (Table 1). Step-by-step methods are provided to produce publication quality genome images using the two command-line only applications CGView Comparison Tool and Circos, as well as the graphical-user interface application Easyfig.

2   Resources

This section lists some of the applications and resources currently available for producing linear and or circular genome diagrams along with a brief description. A summary of the features for each program is presented (Table 1). The choice of tool depends firstly upon whether a simple genome map is required or visualization and organization of complex comparative data from multiple sources (**Note 2**). An additional consideration is whether a graphical user interface is preferred over manipulating data files or writing scripts.

BLAST Ring Image Generator (http://brig.sourceforge.net)

The BLAST Ring Image Generator (BRIG) is a versatile, easy-to-use, cross-platform (Windows, Mac, and Unix) application that can be used to compare large numbers of genomes and display in a circular image [2]. Via a user-friendly graphical user interface (GUI), BRIG performs all file parsing and BLAST comparisons automatically, using a locally installed copy of BLAST [3] to perform comparisons between sequences and CGView to render a circular diagram in raster (JPEG, PNG) or vector (SVG) format [2]. The configuration and generation of diagrams is a step by step process where a reference sequence and one or more comparison genomes or sequences are selected and assigned to rings by the user and subjected to a user-declared BLAST [3] analysis (Fig. 1a). The presence or absence of BLAST hits to the sequences in the reference genome are displayed in concentric rings, with colours selected for each comparison genome graduated based upon the percent

Table 1
Applications for creating genome diagrams

Application Type	CGView WS, CMD	CCT CMD	GView GUI	Gview Server WS	GenomeDiagram API	BRIG GUI	Circos CMD	OGDraw WS, CMD	DNA plotter WS, GUI	GenomeVx WS	Easyfig GUI, CMD
Input formats[a]											
FASTA	+	+	+	−	+[b]	+	−[c]	−	+	−	+
Multi-FASTA	−	+	−	−	+[b]	+	−[c]	−	+	−	+
GenBank	+	+	+	+	+[b]	+	−[c]	+	+	+	+
GFF	+	+	+	−	+[b]		−[c]	−	+	−	−
EMBL	+	+	+	+	+[b]	+	−[c]	−	+	−	+
SAM	−	−	−	−	−	+	−	−	−	−	−
Output formats											
TIF	−	−	−	−	+	−	−	+	−	−	−
PNG	+	+	+	+	+	+	+	+	+	−	−
JPG	+	+	+	+	+	+	−	+	+	−	−
BMP	−	−	−	−	+	−	−	−	+	+	+
SVG/SVGZ	+	+	+	+	+	+	+	−	+	−	+
PS/EPS	−	−	−	−	+	−	−	+	−	−	−
PDF	−	−	+	−	+	−	−	−	−	+	−
Features											
GC content	+	+	+	+	+	+	+	+	+	−	+
GC skew	+	+	+	+	+	+	+	−	+	−	+
Add custom features	+	+	+	+	+	+	+	−	+	+	+
Add analysis data	+	+	+	+	+	+	+	−	+	−	+

(continued)

Table 1
(continued)

Application Type	CGView WS, CMD	CCT CMD	GView GUI	Gview Server WS	GenomeDiagram API	BRIG GUI	Circos CMD	OGDraw WS, CMD	DNA plotter WS, GUI	GenomeVx WS	Easyfig GUI, CMD
Comparison of genomes	+[d]	+	−	+	+	+	+	−	−	−	+
%id and e-value filtering	+	+	−	+	+	+	+	−	−	−	+
Diagram type											
Linear images	−	−	+	+	+	−	−	−	+	−	+
Circular images	+	+	+	+	+	+	+	+	+	+	−
BLAST programs											
BLASTN	+	+	−	+	−	+	−[e]	−	−	−	+
BLASTP	−	+	−	+	−	+	−[e]	−	−	−	−
TBLASTX	+	+	−	+	−	+	−[e]	−	−	−	+
BLASTX	+	+	−	+	−	+	−[e]	−	−	−	−
TBLASTN	−	+	−	+	−	+	−[e]	−	−	−	−

CMD command-line, *GUI* graphical user interface, *WS* web-service, *API* application-programming interface

[a]Different file formats can be easily converted to other formats using EMBOSS Seqret (http://www.ebi.ac.uk/Tools/sfc/emboss_seqret/), Artemis [8] or custom scripts, e.g., GFF files can be converted to GenBank, EMBL or FASTA format

[b]Functions to read in different flat file formats, perform analyses or read analysis data are conferred programmatically using an appropriate python script

[c]Input data must be parsed into appropriate files for Circos

[d]A maximum three comparison genomes can be processed by CGView Server

[e]Circos does not run or analyse BLAST results natively. Instead data from any program can be displayed provided it has been converted to an appropriate format for Circos to parse

A.

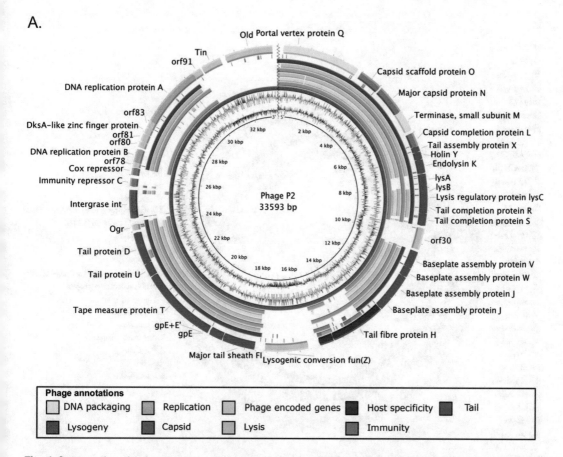

Fig. 1 Comparative circular genome maps generated using BRIG. (**a**) Comparative circular genome map of P2-like phages. Enterobacteria phage P2 [AF063097; [4]] was used as the central reference sequence and the innermost rings show GC content (*black*) and GC skew (*purple/green*). The next rings shows BLASTn comparisons of seven P2-like phage genomes against the P2 genome (L-413C [AY251033; [5]], 186 [U32222], PSP-3 [AY135486; [6]], WPhi [AY135739], FSL SP-004 [KC139521; [7]], fiAA91-ss [KF322032; [8]]), with the genomes of other tailed phages (Fels-2 [NC_010463; [9]], phiCTX [AB008550; [10]], T4 [AF158101; [11]], Mu [AF083977; [12]], T7 [V01146; [13]] and lambda [J02459; [14]]) collapsed into a single ring. The BLASTn hits are colored on a gradient according to % identity as shown in the key (*right*). The outermost ring shows the CDSs of phage P2 as arrows, color-coded according to function (*bottom key*). (**b**) Comparative circular genome map of *Pseudomonas* phage assembled genomes and unassembled sequencing data. The central reference sequence is the complete genome of *Pseudomonas* phage phiPsa374 (KJ409772; [15]) with the colored rings showing BLASTn hits to a draft genome assembly (contigs > 2 kb) of the sequencing reads of phiPsa440 (PRJNA236447; [15]), with the complete genomes of other *Pseudomonas* phages (JG004 [GU988610; [16]], PAK_P1 [KC862297; [17]], PAK_P2 [KC862298] and PaP1 [HQ832595; [18]]) and other tailed phages (as above) collapsed into a single ring each. The sequencing reads of phiPsa374 and phiPsa440 were mapped to the complete phiPsa374 genome using BWA and the coverage of the mapped reads (sam format) is shown, as per the key (*far right*). The read mapping data highlights the terminal redundancy at the genome ends of phiPsa374 (*red graph*; [15]) and the misassembly in the draft genome of phiPsa440 (*blue graph*). The outermost ring shows the CDSs of phiPsa374 as *grey arrows*

B.

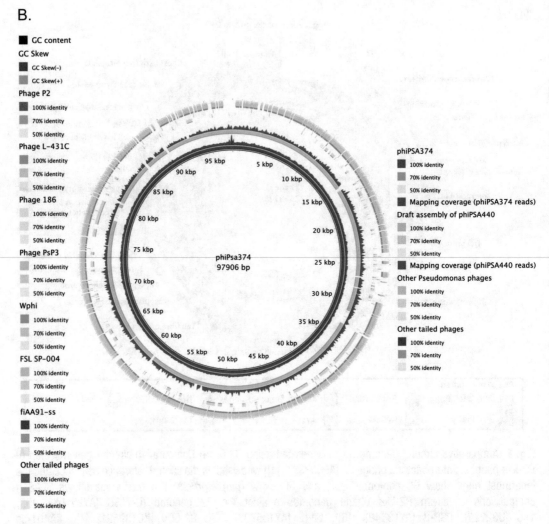

Fig. 1 (continued)

identity of BLAST hits (**Note 3**). Regions of interest, custom labels and additional analyses can be added to the diagram by providing custom annotations either by entering each entry manually or by providing the information in a tab-delimited text, GenBank, EMBL, or multi-FASTA format file. Additional diagram elements such as the height and size of the final image, features, ticks, and labels can also be configured. The configuration settings for each BRIG project can be saved as a template for future use. In addition to visualizing complete genome comparisons, BRIG can also display the presence/absence/truncation/variation of genes in comparison genomes with a user-defined multi-FASTA file of genes as the input reference, as well as visualizing comparisons with draft genomes and unassembled sequence data in SAM-formatted read-mapping files (Fig. 1b).

CGView (http://wishart.biology.ualberta.ca/cgview)

CGView is available as either a web-service or command line application for Unix/Linux (requires proprietary sun-java6-jdk package) [19]. The CGView server produces circular genome diagrams in PNG format, where BLAST results for up to three comparison sequences, or sets of sequences in FASTA format, are displayed as concentric rings. Comparison sequences may be analyzed using the BLAST programs BLASTn, tBLASTx, and BLASTx with options for controlling the query split size and overlap. The BLAST hits for each comparison sequence can be filtered by specifying cutoff thresholds for percentage identity and alignment length and displayed with partial opacity to enable the identification of overlapping hits. For tBLASTx and BLASTx analysis, the resultant hits can be displayed by reading frame. Maps can display GC skew and GC content plotted as the deviation from the average of the entire sequence as well as additional feature and analysis data provided from optional General Feature Format (GFF) files. Finally, zoomed images centred upon a specific base can be produced to illustrate a region of interest in greater detail.

CGView Comparison Tool (http://stothard.afns.ualberta. ca/downloads/CCT)

The CGView Comparison Tool (CCT) is a command line application that retains and expands the functionality of CGView by enabling the comparative analysis and visualization of large numbers of sequences [20]. CCT can be installed and configured manually, or downloaded as a virtual machine (compatible with Windows, Mac, and Unix) of the Ubuntu Linux operating system with CCT and all dependencies already installed. A reference genome or sequence can be supplied in FASTA, GenBank or EMBL formats. Sequences for comparison can be supplied as GenBank or EMBL format or as multi-FASTA files consisting of nucleotide (.fna) or protein (.faa) sequences. Comparisons are performed using a locally installed copy of BLAST+ and hits to the reference sequence in the comparison genomes can be drawn with a height proportional to the percent identity of the hit or a colored with a gradient scale. Additional features and analyses, for example the location of conserved domains and expression data, can be displayed by providing data as GFF files. Several utility scripts are supplied with CCT that allow sequences of interest to be downloaded directly from NCBI (GenBank). Diagrams can be produced in several sizes in PNG, JPEG, SVG and SVGZ formats.

Circos (http://circos.ca)

Circos is a highly flexible command line application written in Perl and can be run on Windows, Unix, and Mac [21]. Data are supplied to Circos using GFF-style data tables and the appearance of the map and associated elements are controlled by editing Apache-like configuration files and images are rendered in PNG and SVG format. Circos is ideal for illustrating positional relationships such as synteny, insertion, deletions and rearrangements between genomes (**Note 4**). These positional relationships are represented by links displayed as connecting lines or ribbons, defined between pairs of positions. Data from analyses can be displayed in 2D tracks as highlights, heatmaps, tiles, scatter, line, and histogram plots. This inherent flexibility allows Circos to display multivariate data at several levels of detail. However, it should be noted that Circos does not perform any analysis nor is it able to read in genome flat files natively. Instead, sequence and analysis data must first be converted into a format parsable by Circos.

DNAPlotter   (https://www.sanger.ac.uk/science/tools/dnaplotter)

DNAPlotter is an interactive application for Windows, Unix and Mac, which allows the generation of linear and circular DNA maps in raster or vector formats [22]. DNAPlotter is included within the Artemis annotation tool [23] but can also be downloaded as a standalone program or executed as a Java Webstart application. DNAPlotter implements the Artemis libraries to filter features, which can then be separated into individual tracks defined using the track manager. Graphs of GC content and GC skew may also be displayed and additional flat files containing feature information in GenBank, EMBL, or GFF format may be read in and displayed on separate tracks.

Easyfig (http://mjsull.github.io/Easyfig/)

Easyfig is an application available for Windows, Unix and Mac, that allows the creation of linear diagrams of one or more genomes and BLAST comparisons. It has an easy-to-use graphical user interface or can be run via command-line [24]. Easyfig can plot the feature objects gene, coding sequence (CDS), tRNA or any other user-defined features in the input annotation file (GenBank or EMBL format). These features can be colour coded within Easyfig or by adding the feature qualifier "/colour=" to gene features in the annotation flat file, a task easily performed using Artemis [23]. Comparisons between two or more genomes can be performed using BLASTn or tBLASTx from within Easyfig (Fig. 2). BLAST can be downloaded from within Easyfig if not already installed. BLAST hits are displayed as cross-links between genomes, colored with a gradient scale of percent identity. Users can define the length, expect value and percent identity of BLAST hits to be rendered in the comparison diagram. Subregions may be selected as an alternative to entire genomes and graphs can also be added to the figure (GC content, GC skew and user-defined custom graphs, e.g., transcriptomic data or sequence read coverage). Easyfig can render images in either BMP or SVG format.

GenomeDiagram (http://biopython.org/wiki/Download)

GenomeDiagram is a command-line module packaged as part of the Biopython distribution, designed to display genomes and comparative genomic data as linear or circular diagrams [27]. Maps

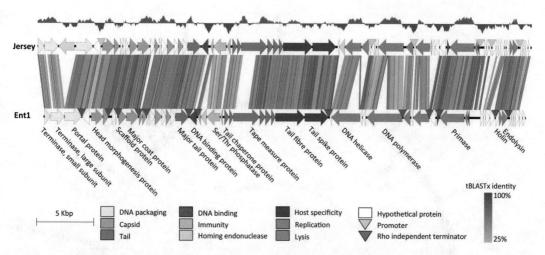

Fig. 2 Comparative linear genome map of *Salmonella* siphoviruses Jersey [KF148055; [25]] and vB_SenS-Ent1 [HE775250; [26]] generated using Easyfig. The genome annotations are color-coded according to function, as shown in the key. Regions of grey between the two genomes depict sequence similarity, shaded according to % tBLASTx identity as shown in the key (*bottom right*). The graph above Jersey shows variation above (*red*) and below (*blue*) the average GC content (49.97%; *black line*)

of individual genomes and genome comparisons are constructed using python scripts. Genome flat files can be loaded and parsed using the SeqIO and SeqFeature modules of Biopython. Similarly, files containing comparative data can be parsed from within the script and data used to either colour individual tracks or produce cross-links illustrating regions of sequence homology. While GenomeDiagram requires the knowledge (and patience) to code an appropriate script, the resulting images can be very powerful. Images produced by GenomeDiagram can be output in a variety of vector and raster formats. A tutorial for creating genome maps is available online (http://biopython.org/DIST/docs/tutorial/Tutorial.html#htoc212).

GenomeVx (http://wolfe.ucd.ie/GenomeVx)

GenomeVx is a webservice which renders simple circular genome diagrams in PDF format using CDS, tRNA and rRNA features extracted from GenBank flat files or entered manually [28]. Features can be colored automatically or by using a vector graphics editing program (Fig. 3). Custom features can be entered manually and added to one or more inner tracks.

GView (https://www.gview.ca/wiki/GView/WebHome)

GView is a GUI application for Windows, Unix and Mac that allows viewing and examining prokaryotic genomes in a circular or linear context [30]. GView can read standard sequence file formats (EMBL, GenBank and GFF) and additional annotations can be optionally added in GFF format. Additional customization on genome maps can be also made in Gview, using the "Genome

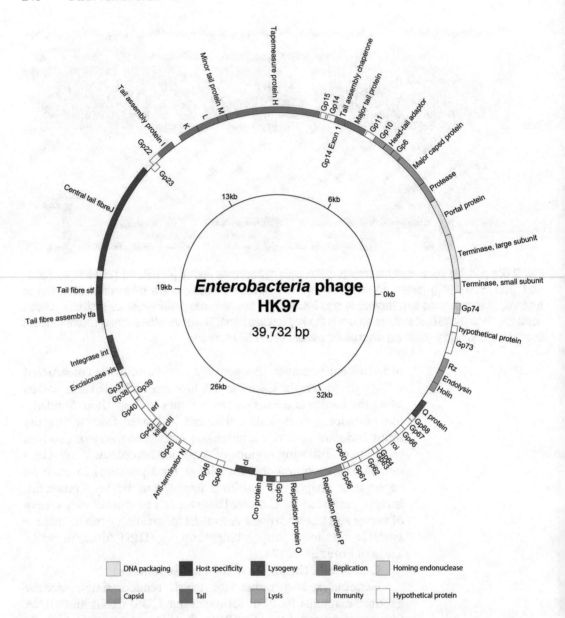

Fig. 3 Circular genome map of siphovirus HK97 [NC_002167; [29]] generated using GenomeVx. Coding sequences are labelled with the annotated product and color-coded according to their functional module. The colour code is provided at the *bottom* of the diagram

Style Sheet" (GSS) format and these settings can be saved for later use. Diagrams can be output in raster or vector formats.

GView Server (https://server.gview.ca)

The GView Server expands upon the functionality of GView by providing a web service front-end for performing comparative analyses between genomes. A range of BLAST analyses are available allowing the preparation of diagrams illustrating sequences representing the core, unique, accessory and pan-genome. Once the job has been completed by the server, diagrams can be viewed immediately in either linear or circular layout by launching the GView Webstart application or alternatively, the results can be downloaded with or without the GView executable file. The BLAST results table can be downloaded separately in Excel or comma-separated value (csv) text formats.

OrganellarGenomeDRAW   (http://ogdraw.mpimp-golm.mpg.de)

OrganellarGenomeDRAW (OGDRAW) is available as a web service and a command line application for Unix/Linux platforms [31]. While optimized for the display of organellar genomes it can be used to produce circular maps of bacteriophage genomes (Fig. 4). The output may be customized by creating a configuration file, allowing custom feature classes to be defined. In addition to GC content graphs, cutting sites for selected restriction enzymes and transcriptomic data can also be displayed. OGDraw renders maps in raster and vector formats.

3   Implementation and Use

This section describes the step-by-step methods for producing images using the two command-line only applications CGView Comparison Tool and Circos, which create different types of circular images, plus the GUI application Easyfig that creates linear diagrams. For the other applications described in this chapter, detailed instructions in the manuals and tutorials are available on their respective websites.

3.1 CGView Comparison Tool

1. This method describes the steps required to produce a comparative genome diagram of members of the myovirus genus *Viunavirus* using *Salmonella* phage Vi01 as the reference genome (Fig. 5). The CCT website has extensive guidance for use and a number of tutorials for creating maps.

2. Edit the cgview_comparison_tool/lib/scripts/cgview_xml_builder.pl to add additional colors for the individual comparison genomes:

```
Line 110 blastColors => [
       111 "rgb(139,0,0)", #dark red
       112 "rgb(255,140,0)", #dark orange
       113 "rgb(0,100,0)", #dark green
       114 "rgb(50,205,50)", #lime green
       115 "rgb(0,0,139)", #dark blue
       116 "rgb(106,190,205)", #light blue
       117 ]
```

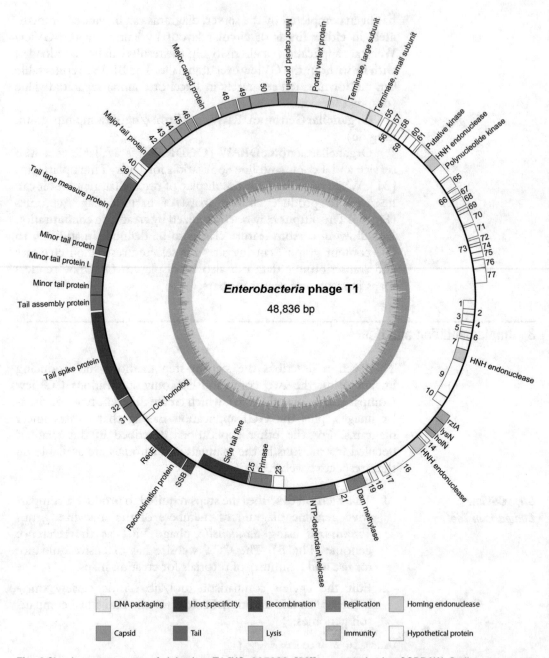

Fig. 4 Circular genome map of siphovirus T1 [NC_005833; [32]] generated using OGDRAW. Coding sequences are labeled with the annotated product and color-coded according to their functional module. The color code is provided at the *bottom* of the diagram

3. Create a new CCT project containing all required directories and configuration files using the command:

```
cgview_comparison_tool.pl -p Viunavirus
```

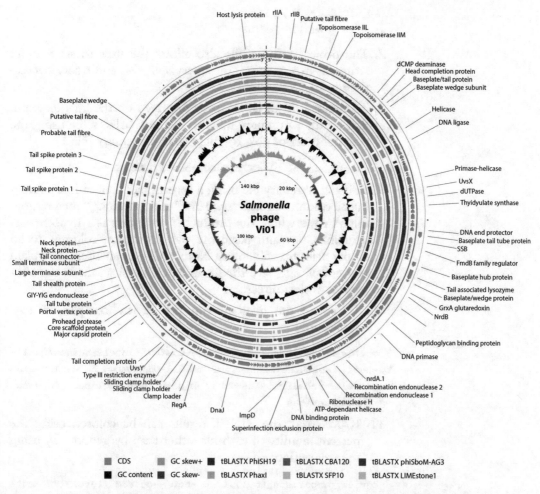

Fig. 5 Comparative circular genome map of the myovirus genus *Viunavirus* prepared using CGView Comparison Tool. Salmonella phage Vi01 [NC_015296; [33]] was employed as the reference sequence and tBLASTx (*E*-value 0.01) used for comparing sequence similarity against PhiSH19 [NC_019530 [34]], PhaxI [NC_019452; [35]], CBA120 [NC_016570; [36]], SFP10 [NC_016073], phiSboM-AG3 [NC_013693; [37]] and LIMEstone1 [NC_019925; [38]]. GC content is depicted in *black* while positive and negative GC skew is denoted by *green* and *purple*, respectively

4. Copy the genome to act as the reference to the reference_genome folder

5. Copy the comparison sequence files to the comparisons folder

6. Edit the project_settings.conf file so that a tBLASTx comparison is performed between the reference and query genomes, GC content and GC skew are displayed on the map and a divider is drawn to indicate that the genomes are linear:

```
query_source = trans
database_source = trans
cog_source = none
draw_gc_content = T
draw_gc_skew = T
```

```
draw_divider = T
map_size = small, medium
```

7. The project settings file also allows the user to set a more stringent expect value, the query split size and query overlap used in BLAST searches.

8. To create a list of labels to be displayed on the final diagram, first create a summary table from the GenBank file using the perl utility script gbk_to_tbl.pl (Supplementary File 1):

```
$ perl gbk_to_tbl.pl <FILENAME.gbk> FILENAME.txt
```

9. Edit the summary file using a spreadsheet application, removing all columns except "seqname" and "product," then removing all rows where the product is described as a hypothetical protein. Copy and paste the two remaining columns to an empty plain text file, using Notepad++ or an alternative plain text editor, and save the file as labels_to_show.txt in the Viunavirus project directory.

10. To create the diagram run the cgview comparison tool application:

```
$ cgview_comparison_tool.pl -t --custom 'labelPlacementQuali-
ty=best labelLineThickness=2 maxLabelLength=250 useInnerLa-
bels=false labelFontSize=20 tick_density=0.25 labels_to_show'
-p Viunalikevirus
```

11. If preferred, the BLAST results can be colored using the percent identity of each hit rather than by genome by using the -cct option:

```
$ cgview_comparison_tool.pl -t -cct --custom 'labelPlacement-
Quality=best labelLineThickness=2 maxLabelLength=250 useIn-
nerLabels=false labelFontSize=20 tick_density=0.25
labels_to_show' -p Viunalikevirus
```

12. Redraw the map in SVG format:

```
$ redraw_maps.sh -p Viunavirus -f svg
```

13. Make any final adjustments to the labels and position of the colour key in a vector graphics editing application.

3.2 Circos

1. This method describes the steps to create a diagram illustrating a tBLASTx comparison of phages HK022 and HK97 to Lambda using Circos version 0.67 (Fig. 6). The configuration and data files required to produce this figure are available as Supplementary material (Supplementary File 2).

2. Create a new directory to contain the files for the Circos project. Create the karyotype.txt file. The karyotype file defines the genomes with data supplied in the format "chr - ID, Label,

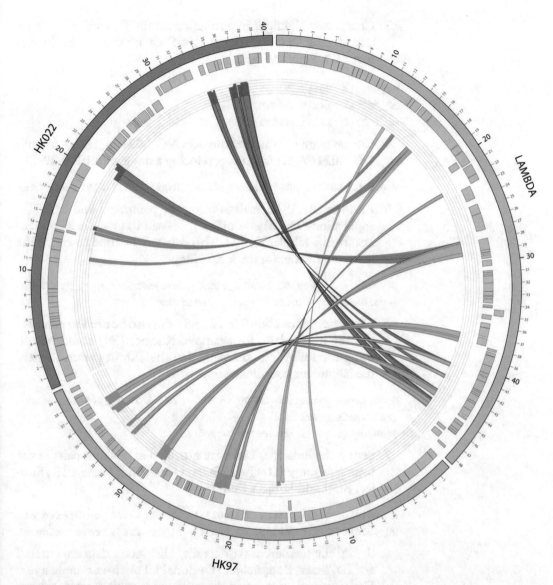

Fig. 6 Circos map depicting tBLASTx alignment results of HK97 [NC_002167; [29]] and HK022 [NC_002166; [29]] relative to Lambda [NC_001416; [14]] with an *E*-value threshold of 0.01. The coloured segments on the *outer ring* depict the genomes of Lambda (*Green*), HK97 (*Blue*) and HK022 (*Red*) respectively. The numbered scale indicates the genome size in kb. Hatch marks (*grey*) on the *inner ring* indicate the location and coding strand of genes. The coloured *ribbons* depict conservation and orientation of translated sequences ≥50 bp relative to Lambda. The *ribbons* terminate at a histogram track displaying percent identity

Start, End and Colour". The ID defined in the karyotype file is used to identify the chromosome in all other data files.

```
chr – NC_001416  Lambda  1 48502 green
chr – NC_002167  HK97   1 39732 blue
chr – NC_002166  HK022  1 40751 red
```

3. Create two highlight files to represent the forward and reverse strand gene features, respectively. Data are entered in the format "ID, Start and End".

```
NC_001416  20147  20767
NC_002167  14715  14975
NC_002166  13751  13972
```

4. Concatenate the query sequences NC_002166_HK97.fna and NC_002167_HK022.fna producing a single multi-FASTA file:

```
$ cat NC_002166_HK97.fna NC_002167_HK022.fna > HK97_HK022.fna
```

5. Run the tBLASTx analysis using the command line BLAST+ application. The flag '-outfmt 7' is used to save the alignment results in a tabular format. This data will be used to create the links and histogram track data files:

```
$ tblastx -query NC_001416_Lambda.fna -subject HK97_HK022.fna
-evalue 0.01 -outfmt 7 -out tblastx.txt
```

6. Other programs aside from BLAST can also be used to generate the alignment data, for example, Nucmer [39] could be run using the following command and the output parsed to yield the alignment coordinates:

```
$ nucmer --maxmatch -b 200 -g 90 -c 65 -l 20 NC_001416_Lambda.
fna Lambda_query.fna
$ show-coords -T out.delta > delta.txt
```

7. Set up the links file. Links are provided as position pairs in the format "Query ID, Query Start, Query End, Subject ID, Subject Start and Subject End":

```
NC_001416  44925  46088  NC_002166  37006  38169  color=red_a1
NC_001416  27518  29125  NC_002167  21024  22631  color=blue_a1
```

8. Set up the histogram.txt data file. Histogram data are entered as "ID, Start, End, Score, [Options]". For this example a text file is created containing the subject id, subject start, subject end, and % identity columns of the tBLASTx results file in addition to a column defining the color of the link.

```
NC_002166  37006  38169  87.37  fill_color=dred
NC_002167  21024  22631  92.54  fill_color=dblue
```

9. Set up the main configuration file circos.conf and the additional configuration files ideogram.conf, ideogram.position.conf, ideogram.label.conf, and ticks.conf saving each one in the Circos project directory. These files are included in the Supplementary material (Supplementary File 2) with comments added to explain the function of variables.

10. Run Circos to generate the genome comparison map using the command:

```
$ circos -conf circos.conf
```

3.3 Easyfig

1. This method describes the steps required to produce a comparative linear tBLASTx genome diagram between *Salmonella* siphoviruses Jersey [KF148055] and vB_SenS-Ent1 [HE775250] (Fig. 2). The Easyfig GUI can be downloaded for free from the website, along with a manual that has extensive guidance for use as well as a number of tutorials for creating figures. The annotation flat files required to produce this figure are available as Supplementary material (Supplementary File 3). Prior to starting this walkthrough, ensure that a local copy of BLAST is installed. If BLAST is not installed locally, open Easyfig and from the menu bar, select Blast → Download Blast automatically.

2. Launch Easyfig. On the main screen, click on the Add feature file button, located underneath the "Annotation files" box. In the newly opened window, navigate to the unzipped Supplemental_file3 folder.

3. Select Jersey.gbk and click on the open button.

4. Click on the Add feature file button again, navigate to the Supplemental_file3 folder, select Ent1.gbk and click on the open button.

5. Click on the Generate tblastx Files button to generate the tBLASTx comparison file. Ensure that the pop up window shows the Supplemental_file3 folder and click choose. This closes this pop up window and initiates tBLASTx.

6. Observe the yellow box on the main screen. Proceed to the next step only when it shows "Performing tblastx...complete".

7. From the menu bar, select Image → Figure.

8. From the Figure Options, select "centre" for Alignment of genomes and in the Legend Option, enter "5000" for Length of scale legend. Ensure that the checkbox for Draw Blast identity legend is checked. Keep the rest of the options as default. Click close once changes are made.

9. From the menu bar, select Image → Annotation.

10. Key in "regulatory" in the text box under misc_feature and ensure that only this check box and the check box next to CDS is checked. There is no need to select colors as they are already defined in the genome flat file. Select "arrow" and "pointer" for CDS and regulatory features respectively. Click close once changes are made.

11. From the menu bar, select Image → Graph.

12. Select GC content from the graph drop down menu. Change the value of Step size to "1" and Window size to "500". Click close once changes are made.

13. On the main screen, click on the Save As button. Ensure that the pop up window shows the Supplemental_file3 folder. Key in "Easyfigexample" as the file name to save as and click on Save to close this window.

14. In the main screen, click on Create Figure. Once the image has been rendered, it can be found in the Supplemental_file3 folder.

4   File Formats

Two dimensional images are stored as either raster or vector graphics formats. Raster image formats utilize a dot matrix data structure and include bitmap (.bmp) portable network graphics (.png), joint photographic experts (.jpg) and tagged image file (.tiff) formats. Because raster image formats are made up of fixed set of dots (pixels) they are resolution-dependent and cannot be increased in size without a concomitant loss of quality.

In contrast, vector image formats, such as extended postscript (.eps) and scalable vector graphic (.svg), store images as paths defined by mathematic equations. This feature makes vector graphics resolution independent and allows images or individual objects to be scaled, moved, rotated or edited without any degradation of image quality. A further advantage is that images rendered in this format can be easily edited using vector graphics image editing software (see below). For publication purposes, once vector images have been scaled and manipulated to the user's requirements they can be rasterized into tiff format files at an appropriate resolution for print (usually 300–600 ppi).

5   Image Editing Software

Images in vector format (EPS, PS, PDF or SVG) can be easily modified by using applications such as proprietary Adobe Illustrator (http://www.adobe.com) or open-source Inkscape (https://inkscape.org). Further enhancements performed using such applications might include the movement, alteration, scaling, rotation, addition, or subtraction of labels and elements. A wide range of tutorials for using Adobe Illustrator and InkScape are available at https://helpx.adobe.com/illustrator/tutorials.html and https://inkscape.prg/en/learn/tutorials, respectively. Images in raster format (BMP, PNG, GIF, JPEG, or TIFF) can be adjusted using applications including proprietary Adobe Photoshop (www.adobe.com/Photoshop) or CorelDraw (http://www.coreldraw.com), or open-source GIMP (http://www.gimp.org).

6   DPI, PPI and Image Size

The pixel dimensions of a digital image are an absolute value. For example, an image with dimensions of 3000 × 3000 pixels would allow for a 10″ by 10″ figure at 300 pixels per inch (PPI). Most of the applications described here allow control of the output pixel size of the image which can be matched to the size specifications in the journal guidelines. Dots per inch (dpi) refers to the resolution of the printing device. It is important to remember that an image displayed on the screen at 100% does not represent the actual physical size of the printed image.

To change the output resolution of an image without altering the physical size in Adobe Photoshop select the Image → Image size menu option. In the dialog window check the "resample image" box and enter a new resolution. The resolution of the image in PPI has now been set to the specified value leaving the pixel dimensions of the image unchanged. To change the image output dimensions select the Image → Image size menu option. In the dialog window select the "resample image" and the "constrain proportions" check boxes and alter the width to the required value by entering data into the pixel dimensions or document size text boxes.

7   Notes

1. In many cases, an additional layer of flexibility is provided by installing the command line versions of the programs discussed in this chapter on an appropriate Linux distribution. BioLinux 8, a dedicated open source bioinformatics platform based on Ubuntu Linux 14.04 LTS base, developed and maintained by the UK Natural Environment Research Council (NERC) is particularly recommended [40]. BioLinux is distributed with a large number of preinstalled bioinformatics packages including Artemis [23], ACT [41], BLAST+, MUMMER3 [39] in addition to BioPerl [42], Biopython [43], and R [44]. UNIX/Linux distributions can also be run from within Windows or Mac OS using virtual machines such as Oracle VM VirtualBox (https://www.virtualbox.org), VMware (http://www.vmware.com) or Parallels (http://www.parallels.com).

2. All of the visualization tools allow for the display of unpublished genomes, provided that the annotated sequences are available as a flat file in GenBank or EMBL format.

3. The choice of color palette is an important element in the design of data graphics. Particularly recommended are Brewer palettes, manually defined selections of colors for sequential, diverging and qualitative data (http://colorbrewer.org). Martin Krzywinski's website has an excellent section on Brewer palettes with swatch files available for loading these palettes into Adobe applications (http://mkweb.bcgsc.ca/brewer).

4. It is important to note that applications which display genome comparisons as concentric rings relative to a reference genome do not serve to illustrate syntenic or positional relationships between the genomes. Additionally, sequences which are present in the comparison genomes but absent in the reference will not be displayed.

Acknowledgments

D. Turner is a Lecturer at the University of the West of England, Bristol, UK. D. Reynolds is a Professor of Health and Environment at the University of the West of England, Bristol, UK. J. M. Sutton is Scientific Leader, Healthcare, Biotechnology and Technology Development Group at Public Health England, Porton Down, UK. E. M. Sim is Research Associate at the ithree institute, University of Technology Sydney, Australia. N. K. Petty is a Senior Lecturer and Microbial Genomics Group leader at the ithree institute, University of Technology Sydney, Australia.

References

1. Tufte ER (2001) The visual display of quantitative information. Graphics Press, Cheshire, CT

2. Alikhan N-F, Petty NK, Ben Zakour NL, Beatson SA (2011) BLAST Ring Image Generator (BRIG): simple prokaryote genome comparisons. BMC Genomics 12(1):402

3. Altschul SF, Gish W, Miller W, Myers EW, Lipman DJ (1990) Basic local alignment search tool. J Mol Biol 215(3):403–410

4. Nilsson AS, Haggård-Ljungquist E (2006) The P2-like bacteriophages. The bacteriophages, 2nd edn. Oxford University Press, New York, NY

5. Garcia E, Chain P, Elliott JM, Bobrov AG, Motin VL, Kirillina O, Lao V, Calendar R, Filippov AA (2008) Molecular characterization of L-413C, a P2-related plague diagnostic bacteriophage. Virology 372(1):85–96

6. Bullas LR, Mostaghimi AR, Arensdorf JJ, Rajadas PT, Zuccarelli AJ (1991) Salmonella phage PSP3, another member of the P2-like phage group. Virology 185(2):918–921

7. Moreno Switt AI, Orsi RH, den Bakker HC, Vongkamjan K, Altier C, Wiedmann M (2013) Genomic characterization provides new insight into Salmonella phage diversity. BMC Genomics 14:481–481

8. Allué-Guardia A, Imamovic L, Muniesa M (2013) Evolution of a self-inducible cytolethal distending toxin type V-encoding bacteriophage from Escherichia coli O157:H7 to Shigella sonnei. J Virol 87(24):13665–13675

9. McClelland M, Sanderson KE, Spieth J, Clifton SW, Latreille P, Courtney L, Porwollik S,

Ali J, Dante M, Du F, Hou S, Layman D, Leonard S, Nguyen C, Scott K, Holmes A, Grewal N, Mulvaney E, Ryan E, Sun H, Florea L, Miller W, Stoneking T, Nhan M, Waterston R, Wilson RK (2001) Complete genome sequence of *Salmonella enterica* serovar Typhimurium LT2. Nature 413 (6858):852–856

10. Nakayama K, Kanaya S, Ohnishi M, Terawaki Y, Hayashi T (1999) The complete nucleotide sequence of phiCTX, a cytotoxin-converting phage of *Pseudomonas aeruginosa*: implications for phage evolution and horizontal gene transfer via bacteriophages. Mol Microbiol 31(2):399–419

11. Miller ES, Kutter E, Mosig G, Arisaka F, Kunisawa T, Rüger W (2003) Bacteriophage T4 genome. Microbiol Mol Biol Rev 67 (1):86–156

12. Morgan GJ, Hatfull GF, Casjens S, Hendrix RW (2002) Bacteriophage Mu genome sequence: analysis and comparison with Mu-like prophages in *Haemophilus*, *Neisseria* and *Deinococcus*. J Mol Biol 317(3):337–359

13. Dunn JJ, Studier FW, Gottesman M (1983) Complete nucleotide sequence of bacteriophage T7 DNA and the locations of T7 genetic elements. J Mol Biol 166(4):477–535

14. Sanger F, Coulson AR, Hong GF, Hill DF, Petersen GB (1982) Nucleotide sequence of bacteriophage λ DNA. J Mol Biol 162 (4):729–773

15. Frampton RA, Taylor C, Holguin Moreno AV, Visnovsky SB, Petty NK, Pitman AR, Fineran PC (2014) Identification of bacteriophages for biocontrol of the kiwifruit canker phytopathogen *Pseudomonas syringae* pv. *actinidiae*. Appl Environ Microbiol 80(7):2216–2228

16. Garbe J, Bunk B, Rohde M, Schobert M (2011) Sequencing and characterization of *Pseudomonas aeruginosa* phage JG004. BMC Microbiol 11:102

17. Debarbieux L, Leduc D, Maura D, Morello E, Criscuolo A, Grossi O, Balloy V, Touqui L (2010) Bacteriophages can treat and prevent *Pseudomonas aeruginosa* lung infections. J Infect Dis 201(7):1096–1104

18. Lu S, Le S, Tan Y, Zhu J, Li M, Rao X, Zou L, Li S, Wang J, Jin X, Huang G, Zhang L, Zhao X, Hu F (2013) Genomic and proteomic analyses of the terminally redundant genome of the *Pseudomonas aeruginosa* phage PaP1: establishment of genus PaP1-like phages. PLoS One 8(5):e62933

19. Grant JR, Stothard P (2008) The CGView Server: a comparative genomics tool for circular genomes. Nucleic Acids Res 36(Web Server issue):W181–W184

20. Grant J, Arantes A, Stothard P (2012) Comparing thousands of circular genomes using the CGView Comparison Tool. BMC Genomics 13(1):202

21. Krzywinski MI, Schein JE, Birol I, Connors J, Gascoyne R, Horsman D, Jones SJ, Marra MA (2009) Circos: an information aesthetic for comparative genomics. Genome Res 19:1639–1645

22. Carver T, Thomson N, Bleasby A, Berriman M, Parkhill J (2009) DNAPlotter: circular and linear interactive genome visualization. Bioinformatics 25(1):119–120

23. Rutherford K, Parkhill J, Crook J, Horsnell T, Rice P, Rajandream MA, Barrell B (2000) Artemis: sequence visualization and annotation. Bioinformatics 16(10):944–945

24. Sullivan MJ, Petty NK, Beatson SA (2011) Easyfig: a genome comparison visualizer. Bioinformatics 27(7):1009–1010

25. Anany H, Switt AM, De Lappe N, Ackermann H-W, Reynolds D, Kropinski A, Wiedmann M, Griffiths M, Tremblay D, Moineau S, Nash JE, Turner D (2015) A proposed new bacteriophage subfamily: "Jerseyvirinae". Arch Virol 160:1021

26. Turner D, Hezwani M, Nelson S, Salisbury V, Reynolds D (2012) Characterization of the *Salmonella* bacteriophage vB_SenS-Ent1. J Gen Virol 93(Pt 9):2046–2056

27. Pritchard L, White JA, Birch PRJ, Toth IK (2006) GenomeDiagram: a python package for the visualization of large-scale genomic data. Bioinformatics 22(5):616–617

28. Conant GC, Wolfe KH (2008) GenomeVx: simple web-based creation of editable circular chromosome maps. Bioinformatics 24 (6):861–862

29. Juhala RJ, Ford ME, Duda RL, Youlton A, Hatfull GF, Hendrix RW (2000) Genomic sequences of bacteriophages HK97 and HK022: pervasive genetic mosaicism in the lambdoid bacteriophages. J Mol Biol 299 (1):27–51

30. Petkau A, Stuart-Edwards M, Stothard P, Van Domselaar G (2010) Interactive microbial genome visualization with GView. Bioinformatics 26(24):3125–3126

31. Lohse M, Drechsel O, Bock R (2007) OrganellarGenomeDRAW (OGDRAW): a tool for the easy generation of high-quality custom graphical maps of plastid and mitochondrial genomes. Curr Genet 52(5-6):267–274

32. Roberts MD, Martin NL, Kropinski AM (2004) The genome and proteome of coliphage T1. Virology 318(1):245–266

33. Pickard D, Toribio AL, Petty NK, van Tonder A, Yu L, Goulding D, Barrell B, Rance R, Harris D, Wetter M, Wain J, Choudhary J, Thomson N, Dougan G (2010) A conserved acetyl esterase domain targets diverse bacteriophages to the vi capsular receptor of *Salmonella enterica* Serovar Typhi. J Bacteriol 192(21):5746–5754

34. Hooton S, Timms A, Rowsell J, Wilson R, Connerton I (2011) *Salmonella* Typhimurium-specific bacteriophage PhiSH19 and the origins of species specificity in the Vi01-like phage family. Virol J 8(1):498

35. Shahrbabak SS, Khodabandehlou Z, Shahverdi AR, Skurnik M, Ackermann H-W, Varjosalo M, Yazdi MT, Sepehrizadeh Z (2013) Isolation, characterization and complete genome sequence of PhaxI: a phage of *Escherichia coli* O157 : H7. Microbiology 159 (Pt 8):1629–1638

36. Kutter E, Skutt-Kakaria K, Blasdel B, El-Shibiny A, Castano A, Bryan D, Kropinski A, Villegas A, Ackermann H-W, Toribio A, Pickard D, Anany H, Callaway T, Brabban A (2011) Characterization of a ViI-like phage specific to *Escherichia coli* O157:H7. Virol J 8(1):430

37. Anany H, Lingohr E, Villegas A, Ackermann H-W, She Y-M, Griffiths M, Kropinski A (2011) A *Shigella boydii* bacteriophage which resembles *Salmonella* phage ViI. Virol J 8 (1):242

38. Adriaenssens EM, Van Vaerenbergh J, Vandenheuvel D, Dunon V, Ceyssens P-J, De Proft M, Kropinski AM, Noben J-P, Maes M, Lavigne R (2012) T4-related bacteriophage LIMEstone isolates for the control of soft rot on potato caused by '*Dickeya solani*'. PLoS One 7(3):e33227

39. Kurtz S, Phillippy A, Delcher AL, Smoot M, Shumway M, Antonescu C, Salzberg SL (2004) Versatile and open software for comparing large genomes. Genome Biol 5(2):R12

40. Field D, Tiwari B, Booth T, Houten S, Swan D, Bertrand N, Thurston M (2006) Open software for biologists: from famine to feast. Nat Biotechnol 24(7):801–803

41. Carver TJ, Rutherford KM, Berriman M, Rajandream M-A, Barrell BG, Parkhill J (2005) ACT: the Artemis comparison tool. Bioinformatics 21(16):3422–3423

42. Stajich JE, Block D, Boulez K, Brenner SE, Chervitz SA, Dagdigian C, Fuellen G, Gilbert JGR, Korf I, Lapp H, Lehväslaiho H, Matsalla C, Mungall CJ, Osborne BI, Pocock MR, Schattner P, Senger M, Stein LD, Stupka E, Wilkinson MD, Birney E (2002) The bioperl toolkit: perl modules for the life sciences. Genome Res 12(10):1611–1618

43. Cock PJA, Antao T, Chang JT, Chapman BA, Cox CJ, Dalke A, Friedberg I, Hamelryck T, Kauff F, Wilczynski B, de Hoon MJL (2009) Biopython: freely available Python tools for computational molecular biology and bioinformatics. Bioinformatics 25(11):1422–1423

44. R Core Team (2017) R: a language and environment for statistical computing. http://www.R-project.org

Part IV

Bacteriophage Genetics

Chapter 19

Transposable Bacteriophages as Genetic Tools

Ariane Toussaint

Abstract

Phage Mu is the paradigm of a growing family of bacteriophages that infect a wide range of bacterial species and replicate their genome by replicative transposition. This molecular process, which is used by other mobile genetic elements to move within genomes, involves the profound rearrangement of the host genome [chromosome(s) and plasmid(s)] and can be exploited for the genetic analysis of the host bacteria and the in vivo cloning of host genes. In this chapter we review Mu-derived constructs that optimize the phage as a series of genetic tools that could inspire the development of similarly efficient tools from other transposable phages for a large spectrum of bacteria.

Key words Bacteriopage Mu, Transposable phages, Mini-Mu, Mini-muduction, In vivo gene cloning

1 Introduction

Mu (http://viralzone.expasy.org/all_by_species/507.html) is the paradigm of a large family of phages and prophages that replicate their genome by replicative transposition, hence their name "transposable phages." They are also similar in genome length (35–40 kbp), genetic organization and protein content. It was therefore proposed that they should be organized into a new taxonomic family "Saltoviridae" (from the Salto skating figure/jump; [1]). At least 26 transposable phages have now been described and their genomes sequenced. They infect a very wide range of Gram-negative bacteria. In addition, many transposable prophages have been identified in bacterial genome sequences, whether Gram⁻ [2] or Gram⁺ [3].

Mu was discovered by A.L. Taylor in 1963 [4]. He readily noticed that, as a result of its promiscuous integration, the phage induces insertion mutations, which were soon shown to be strongly polar (*see* for example [5]).

The powerful potential of Mu as a genetic tool became obvious when major host chromosome rearrangements (inversions, duplications, deletions of adjacent genes, replicon fusions, and

Martha R.J. Clokie et al. (eds.), *Bacteriophages: Methods and Protocols, Volume 3*, Methods in Molecular Biology, vol. 1681,
https://doi.org/10.1007/978-1-4939-7343-9_19, © Springer Science+Business Media LLC 2018

transpositions of host DNA segments) were recognized as an intrinsic part of the unique mode of Mu replication by replicative transposition ([6] and references therein for a historical review of Mu transposition). Many useful genetic tools were derived from Mu, devised to avoid killing while retaining the capacity to induce host chromosomal rearrangements and combining Mu properties with selective markers, plasmid origins of replication, reporter genes and other features that make them useful for all sorts of genetic manipulations. At first view such tools, which only exploit in vivo processes (conjugation and viral infection) may appear out of date in face of new generation sequencing and the variety of systems available for targeted mutagenesis, such as CRISPR derived tool kits (*see* for instance [7]). However, these latter procedures involve transformation, which in many cases remains much less efficient than conjugation or infection to introduce DNA in a cell.

In this chapter, as an introduction to three methodological chapters describing specific Mu-based methods (gene mapping and in vivo cloning, Mu-printing and methods based on in vitro assembled Mu transposition complexes), we shall overview how the interactions of Mu with its host DNA (chromosome, other pro-phages, and plasmids) has been exploited to mutate and move genes within and between bacterial cells and to study gene expression and function. These methods could be easily developed with other transposable phages, to be used in a variety of important bacterial species, which are not susceptible to Mu infection.

2   Mu in Brief

Mu is a temperate phage that infects *E coli* and other Enterobacteria. Mu viral particles contain a ~40 kb linear double stranded DNA consisting of the Mu 38 kbp dsDNA genome, flanked by variable host DNA sequences (variable ends or VE DNA) 100–150 bp on one side defined as left, and 1–1.5 kbp on the right end. VE DNA is the result of the packaging of viral DNA by a regular full head mechanism, which initiates at the *pac* site located near the left end and proceeds on randomly integrated Mu copies, spread through the host genome as a result of the transpositional replication/replicative transposition mode of Mu DNA amplification (http://viralzone.expasy.org/all_by_species/4017.html for an illustration of the process [1].

The Mu genome consists of seven functional blocks (Fig. 1 and http://viralzone.expasy.org/all_by_species/4356.html, for more details). Two early transcriptional units drive the synthesis of (1) the lysogenic repressor Repc, and (2) the lytic repressor MuNer, the MuA transposase, the MuB transposition activator required for integration and replication and the "semi-essential region." Most of the genes in this SE region are dispensable for viral development,

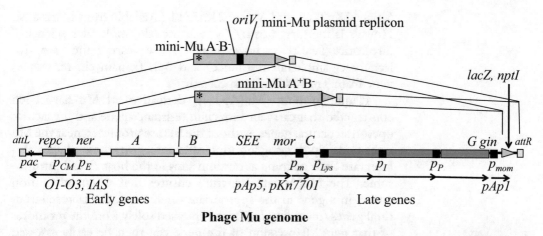

Fig. 1 Genetic and physical map of Mu and mini-Mu's. See text for details on genes promoters (*filled squares*). The position of selectable markers (ampicillin and kanamycin for pAp1, pAp5, and Kn7701 respectively) and fusion reporter genes (*lacZ* and *nptI*) are indicated. Mini-Mu with or without *A* and/or *B* genes and a plasmid origin contain the repressor gene *repc*, the *attL* and *attR* MuA transposase binding sites, and the operator/*IAS* region. *SEE* semi-essential early region. VE ends are not indicated

and of unknown function. The last gene product from the operon is MuMor. It activates transcription of the middle operon (absent from most "Saltoviridae") that consists of the single *C* gene. MuC activates the four late transcriptional units from the p_{lys}, p_I, p_P and p_{mom} promoters, which control the synthesis of the lysis, head, tail and fiber genes and the host restriction evasion system Mom.

Repc, MuNer, and MuA bind to the overlapping operators (*O1–O3*, *see* Fig. 1) and internal transposition enhancer (*IAS*), thereby regulating the lytic-lysogeny switch (Repc binding to the operators), early transcription (MuNer binding to the operators) and the early steps of transpososome assembly (Repc and MuA binding to the *IAS*; for details *see* http://viralzone.expasy.org/all_by_species/4516.html).

The *mom* gene product modifies a fraction of the adenines residues in Mu and host DNA (http://viralzone.expasy.org/all_-by_species/4277.html), such that Mu DNA is partially resistant to several restriction enzymes both in vivo and in vitro.

3   Mu as a Genetic Tool

3.1   Thermoinducible Mu and Mu Derivatives with Selectable Markers

The original Mu phage cannot be induced by UV or mitomycin C; hence the most widely used mutant of Mu is the thermoinducible Mu*cts*62 [8]. To make the selection of lysogens more direct, drug resistances and other selectable genes have been introduced in plaque-forming derivatives, Mu*cts*62Apl and Mu*cts*62Ap5 [9] and

Mu*cts*62Kn7701 and Mu*cts*62Kn7711 (available from Martha M. Howe's laboratory; https://www.uthsc.edu/molecular_sciences/directories/faculty/m_howe.php). These carry the *bla* (β-lactamase, ampicillin resistance) and *nptl* (kanamycin resistance) gene from Tn*l* and Tn*5*, respectively.

Plaque forming [Mu(Ap,lac)] derivatives of Mu have been constructed that carry an ampicillin resistance gene and the lactose operon structural genes, without the lactose promoter, near the Mu right end. Upon lysogenization after infection, these phages also integrate their genome at random sites in the host *E. coli* chromosome. The Mu(Ap,lac) structure ensures that when integration occurs in a gene in the appropriate orientation, the lactose structural genes (including *lacZ*) are expressed solely from the promoter of that gene. Expression of the gene can then be easily assessed through β-galactosidase measurement [10].

Because of the very low selectivity for target site selection during Mu integration/transposition (e.g., [11–14]) the phage, which contains several transcriptional terminators, can be used to isolate defective, strong polar mutations in most if not any gene in the chromosome of any bacterial strain susceptible to Mu infection. About 1% auxotrophs are recovered after infection of *Escherichia coli* [4] with around 10^7 phages. Mu insertional mutations can be screened, selected and mapped as any other type of mutation. One should however remember that due to the presence of the prophage, their transfer by conjugation or transduction into a non-lysogenic host provokes induction of the Mu prophage. This may not only decrease the efficiency of transfer but also lead to the appearance of secondary Mu-induced mutations in the new host. Such drawbacks can be eliminated by using, instead of the phage, Mu disarmed derivatives, the mini-Mu's (see more below).

3.2 Expanding Mu Host Range

Originally isolated from *E. coli* K12, and almost exclusively studied in this species, Mu is also infecting and developing in other enterobacteria. This results in part from the presence on the Mu 38 kb dsDNA genome, of an invertible DNA segment, the G region, that contains two sets of tail fiber and fiber chaperon genes. Each set is expressed in one orientation [G(+) or G(−)]. Inversion is controlled by the phage encoded Gin invertase (reviewed in [15], *see* http://viralzone.expasy.org/all_by_species/4277.html for an illustration). Some enterobacteria are naturally sensitive to Mu virions with either the G(+) or G(−)-directed host range, Mu-sensitive *Shigella*, *Salmonella enterica* Typhimurium, *Salmonella* Typhi, *Salmonella* Montevideo, *Serratia marcescens*, *Citrobacter freundii*, *Enterobacter*, *Klebsiella pneumoniae*, and *Erwinia* have been found either among natural isolates [4, 16–18] or in mutant derivatives. For example, *S. enterica* and *K. pneumoniae* that carry a deletion extending from the *his* locus become Mu-sensitive (e.g., [17, 19]).

In resistant strains, broad host range plasmids (e.g., IncP plasmids RP4, RK2, R68) containing a Mucts62 prophage have been used to introduce the phage by mating [20–22]. Transfer to different bacterial species occurs at widely different frequencies. In enterobacteria such as *Erwinia stuartii*, *Klebsiella aerogenes*, and *K. pneumoniae* transfer occurs at 10^{-3}–10^{-5} per input donor [23–25] and transconjugants that acquire the plasmid produce 10^8 phage/ml upon induction at 42 °C. In other bacterial species transfer occurs at a very low frequency (10^{-7}–10^{-8} per input donors from *E. coli* to *A. tumefaciens*, *Ralstonia solanacearum*, *K. pneumoniae*, and *Sinorhizobium meliloti* or *Rhizobium leguminosarum* for instance; [20, 21, 25]) and the plasmids in the few recovered transconjugants are largely deleted. Nevertheless, Mu appears to transpose to a certain extent in *Agrobacterium* [21].

Although such attempts to introduce Mu in nonsensitive hosts may still be considered, the presence of a very efficient host range variation system, the diversity generating retroelement DGRE ([26] and references therein), on several "Saltoviridae" genomes offers an interesting perspective for developing broad host range derivatives. In addition, it seems that using the variety of "Saltoviridae" available within a very large spectrum of host species, including Firmicutes, combined with new in vitro recombination technologies available, would provide more straightforward ways to design "Salto" and "mini-salto" derivatives targeting any bacterium of interest.

3.3 Mu-Mediated Gene Transfer by Conjugation; Polarized and Unpolarized Transfer

It was by mating an *E. coli*/F*lac* infected by Mu with a polyauxotrophic *E. coli* recipient, that van de Putte and Gruijthuijsen ([27] first observed that all the donor chromosomal markers were transferred to the recipient at the same high frequency (10^{-4}) even when the donor was RecA$^-$. It is now obvious that upon infection of the donor, the Mu genome integrates at random into the host chromosome or the F so that replicon fusion occurs between the chromosome and the plasmid. The donor population becomes a mixture of "Hfr" with F integrated at random positions, flanked by two Mu prophages in the same orientation The overall probability of any gene to be near the F is equal, leading to equal transfer efficiencies for all genes.

Transfer from the donor can also be polarized by different means. If a Rec$^+$ strain carries a Mu prophage at a fixed position in the chromosome and in the F, homologous recombination between the two prophages generates a unique type of "Hfr" with a transfer polarity directed by the respective orientations of the two Mu prophages [28]. Alternatively, Hfr's can be selected, in Rec$^+$ or in Rec$^-$ bacteria, by different means such as for instance selection at high temperature of cells that keep a plasmid thermosensitive for replication after infection with Mu. Each selected clone has integrated the plasmid at a given chromosomal location flanked by two Mu copies in the same orientation as a result of replicon fusion [27]. Using broad host range IncP plasmids or their derivatives with thermosensitive replication allows the use of these strategies in a large spectrum of bacterial species. In methodological Chapter 21, van Gijsegem provides more details about Mu-mediated gene transfer (concentrating on the more efficient mini-Mu derivatives see below) and the possibility to exploit this property for in vivo gene cloning,

3.4 Mu-Based Suicide Vectors

Efficient transposon mutagenesis requires efficient introduction of the transposon into the desired strain, followed by the elimination ("suicide") of the transposon vector.

Mu inserted in a plasmid can be the killing actor of the suicide if induced when introduced in a new host. Killing can result from expression of some lethal phage function (e.g., Kil, Lys) or/and replication. Plasmid pJB4JI [29] has been most popular and a powerful tool in many instances despite the fact that its transfer and suicide/Tn5 transposition efficiencies are very strain dependent. pJB4JI derives from pPHI, which encodes resistance for chloramphenicol, gentamycin spectinomycin and streptomycin by insertion of a Muc^+::Tn5. pJB4JI has been transferred and successfully used in a large series of strains including *Caulobacter crescentus* [30], *Acinetobacter* sp. [31], *Azotobacter vinelandii* [32], *Aeromonas hydrophila* [33], e.g., fast- and slow-growing *Rhizobium* (e.g., [34, 35]), *A. tumefaciens* (e.g., [36, 37]), several *Erwinia* species (e.g., [38, 39]).

4 Mu-Derived Genetic Tools

4.1 Mini-Mu's

To optimize the potential of Mu as a genetic tool, it was essential to reduce its killing effects, while keeping its promiscuous integration and capacity to promote chromosomal rearrangements.

To function as a transposon, a Mu element requires only the MuA transposase and its cognate binding sites at Mu ends, *attL* and *attR*. However the MuB protein and *IAS* enhancer significantly stimulate transposition and associated genome rearrangements. MuA and MuB proteins can be provided *in cis* or *in trans* by a prophage or by a suitable plasmid construct. The optimal combination is of course a mini-Mu with all transposase binding sites (*attL*, *attR*, *IAS*) but deleted from all the phage coding genes including *A* and *B*, and a complementing plasmid that can be eliminated once the mini-Mu has integrated in its new host, which prevents the occurrence of secondary transposition events (recently reviewed in [40]).

Alternatively, the disarmed mini-Mu can be complemented by a complete helper Mu phage. In this case, all phage structural and lysis proteins are provided by the helper, in addition to the transposition proteins. The mini-Mu replicates and is packaged into viral particles. Successful packaging requires that the Mu-derived element be <40 kb long to fit into the phage head, and carry a functional *pac* site (*see* Fig. 1). As a result of full head packaging, any internal deletion of phage DNA will be compensated by an equivalent lengthening of the right VE.

The most useful mini-Mu's that retain *attL*, *attR*, *IAS*, and the *pac* site and are deleted for the phage lytic and structural functions, carry a selection marker (pAp1 or Kn7701 see above) and the thermosensitive allele of the phage lytic repressor Rep*cts*62.

These can be easily propagated as viral particles formed upon thermal induction of Mu*cts*62-mini-Mu*cts*62 double lysogens [41–43]. The lysates obtained contain about 1% mini-Mu containing virions, the genome of which consists of the mini-Mu with a long VE at its right end (40 kbp minus the length of the mini-Mu). In some cases this VE contains the left end of the complementing Mu*cts*62 (which happened to transpose in the appropriate orientation near a mini-Mu) including the *A* (and eventually the *B*) gene, or a second mini-Mu in the same or opposite orientation (*see* Fig. 2).

Single mini-Mu lysogens (and among them mini-Mu induced host mutations) can be isolated upon infection (usually at low temperature because of the presence of the *cts* allele) with a mixed Mu-mini-Mu lysate. They are recovered among clones that carry the mini-Mu selectable marker (usually antibiotic resistance) but do not produce phage. This works, though at a slightly lower

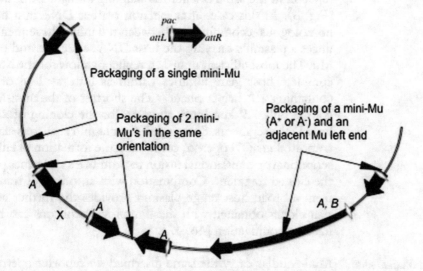

Fig. 2 DNA packaging after induction of a Mu*cts*-mini-Mu*cts* double lysogen. Several mini-Mu and one Mu are represented with *arrows* pointing toward the Mu right end. *White diamonds* represent *pac* sites. Phage particles that contain two mini-MuA+ in the same orientation are those that will promote mini-Muduction of the bacterial segment *X* when infecting a new host. The particle that contains the mini-Mu and the left end of a helper Mu, can deliver and integrate the mini-Mu into a new host

frequency, even when the mini-Mu does not carry the *A* gene, promoted by those mini-Mu residing on a DNA with a helper Mu left end containing the *A* (and possibly *B*) gene.

4.1.1 Mini-Muduction

In mixed Mu mini-Mu lysates, it is not uncommon that two copies of the mini-Mu flanking host DNA, in similar or opposite orientation, fit within a 40 kbp segment of DNA [42] packaged into a virion. When ejected into a new host, provided the mini-Mu's are A+ and in the same orientation, the whole mini-Mu-host DNA-mini-Mu segment can be integrated by MuA mediated transposition using *attL* from one mini-Mu and *attR* from the second so that the new host acquires DNA from the former one. This process, called mini-Muduction [44] does not require homologous recombination and allows for transferring DNA between hosts of different species [45].

4.2 Mini-Mu Plasmid Replicons and In Vivo Construction of Gene Libraries

Mini-Mu*cts* plasmid replicons carry, between Mu ends, a high- or low-copy-number plasmid origin of replication and plasmid functions required for replication and maintenance (e.g., from the high-copy-number plasmid pMB1 or P15A or the low-copy-number broad-host-range plasmid pSa). They also carry selectable genes for resistance to kanamycin, chloramphenicol, or streptomycin/spectinomycin. Just like any other mini-Mu*cts*, they can be complemented by a helper Mu*cts* to transpose at high frequency and be packaged. Viral particles are generated, some of which contain a 40 kbp DNA segment consisting of two copies of the mini-Mu replicon in the same orientation flanking a fragment of host DNA (Fig. 3). In this case, after ejection of that DNA in a new host, homologous recombination between the mini-Mu sequences, produces a plasmids carrying the host DNA sequences and one mini-Mu. The most efficient mini-Mu replicons allow for the isolation of complete host gene libraries within as little as 1 μl of a lysate containing 10^6 helper phages. The shortest of the mini-Mu replicons is only 7.9 kbp long, allowing for the cloning of 23.2-kbp-long DNA segments. Some mini-Mu plasmid replicons also carry a truncated *lacZYA* operon, combining the formation of either transcriptional or translational fusion to promoters and genes present in the cloned fragment. Combination with an origin of transfer *oriT* from a broad host range plasmid provides the further advantage that clones obtained with these mini-Mu replicons can be mobilized by conjugation [46, 47].

4.3 Mini-Mudlac and Other Reporter Fusion Generating Derivatives

Mini-Mudlac carry the same disarmed *lac* reporter operon as Mu (Ap,Lac) described above. They can be grown into phage particles as other mini-Mu's and when they insert in the appropriate orientation in a gene of a new infected host, the *lac* operon is expressed from that gene promoter. Depending on the site of insertion, translation initiated at the gene start codon will either proceed

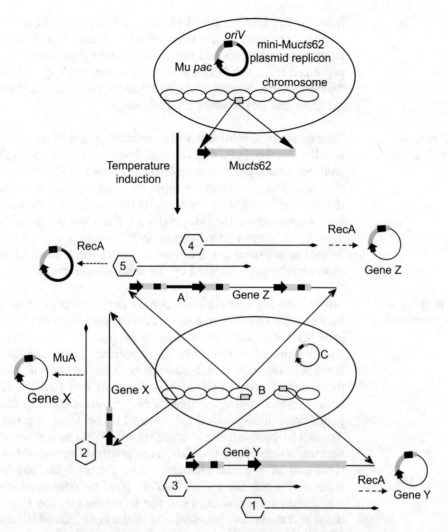

Fig. 3 In vivo cloning with mini-Mu-replicon. The starting strain contains a plasmid with the mini-Mu*cts* replicon and a helper Mu*cts* in the chromosome. Grey blocks: Mu DNA. Black arrow: Mu *pac* site, indicating the start and direction of headful packaging. Black square: plasmid vegetative origin of replication *oriV*. Bold line: plasmid DNA. Thin line: chromosomal DNA. After induction at 42 °C, the first replication of the mini-Mu leads to replicon fusion between the plasmid and the chromosome (A). Two mini-Mu replicons in the same orientation flank the rest of the plasmid DNA. The first replication of the helper Mu provoked here an excision/ deletion (B), splitting the chromosome in two pieces, as did a further replication of the mini-Mu generating a plasmid with one copy of the mini-Mu replicon and host DNA (C). Mu and mini-Mu replicon copies accumulate in the host chromosome, next to various genes (X, Y, Z), in either orientation (shown as blown ups). Heads synthesized by the helper Mu recognize all *pac* sites, and package e.g. a Mu (1), a mini-Mu replicon with a long right VE (here containing gene X, 2), a mini-Mu replicon and a nearby Mu left end (3) or a second mini-Mu replicon in the same orientation (4). This generates particles with intervening host DNA (genes Y or Z), or two mini-Mu replicons and the intervening plasmid DNA (5). Upon ejection of these packaged DNA into a new host (dashed arrows), homologous recombination between two mini-Mu replicons regenerates a plasmid with host DNA (3 and 4). Genes Y and Z have thus been cloned. If MuA is present, replicative transposition associated excision/deletion could also regenerate a plasmid with host DNA (2, cloning of gene X)

through the *lacZ* gene and produce a fusion protein with the N-terminal segment of the gene of interest and β-galactosidase, or encounter a stop codon. In that case genuine β-galactosidase will be produced from the *lacZ* start codon, corresponding to a so called "gene fusion". Several types of mini-Mudlac were engineered to generate one or the other type of fusion.

4.3.1 Gene Fusion Generating Derivatives

Various types of mini-Mu with one among several reporter genes, usually *lac*, *lux* [48, 49] or *nptI* (neomycin/kanamycin/G-418 aminoglycoside phosphotransferase resistance [50]) have been constructed. The promoterless reporter gene/operon is always near the Mu right-end, which was engineered to allow for transcription to proceed across the Mu end into the reporter gene/operon. Translation stops and reinitiates at the reporter gene start codon so that he reporter gene/operon is under the control of the promoter of the gene targeted by the mini-Mu insertion.

4.3.2 Protein Fusion Generating Derivatives

Other mini-Mu fusion elements were designed to generate protein fusions with the *lacZ* and *nptI*-encoded proteins [48]. These two proteins were selected because they remain active when extra amino-terminal amino acids are appended. These Mu derivatives have, near the right-end, truncated gene fragments, which carry no promoter, no translation-initiation region and lack the first few codons, which are dispensable for enzymatic activity. The truncated genes were incorporated into Mu, 117-bp from the right-end. These 117-bp contain no nonsense triplet in the correct phase so that transcription and translation can proceed from outside into the right-end of the Mu element across the 117 bp and into the truncated gene. The fusion LacZ or NptI proteins produced contain amino acids coming from the translation of the 117-bp segment and the amino-terminal end of the gene into which the mini-Mu element has inserted. The generation of this type of fusion requires that the Mu insertion occurs in the appropriate orientation and phase within a translated structural gene. Sometimes, however, one out-of-phase insertion produces an active protein, if translation from the inserted gene terminates and subsequently reinitiates in the correct phase. Furthermore, at a low frequency, translation can "slip" and change phase or be terminated or initiated from a nonstandard sequence. Such low levels of gene products can be readily detected with the sensitive assays for β-galactosidase.

Each of the *lacZ* and *nptI* systems have their distinct advantages, such as the extensive genetic techniques and assays for β-galactosidase and the versatile selection of aminoglycoside resistance for *nptI*, which can be carried out in many organisms, including bacteria, yeast [51], mammalian tissue-culture cells [52], and whole plants (e.g., [53, 54]), Hybrid proteins generated by this type of fusion have many uses such as studies of translation regulation, binding of the protein to nucleic acids, raising of antibodies, localization of the polypeptide on or through membranes.

5   Other Mu-Derived Genetic Tools

5.1   Hybrids Between Mu and Other Phage Genomes

Mu derivatives adapted to infect particular bacterial species, e.g., *S. typhimurium*, have been isolated (reviewed in [55]).

These hybrids between Mu and *Salmonella* temperate and generalized transducing phage P22, carry two-thirds of the P22 genome sandwiched between the ends of Mu. Insertions of these elements in the *Salmonella* chromosome generate P22 prophages that cannot excise. Upon UV induction (which inactivates the P22 repressor), the prophage replicates in situ, resulting in the amplification of neighboring regions of the chromosome. In an induced lysate of a Mud-P22 lysogen, the P22 moiety provides the machinery (P22 packaging *pac* site, head proteins including TerL-TerS terminase) for processive headful packaging, in one direction, from the P22 *pac* site. Each packaging produces three contiguous headful of adjacent DNA, each of which contains a DNA molecule corresponding to several minutes of chromosomal DNA adjacent to the site of prophage insertion and is assembled into a phage particle. Ordered libraries of the *S*. Typhimurium chromosome can be generated as lysates from a representative set of Mud-P22 insertions. This type of approach should be applicable to a variety of other bacteria, provided they are sensitive to a characterized temperate phage [56].

5.2   Mini-Mu Based Gene Amplification

Kurahashi et al. (cited in [39]) were the first to report the use of mini-Mu's to amplify a particular metabolic pathway (L-threonine-overproduction in *E. coli*). They used a dual-system where a disarmed mini-Mu containing the L-threonine metabolic pathway is inserted in the *E. coli* chromosome and amplified by replicative transposition using transient complementation by a plasmid producing MuA and MuB. The level of expression of the metabolic pathway carried by the mini-Mu increases proportionally with the mini-Mu copy number although it is influenced by the location of the mini-Mu's in the chromosome ([40] and references therein).

5.3   Mu Derivatives for In Vitro-Based Transposition Applications

Once the Mu DNA segments and proteins required for Mu transposition had been identified by genetic analysis, purification of the MuA transposase and MuB transposition activator allowed for setting up in vitro assays using appropriate mini-Mu's and target plasmid DNA [57]. Within 10 years the chemistry of the transposase-mediated reaction (hydrolysis of a specific phosphodiester bond at each Mu terminal nucleotide exposing 3'-OH ends, which then attack target DNA at staggered positions 5 bp apart), confirmed a model originally proposed by Shapiro [58]. Both steps are direct phosphoryl-transfer reactions that do not involve covalent protein-DNA intermediates. MuA, Mu *attL*, *attR*, and *IAS* are the minimal requirements for the assembly of a functional transpososome that can be introduced into a host of interest by electroporation [59, 60]. The method, as well as other in vitro based applications of mini-Mu-transposase complexes, is described in detail in Chapter 20.

5.3.1   MuDel

MuDel is one example of a tool exploiting in vitro mini-Mu transposition (*see* Chapter 20, for more details) for the isolation of small, targeted deletions of a fixed number of residues in proteins of interest [61, 62]. Several applications were developed based on the same MuDel, for example to replace trinucleotide deletions at random positions in a target gene with a randomized trinucleotide sequence donated by various DNA cassettes, allowing for the isolation of numerous targeted substitutions in proteins of interest [62].

6   Avoiding Mu to Become a Nuisance

As mentioned earlier, transfer of a Mu prophage, whether c^+ or *cts*, by conjugative transfer or generalized transduction, provokes its induction (zygotic induction). As a result, Mu lysogens, including mutants resulting from prophage insertion, are often recovered in the recipient progeny. In addition, experiments that involve the mixing of cultures of a Mu lysogen and a nonlysogenic strain, often result in the lysogenization of the latter, unless it is resistant to the phage. This probably results from the spontaneous induction of a few lysogens, even in a priori noninducing growth conditions. Although quite a lot of information is available on the Mu Repc protein (especially Rep*cts*62), on its role in combination with host proteins in the lysis-lysogeny switch ([63] and references therein), the exact conditions that lead to Mu wild type induction still remain to be elucidated. Muc^+ and Mu*cts*62 prophages are for instance induced in stationary phase, so that phages can be recovered upon dilution of an overnight culture in fresh broth and growth for a couple of hours ([63, 64] and references therein). This has to be kept in mind when using Mu lysogens, even in experiments where Mu properties are not directly exploited. Ferrières and coworkers [65] reported such a case that went overlooked for years. They used a suicide system, which was designed by Pühler's laboratory [66] to isolate random transposon insertions in various bacterial species. It involves a donor strain with a broad host range suicide vector. This plasmid is complemented for replication by a replicase gene located in the chromosome. Upon transfer in a new host (devoid of the replicase) the plasmid is unable to maintain, allowing for the selection of cells that have integrated the transposon into their chromosome. Muc$^+$ was used to construct the donor strains (known as SM10 λpir and S17-1 λpir), remained there as a prophage and, it turns out, transfers to *E. coli* recipients at high frequency both by conjugation and infection. This may seriously compromise the correct characterization of the isolated transposon-induced mutants. New Mu-free donors were derived from SM10 λpir and S17-1 λpir that overcome that problem [65].

7 Conclusions

Mu is one among many transposable elements that can be used as genetic tools. Through the years a large number of disarmed Mu derivatives have been isolated that turned out very useful in a growing number of bacterial species. The most recent ones (reviewed in [40] for in vivo methods, Chapter 20 for in vitro methods) have several advantages over mini-Mu's isolated earlier. First, they restrict and stabilize as much as possible the number of transposition events in the recipient cell. Second, they do not rely on infection and hence are applicable in a larger number of species of interest. One limitation of these methods, however, is that they rely on transformation for moving DNA into new hosts, which may turn limiting in some bacterial species. It seems, therefore that combining the controlled transposition capabilities of these new generation constructs with the old mode of moving DNA between strains by conjugation or infection (the later by using complementation to generate viral particles, which deliver the mini-Mu), in particular with the variety of available "Saltoviridae," should still provide many efficient protocols for studying an ever larger spectrum of bacterial species.

Acknowledgments

I thank Cold Spring Harbor Laboratory Press for permitting to reproduce part of a chapter that they originally published in Phage Mu in 1987. I dedicate this chapter to the memory of Ahmad Bukahri, Ditmar Kamp, Malcolm Casadaban, and Neville Symonds who all decisively contributed to the understanding of Mu biology and to the development of Mu-based genetic tools. The reference list in this review is obviously incomplete, and I deeply apologize to those whose work has not been cited.

References

1. Hulo C, Masson P, Le Mercier P, Toussaint A (2015) A structured annotation frame for the transposable phages: a new proposed family "Saltoviridae" within the Caudovirales. Virology 477:155

2. Lima-Mendez G, Toussaint A, Leplae R (2011) A modular view of the bacteriophage genomic space: identification of host and lifestyle marker modules. Res Microbiol 162:737–746

3. Toussaint A (2013) Transposable Mu-like phages in Firmicutes: new instances of divergence generating retroelements. Res Microbiol 164:281–287

4. Taylor AL (1963) Bacteriophage-induced mutation in Escherichia coli. Proc Natl Acad Sci U S A 50:1043–1051

5. Silverman M, Simon M (1973) Genetic analysis of bacteriophage Mu-induced flagellar mutants in Escherichia coli. J Bacteriol 116:114–122

6. Harshey RM (2012) The Mu story: how a maverick phage moved the field forward. Mob DNA 3:21

7. Sander JD, Joung JK (2014) CRISPR-Cas systems for editing, regulating and targeting genomes. Nat Biotechnol 32:347–355

8. Vogel JL, Zhu Juan Li, Howe MM, Toussaint A Higgins NP (1991) Temperature sensitive mutations in bacteriophage Mu c repressor define a DNA binding site. J.Bacteriol 173:6568–6577

9. Leach D, Symonds N (1979) The isolation and characterisation of a plaque-forming derivative of bacteriophage Mu carrying a fragment of Tn3 conferring ampicillin resistance. Mol Gen Genet 172:179–184

10. Casadaban MJ, Cohen SN (1979) Lactose genes fused to exogenous promoters in one step using a Mu-lac bacteriophage: in vivo probe for transcriptional control sequences. Proc Natl Acad Sci U S A 76:4530–4533

11. Castilho BA, Casadaban MJ (1991) Specificity of mini-Mu bacteriophage insertions in a small plasmid. J Bacteriol 173:1339–1343

12. Haapa-Paananen S, Rita H, Savilahti H (2002) DNA transposition of bacteriophage Mu. A quantitative analysis of target site selection in vitro. J Biol Chem 277:2843–2851

13. Manna D, Breier AM, Higgins NP (2004) Microarray analysis of transposition targets in Escherichia coli: the impact of transcription. Proc Natl Acad Sci U S A 101:9780–9785

14. Manna D, Deng S, Breier AM, Higgins NP (2005) Bacteriophage Mu targets the trinucleotide sequence CGG. J Bacteriol 187:3586–3588

15. van de Putte P, Goosen N. Trends Genet. 1992 Dec;8(12):457–62. Review. PMID: 1337227

16. van de Putte P, Cramer S, Giphart-Gassler M (1980) Invertible DNA determines host specificity of bacteriophage mu. Nature 286:218–222

17. Faelen M, Mergeay M, Gerits J, Toussaint A, Lefebvre N (1981) Genetic mapping of a mutation conferring sensitivity to bacteriophage Mu in Salmonella typhimurium LT2. J Bacteriol 146:914–919

18. Soberon M, Gama MJ, Richelle J, Martuscelli J (1986) Behaviour of temperate phage Mu in Salmonella typhi. J Gen Microbiol 132:83–89

19. Muller KH, Trust TJ, Kay WW (1988) Unmasking of bacteriophage Mu lipopolysaccharide receptors in Salmonella enteritidis confers sensitivity to Mu and permits Mu mutagenesis. J Bacteriol 170:1076–1081

20. Boucher C, Bergeron B, De Bertalmio MB, Denarie J (1977) Introduction of bacteriophage Mu into Pseudomonas solanacearum and Rhizobium meliloti using the R factor RP4. J Gen Microbiol 98:253–263

21. van Vliet F, Silva B, van Montagu M, Schell J (1978) Transfer of RP4::mu plasmids to Agrobacterium tumefaciens. Plasmid 1:446–455

22. Casadesus J, Ianez E, Olivares J (1980) Transposition of Tn 1 to the Rhizobium meliloti genome. Mol Gen Genet 180:405–410

23. Coplin DL, Frederick RD, Majerczak DR, Haas ES (1986) Molecular cloning of virulence genes from Erwinia stewartii. J Bacteriol 168:619–623

24. Murooka Y, Takizawa N, Harada T (1981) Introduction of bacteriophage Mu into bacteria of various genera and intergeneric gene transfer by RP4::Mu. J Bacteriol 145:358–368

25. Rosenberg C, Bergeron B, Julliot JS, Denarie J (1977) Use of RP4 plasmids carrying bacteriophage Mu insertions in nitrogen fixing bacteria Klebsiella pneumoniae and Rhizobium meliloti. Basic Life Sci 9:411–416

26. Medhekar B, Miller JF (2007) Diversity-generating retroelements. Curr Opin Microbiol 10:388–395

27. van de Putte P, Gruijthuijsen M (1972) Chromosome mobilization and integration of F-factors in the chromosome of RecA strains of E. coli under the influence of bacteriophage Mu-1. Mol Gen Genet 118:173–183

28. Zeldis JB, Bukhari AI, Zipser D (1973) Orientation of prophage Mu. Virology 55:289–294

29. Hirsch PR, Beringer JE (1984) A physical map of pPH1JI and pJB4JI. Plasmid 12:139–141

30. Ely B, Croft RH (1982) Transposon mutagenesis in Caulobacter crescentus. J Bacteriol 149:620–625

31. Singer JT, Finnerty WR (1984) Insertional specificity of transposon Tn5 in Acinetobacter sp. J Bacteriol 157:607–611

32. Contreras A, Maldonado R, Casadesus J (1991) Tn5 mutagenesis and insertion replacement in Azotobacter vinelandii. Plasmid 25:76–80

33. Leung KY, Stevenson RM (1988) Tn5-induced protease-deficient strains of Aeromonas hydrophila with reduced virulence for fish. Infect Immun 56:2639–2644

34. Cen Y, Bender GL, Trinick MJ, Morrison NA, Scott KF, Gresshoff PM, Shine J, Rolfe BG (1982) Transposon mutagenesis in rhizobia which can nodulate both legumes and the non-legume parasponia. Appl Environ Microbiol 43:233–236

35. Noel KD, Sanchez A, Fernandez L, Leemans J, Cevallos MA (1984) Rhizobium phaseoli symbiotic mutants with transposon Tn5 insertions. J Bacteriol 158:148–155

36. Garfinkel DJ, Nester EW (1980) Agrobacterium tumefaciens mutants affected in crown gall tumorigenesis and octopine catabolism. J Bacteriol 144:732–743

37. Kang HW, Wirawan IG, Kojima M (1992) Isolation and genetic analysis of an Agrobacterium tumefaciens avirulent mutant with a chromosomal mutation produced by transposon mutagenesis. Biosci Biotechnol Biochem 56:1924–1928

38. Chatterjee AK, Thurn KK, Feese DA (1983) Tn5-Induced Mutations in the Enterobacterial Phytopathogen Erwinia chrysanthemi. Appl Environ Microbiol 45:644–650

39. Zink RT, Kemble RJ, Chatterjee AK (1984) Transposon Tn5 mutagenesis in Erwinia carotovora subsp. carotovora and E. carotovora subsp. atroseptica. J Bacteriol 157:809–814

40. Akhverdyan VZ, Gak ER, Tokmakova IL, Stoynova NV, Yomantas YA, Mashko SV (2011) Application of the bacteriophage Mu-driven system for the integration/amplification of target genes in the chromosomes of engineered Gram-negative bacteria–mini review. Appl Microbiol Biotechnol 91:857–871

41. Toussaint A, Faelen M, Resibois A (1981) Chromosomal rearrangements induced by mini-Mu and mini-D108: mini review and new data. Gene 14:115–119

42. Resibois A, Toussaint A, van Gijsegem F, Faelen M (1981) Physical characterization of mini-mu and mini-D108. Gene 14:103–113

43. Groisman EA, Casadaban MJ (1987) Cloning of genes from members of the family Enterobacteriaceae with mini-Mu bacteriophage containing plasmid replicons. J Bacteriol 169:687–693

44. Faelen M, Toussaint A, Resibois A. Mol Gen Genet. 1979 Oct 3;176(2):191–7. PMID: 160973

45. Lefebvre N, Toussaint A (1981) Transfer of Salmonella typhimurium and Klebsiella pneumoniae genes in E. coli K12 by mini-muduction. Mol Gen Genet 181:268–272

46. Groisman EA, Castilho BA, Casadaban MJ (1984) In vivo DNA cloning and adjacent gene fusing with a mini-Mu-lac bacteriophage containing a plasmid replicon. Proc Natl Acad Sci U S A 81:1480–1483

47. Groisman EA, Casadaban MJ (1986) Mini-mu bacteriophage with plasmid replicons for in vivo cloning and lac gene fusing. J Bacteriol 168:357–364

48. Castilho BA, Olfson P, Casadaban MJ (1984) Plasmid insertion mutagenesis and lac gene fusion with mini-mu bacteriophage transposons. J Bacteriol 158:488–495

49. Casadaban MJ, Chou J (1984) In vivo formation of gene fusions encoding hybrid beta-galactosidase proteins in one step with a transposable Mu-lac transducing phage. Proc Natl Acad Sci U S A 81:535–539

50. Ratet P, Richaud F (1986) Construction and uses of a new transposable element whose insertion is able to produce gene fusions with the neomycin-phosphotransferase-coding region of Tn903. Gene 42:185–192

51. Jimenez A, Davies J (1980) Expression of a transposable antibiotic resistance element in Saccharomyces. Nature 287:869–871

52. Colbere-Garapin F, Horodniceanu F, Kourilsky P, Garapin AC (1981) Construction of a dominant selective marker useful for gene transfer studies in animal cells. Dev Biol Stand 50:323–326

53. Fraley RT, Rogers SG, Horsch RB, Sanders PR, Flick JS, Adams SP, Bittner ML, Brand LA, Fink CL, Fry JS et al (1983) Expression of bacterial genes in plant cells. Proc Natl Acad Sci U S A 80:4803–4807

54. Herrera-Estrella L, Block MD, Messens E, Hernalsteens JP, Montagu MV, Schell J (1983) Chimeric genes as dominant selectable markers in plant cells. EMBO J 2:987–995

55. Crain JA, Maloy SR (2007) Mud-P22. Methods Enzymol 421:249–259

56. Youderian P, Sugiono P, Brewer KL, Higgins NP, Elliott T (1988) Packaging specific segments of the Salmonella chromosome with locked-in Mud-P22 prophages. Genetics 118:581–592

57. Mizuuchi K (1983) In vitro transposition of bacteriophage Mu: a biochemical approach to a novel replication reaction. Cell 35:785–794

58. Shapiro JA (1969) Mutations caused by the insertion of genetic material into the galactose operon of Escherichia coli. J Mol Biol 40:93–105

59. Lamberg A, Nieminen S, Qiao M, Savilahti H (2002) Efficient insertion mutagenesis strategy for bacterial genomes involving electroporation of in vitro-assembled DNA transposition complexes of bacteriophage mu. Appl Environ Microbiol 68:705–712

60. Pajunen MI, Pulliainen AT, Finne J, Savilahti H (2005) Generation of transposon insertion mutant libraries for Gram-positive bacteria by electroporation of phage Mu DNA transposition complexes. Microbiology 151:1209–1218

61. Jones DD (2005) Triplet nucleotide removal at random positions in a target gene: the tolerance of TEM-1 beta-lactamase to an amino acid deletion. Nucleic Acids Res 33:e80

62. Baldwin AJ, Arpino JA, Edwards WR, Tippmann EM, Jones DD (2009) Expanded chemical diversity sampling through whole protein evolution. Mol Biosyst 5:764–766

63. Ranquet C, Toussaint A, de Jong H, Maenhaut-Michel G, Geiselmann J (2005) Control of bacteriophage mu lysogenic repression. J Mol Biol 353:186–195

64. Lamrani S, Ranquet C, Gama MJ, Nakai H, Shapiro JA, Toussaint A, Maenhaut-Michel G (1999) Starvation-induced Mucts62-mediated coding sequence fusion: a role for ClpXP, Lon, RpoS and Crp. Mol Microbiol 32:327–343

65. Ferrieres L, Hemery G, Nham T, Guerout AM, Mazel D, Beloin C, Ghigo JM (2010) Silent mischief: bacteriophage Mu insertions contaminate products of Escherichia coli random mutagenesis performed using suicidal transposon delivery plasmids mobilized by broad-host-range RP4 conjugative machinery. J Bacteriol 192:6418–6427

66. Simon R, Priefer U, and Pühler A (1983) A broad host range mobilization system for in vivo genetic engineering: transposon mutagenesis in Gram negative bacteria. Biotechnology 1:784–791

Chapter 20

Applications of the Bacteriophage Mu In Vitro Transposition Reaction and Genome Manipulation via Electroporation of DNA Transposition Complexes

Saija Haapa-Paananen and Harri Savilahti

Abstract

The capacity of transposable elements to insert into the genomes has been harnessed during the past decades to various in vitro and in vivo applications. This chapter describes in detail the general protocols and principles applicable for the Mu in vitro transposition reaction as well as the assembly of DNA transposition complexes that can be electroporated into bacterial cells to accomplish efficient gene delivery. These techniques with their modifications potentiate various gene and genome modification applications, which are discussed briefly here, and the reader is referred to the original publications for further details.

Key words DNA transposition technology, Phage Mu, Transpososome

1 Introduction

Bacteriophage Mu transposition is one of the best characterized DNA transposition systems, and it is the first, for which an in vitro reaction was established [1]. Thereafter, a substantially simpler version of the original reaction has been developed, and it requires only a simple buffer and three purified macromolecular components: transposon DNA, MuA transposase, and target DNA [2, 3]. This minimal in vitro reaction has further been modified to yield a variety of elaborate applications, e.g., for DNA sequencing [4], protein engineering [5–10], SNP discovery [11, 12], and constructing of gene targeting vectors [13–16]. The reaction has found its utility also in functional analyses of proteins, genes, and entire genomes [17–23]. With an additional in vivo step, the minimal system can be used for efficient gene delivery not only in a variety of bacteria but also in yeast and mammalian cells [24–28]. The gene delivery technology is characteristically species nonspecific, and it can be used to generate exhaustive insertion mutant libraries for many types of microorganisms, with the latest

Martha R.J. Clokie et al. (eds.), *Bacteriophages: Methods and Protocols, Volume 3*, Methods in Molecular Biology, vol. 1681, https://doi.org/10.1007/978-1-4939-7343-9_20, © Springer Science+Business Media LLC 2018

development widening the scope of the technology also to archaeal species [17]. All of the abovementioned methodologies are critically dependent on the high efficiency and low target site selectivity of the Mu in vitro transposition reaction [2, 29–31], which makes the system ideal for a wide variety of applications.

In this paper we describe general protocols and principles for Mu in vitro transposition reaction as well as the assembly of DNA transposition complexes (Mu transpososomes) and their subsequent electroporation into bacterial cells as an example of the gene delivery methodology.

2   Materials

MuA transposase protein:

MuA can be purified by using the published protocol [32]. The protein preparation is recommended to be flash-frozen and stored at −80 °C. Alternatively, you may obtain MuA from Thermo Fisher Scientific (Waltham, USA). As this product is not frozen, the manufacturer recommends its storage at −20 °C (https://www.thermofisher.com/order/catalog/product/F750).

MuA storage buffer:

25 mM HEPES pH 7.6, 0.1 mM EDTA, 1 mM DTT, 20% (w/v) glycerol, 500 mM KCl. This buffer is recommended for long term storage of MuA preparations made in house.

MuA dilution buffer:

0.3 M NaCl, 25 mM HEPES pH 7.6, 0.1 mM EDTA, 1 mM DTT, 10% (w/v) glycerol.

Use this buffer to dilute MuA from the stock solution directly prior to use. Store at −20 °C.

5× MuA stop:

0.1% bromophenol blue, 2.5% SDS, 50 mM EDTA, 25% Ficoll 400. Store at −20 °C.

Triton X-100:

Prepare 1.25% solution from 10% stock solution by diluting with H_2O directly prior to use.

2× MIX:

50 mM Tris–HCl pH 8.0, 200 μg/ml BSA (bovine serum albumin), 30% (w/v) glycerol. Use high quality molecular biology grade BSA. Store at −70 °C.

1 M HEPES pH 7.6:

Dissolve 13.015 g HEPES (sodium salt) and 11.915 g HEPES (free acid) in 100 ml H_2O. Filter-sterilize. Do not try to adjust additionally the pH of the solution!

5× complex buffer:

750 mM Tris–HCl pH 6.0, 0.125% Triton TX-100, 750 mM NaCl, 0.5 mM EDTA. Filter-sterilize. Store at −20 °C (e.g., in 1 ml aliquots). This buffer is used for transpososome assembly.

Transposon DNA:

Mu transposon DNA can be prepared in house from carrier plasmids by using the published protocol [2] (*see* **Note 1**). Several ready-to-use transposons (called Entranceposons) are available from Thermo Fisher Scientific (Waltham, USA) (https://www.thermofisher.com, search for "entranceposon").

TAE buffer:

To prepare a concentrated (50×) stock solution of TAE dissolve 242 g Tris base in approximately 750 ml deionized water. Add 57.1 ml of glacial acetic acid and 100 ml of 0.5 M EDTA (pH 8.0). Adjust the final volume to 1 l with deionized water.

SOB:

2% Bacto Tryptone, 0.5% Bacto yeast extract, 10 mM NaCl, 2.5 mM KCl. Autoclave. This SOB solution is made without the addition of $MgCl_2$.

SOC:

Add to 100 ml of SOB solution 1 ml of both 2 M $MgCl_2$ and 2 M glucose from stock solutions sterilized by filtration through a 0.22 μm filter.

***E. coli* cells**:

Any standard *E. coli* strain that is suitable for cloning can be used for the methods described.

3   Methods

3.1   General Protocol for Mu In Vitro Transposition Reactions

A standard reaction protocol is described here, but it can be modified in many ways (*see* **Notes 2** and **3**).

Reagent	Volume
2× MIX	12.5 μl
Target DNA (typically 50–500 ng)	typically 1–2 μl
Transposon DNA (0.5 pmol/μl)	1 μl
2.5 M NaCl	1 μl
1.25% Triton X-100 (freshly diluted)	1 μl
0.25 M $MgCl_2$	1 μl
H_2O	up to 24 μl
MuA (220 ng/μl)	1 μl
	Σ 25 μl

1. Dilute MuA into a final concentration of 220 ng/µl with cold MuA dilution buffer. Keep MuA preparation on ice.

2. Assemble the reactions on ice without MuA.

3. Add MuA to start the reaction and incubate at 30 °C, typically for 1 h (*see* **Note 4**).

4. Stop the reaction by the addition of 25 µl of 1% SDS and incubate at room temperature for 30 min. Add 50 µl of water to reduce the salt concentration. Electrotransformation can be done with this preparation using 1 µl aliquots (*see* **Note 5**).

5. Alternatively, for gel analysis, stop the reaction by the addition of 5× MuA stop (e.g., 5 µl sample + 1.5 µl 5× MuA stop). Electrophorese the products using an agarose gel in TAE buffer and, following the run, stain the gel with ethidium bromide to visualize the reaction products.

3.2 General Protocol for the Assembly of Mu Transpososomes and Their Subsequent Electroporation into E. coli

A standard 20 µl reaction protocol is described here, but the reaction can be scaled up to include the reaction volume of 80 µl.

3.2.1 In Vitro Assembly of Mu Transpososomes

Reagent	Volume
5× complex buffer	4 µl
Glycerol	10 µl
H_2O	4 µl
Transposon DNA (1.1 pmol/µl)	1 µl
MuA (400 ng/µl)	1 µl
	Σ 20 µl

1. Adjust the transposon concentration to 1.1 pmol/µl.

2. Dilute MuA into a final concentration of 400 ng/µl with cold MuA dilution buffer. Keep MuA preparation on ice.

3. Assemble the reaction with 5× complex buffer, glycerol, and H_2O at room temperature, transfer the tube on ice.

4. Add transposon to the reaction.

5. Add MuA to start the reaction and incubate at 30 °C, typically for 2 h (*see* **Note 6**).

6. Transpososome assembly can be monitored using native agarose gel electrophoresis. Prepare 2% agarose gel (NuSieve 3:1, Lonza; http://www.lonza.com/) containing 87 µg/ml BSA (Sigma) and 87 µg/ml heparin (Sigma) in TAE buffer. Run the gel using buffer circulation. Electrophorese at 5.3 V/cm for 2 h at 4 °C. Prior to loading, add 0.2 volume of 25% Ficoll 400 to the samples. Stain the gel after the run to visualize stable protein-DNA complexes, i.e., transpososomes.

3.2.2 Electroporation of Mu Transpososomes into E. coli

1. Dilute the Mu transpososome preparation 1:5 or 1:10 with H_2O to reduce the salt concentration onto a level suitable for electroporation.

2. Thaw competent *E. coli* cells on ice (*see* **Note 7**).

3. Add 1 μl of diluted transpososome preparation into 25 μl of electrocompetent cells in a cold tube. Mix gently.

4. Rapidly transfer the mixture into an ice-cold electroporation cuvette (0.1 cm electrode spacing, Bio-Rad).

5. Electroporate immediately using the following pulse settings: voltage 1.8 kV, resistance 200 Ω, and capacitance 25 μF (*see* **Notes 8** and **9**).

6. Add 1 ml SOC (room temperature solution), transfer to a microcentrifuge tube.

7. Incubate at 37 °C by shaking (220 rpm) for 40 min (*see* **Note 10**).

8. Spread the cells onto appropriate selection plates.

Electroporation protocols for other cell types can be found in recent literature (*see* **Notes 11** and **12**).

4 Notes

1. Mini-Mu transposons utilized in in vitro reactions are linear DNA molecules that contain in each of their ends, in an inverted relative orientation, a 50 bp segment from the right end of the phage Mu genome. This so-called R-end segment contains a pair of MuA transposase binding sites. The DNA between the R-ends can be of any origin and modified with regard the needs of each particular application. Also the R-end DNA can be modified, at least to some extent, as restriction sites and translation stop signals have been engineered successfully into these ends to allow downstream processing possibilities. These R-end modifications enable efficient protein engineering applications, by which short insertions [2, 5], deletions [6, 7], single amino acid substitutions [8], or domain additions [9] can be produced.

2. The standard reaction described generates reaction products in amounts sufficient for a majority of applications. However, if a more efficient reaction is needed for the product generation, the concentration of MuA and donor DNA (=transposon) can be increased. The Mu transposition reaction proceeds within the context of the Mu transpososome that contains four molecules of MuA synapsing two transposon ends. Therefore, the stoichiometry between MuA and transposon ends should be kept relatively constant. With our standard transposons (1–2 kb

in length) we have used the stoichiometry of 1 pmol transposon ends (equals to 0.5 pmol of mini-Mu transposon DNA) and 2.7 pmol (220 ng) MuA. MuA is used in a moderate excess, as it binds not only to its binding sites in the R-ends but also sequence nonspecifically along the entire transposon DNA, albeit with a lower affinity. Thus, if longer transposons or more target DNA need to be used, also the MuA concentration in the reaction should be increased. MuA binds to plastic surfaces, e.g., pipette tips and tube walls. Therefore, it is worth minimizing such contacts when working with the enzyme. The volume of the standard reaction is 25 μl, but it can be increased at least up to 100 μl.

3. Make sure that your target DNA does not contain the same selectable marker gene that is present in the transposon DNA that you are using. The use of different selectable markers in the target and donor DNA allows the easy selection of proper integration products.

4. Depending on the transposon DNA, different incubation times may be needed to allow the reaction to proceed into completion. In particular, extended incubation times are needed with long transposons, or if the transposon used contains modified R-ends.

5. The SDS treatment disassembles transpososomes. The protocol described allows a quantitative analysis of reaction products by the use of biological selection. This is, reaction products are transformed or electroporated into *E. coli* and scored as colonies on appropriate selection plates. Transpososomes may also be disassembled using phenol extraction. This is recommended particularly if further downstream processing is included in the application protocol. In short, reaction products from several reactions may be pooled, extracted with phenol and subsequently with chloroform, ethanol-precipitated, and resuspended in a buffer appropriate for further processing.

6. MuA transposase and transposon DNA assemble transpososomes in the absence of divalent metal ions. Under these conditions transpososomes are inactive but can be activated by the addition of Mg^{2+}. Extended complex assembly time may be needed with long transposons, or if the transposon used contains modified R-ends. Transpososomes are stable under the conditions used for the assembly. The transpososome preparation can be flash-frozen under liquid nitrogen and stored at $-80\ ^{\circ}C$ for later use.

7. It is important that the recipient cells used are kept in a solution devoid of Mg^{2+} ions to prevent the activation of transposition chemistry prior to electroporation.

8. Mu transpososomes will encounter Mg^{2+} ions inside the recipient cell and become activated for transposition. Subsequently, transpososomes are able to integrate the delivered transposon DNA into the host chromosome.

9. The protocol has been developed for Genepulser II electroporation apparatus (Bio-Rad). If other brand is used, optimal pulse parameters may differ.

10. During the incubation, the cells will recover from the stress inflicted by freezing and electrical pulse. To avoid cell duplication prior to plating, the incubation time may need to be adjusted depending on the bacterial strain used.

11. The electroporation protocol described has been optimized for gram-negative bacteria [24]. Its further optimization for gram-positive bacteria has been published [25].

12. Electroporation of transpososomes into yeast, mouse ES cells, human HeLa, and human ES cells is also feasible [26].

Acknowledgments

This work was supported by the Academy of Finland (Grant 251168).

References

1. Mizuuchi K (1983) In vitro transposition of bacteriophage Mu: a biochemical approach to a novel replication reaction. Cell 35:785–794

2. Haapa S, Taira S, Heikkinen E, Savilahti H (1999) An efficient and accurate integration of mini-Mu transposons in vitro: a general methodology for functional genetic analysis and molecular biology applications. Nucleic Acids Res 27:2777–2784

3. Savilahti H, Rice PA, Mizuuchi K (1995) The phage Mu transpososome core: DNA requirements for assembly and function. EMBO J 14:4893–4903

4. Haapa S, Suomalainen S, Eerikäinen S, Airaksinen M, Paulin L, Savilahti H (1999) An efficient DNA sequencing strategy based on the bacteriophage Mu in vitro DNA transposition reaction. Genome Res 9:308–315

5. Poussu E, Vihinen M, Paulin L, Savilahti H (2004) Probing the a-complementing domain of E. coli b-galactosidase with use of an insertional pentapeptide mutagenesis strategy based on Mu in vitro DNA transposition. Proteins 54:681–692

6. Poussu E, Jäntti J, Savilahti H (2005) A gene truncation strategy generating N- and C-terminal deletion variants of proteins for functional studies: mapping of the Sec1p binding domain in yeast Mso1p by a Mu in vitro transposition-based approach. Nucleic Acids Res 33:e104

7. Jones DD (2005) Triplet nucleotide removal at random positions in a target gene: the tolerance of TEM-1 b-lactamase to an amino acid deletion. Nucleic Acids Res 33:e80

8. Baldwin AJ, Busse K, Simm AM, Jones DD (2008) Expanded molecular diversity generation during directed evolution by trinucleotide exchange (TriNEx). Nucleic Acids Res 36:e77

9. Edwards WR, Busse K, Allemann RK, Jones DD (2008) Linking the functions of unrelated proteins using a novel directed evolution domain insertion method. Nucleic Acids Res 36:e78

10. Hoeller BM, Reiter B, Abad S, Graze I, Glieder A (2008) Random tag insertions by Transposon Integration mediated Mutagenesis (TIM). J Microbiol Methods 75:251–257

11. Orsini L, Pajunen M, Hanski I, Savilahti H (2007) SNP discovery by mismatch-targeting of Mu transposition. Nucleic Acids Res 35:e44

12. Yanagihara K, Mizuuchi K (2002) Mismatch-targeted transposition of Mu: a new strategy to map genetic polymorphism. Proc Natl Acad Sci U S A 99:11317–11321

13. Vilen H, Eerikäinen S, Tornberg J, Airaksinen MS, Savilahti H (2001) Construction of gene-targeting vectors: a rapid Mu in vitro DNA transposition-based strategy generating null, potentially hypomorphic, and conditional alleles. Transgenic Res 10:69–80

14. Zhang C, Kitsberg D, Chy H, Zhou Q, Morrison JR (2005) Transposon-mediated generation of targeting vectors for the production of gene knockouts. Nucleic Acids Res 33:e24

15. Jukkola T, Trokovic R, Maj P, Lamberg A, Mankoo B, Pachnis V, Savilahti H, Partanen J (2005) Meox1Cre: a mouse line expressing Cre recombinase in somitic mesoderm. Genesis 43:148–153

16. Turakainen H, Saarimaki-Vire J, Sinjushina N, Partanen J, Savilahti H (2009) Transposition-based method for the rapid generation of gene-targeting vectors to produce Cre/Flp-modifiable conditional knock-out mice. PLoS One 4:e4341

17. Kiljunen S, Pajunen MI, Dilks K, Storf S, Pohlschroder M, Savilahti H (2014) Generation of comprehensive transposon insertion mutant library for the model archaeon, Haloferax volcanii , and its use for gene discovery. BMC Biol 12:103

18. Krupovic M, Vilen H, Bamford JK, Kivelä HM, Aalto JM, Savilahti H, Bamford DH (2006) Genome characterization of lipid-containing marine bacteriophage PM2 by transposon insertion mutagenesis. J Virol 80:9270–9278

19. Vilen H, Aalto JM, Kassinen A, Paulin L, Savilahti H (2003) A direct transposon insertion tool for modification and functional analysis of viral genomes. J Virol 77:123–134

20. Kekarainen T, Savilahti H, Valkonen JP (2002) Functional genomics on potato virus A: virus genome-wide map of sites essential for virus propagation. Genome Res 12:584–594

21. Laurent LC, Olsen MN, Crowley RA, Savilahti H, Brown PO (2000) Functional characterization of the human immunodeficiency virus type 1 genome by genetic footprinting. J Virol 74:2760–2769

22. Pajunen M, Turakainen H, Poussu E, Peränen J, Vihinen M, Savilahti H (2007) High-precision mapping of protein–protein interfaces: an integrated genetic strategy combining en masse mutagenesis and DNA-level parallel analysis on a yeast two-hybrid platform. Nucleic Acids Res 35:e103

23. Weber M, Chernov K, Turakainen H, Wohlfahrt G, Pajunen M, Savilahti H, Jantti J (2010) Mso1p regulates membrane fusion through interactions with the putative N-peptide-binding area in Sec1p domain 1. Mol Biol Cell 21:1362–1374

24. Lamberg A, Nieminen S, Qiao M, Savilahti H (2002) Efficient insertion mutagenesis strategy for bacterial genomes involving electroporation of in vitro-assembled DNA transposition complexes of bacteriophage Mu. Appl Environ Microbiol 68:705–712

25. Pajunen MI, Pulliainen AT, Finne J, Savilahti H (2005) Generation of transposon insertion mutant libraries for Gram-positive bacteria by electroporation of phage Mu DNA transposition complexes. Microbiology 151:1209–1218

26. Paatero AO, Turakainen H, Happonen LJ, Olsson C, Palomäki T, Pajunen MI, Meng X, Otonkoski T, Tuuri T, Berry C, Malani N, Frilander MJ, Bushman FD, Savilahti H (2008) Bacteriophage Mu integration in yeast and mammalian genomes. Nucleic Acids Res 36:e148

27. Tu Quoc PH, Genevaux P, Pajunen M, Savilahti H, Georgopoulos C, Schrenzel J, Kelley WL (2007) Isolation and characterization of biofilm formation-defective mutants of Staphylococcus aureus. Infect Immun 75:1079–1088

28. Wu Z, Xuanyuan Z, Li R, Jiang D, Li C, Xu H, Bai Y, Zhang X, Turakainen H, Saris PE, Savilahti H, Qiao M (2009) Mu transposition complex mutagenesis in Lactococcus lactis–identification of genes affecting nisin production. J Appl Microbiol 106:41–48

29. Butterfield YSN, Marra MA, Asano JK, Chan SY, Guin R, Krzywinski MI, Lee SS, MacDonald KWK, Mathewson CA, Olson TE, Pandoh PK, Prabhu A-L, Schnerch A, Skalska U, Smailus DE, Stott JM, Tsai MI, Yang GS, Zuyderduyn SD, Schein JE, Jones SJM (2002) An efficient strategy for large-scale high-throughput transposon-mediated sequencing of cDNA clones. Nucleic Acids Res 30:2460–2468

30. Haapa-Paananen S, Rita H, Savilahti H (2002) DNA transposition of bacteriophage Mu. A quantitative analysis of target site selection in vitro. J Biol Chem 277:2843–2851

31. Mizuuchi M, Mizuuchi K (1993) Target site selection in transposition of phage Mu. Cold Spring Harb Symp Quant Biol 58:515–523

32. Baker TA, Mizuuchi M, Savilahti H, Mizuuchi K (1993) Division of labor among monomers within the Mu transposase tetramer. Cell 74:723–733

Chapter 21

Use of RP4::Mini-Mu for Gene Transfer

Frédérique Van Gijsegem

Abstract

Gene cloning is an invaluable technique in genetic analysis and exploitation of genetic properties of a broad range of bacteria. Numerous in vitro molecular cloning protocols have been devised but the efficiency of these techniques relies on the frequency with which the recombinant DNA can be introduced in the recipient strain. Here, we describe an in vivo gene transfer and cloning technique based on transposable bacteriophage Mu property to rearrange its host genome. This technique uses the broad host range plasmid RP4 carrying a transposable mini-MuA$^+$ derivative and was successfully used as well in enteric as in environmental nonenteric bacteria.

Key words In vivo cloning, Chromosome mobilization, Gene mapping, Transcription unit identification

1 Introduction

Gene transfer and gene cloning are seminal for several genetic analyses such as gene mapping or mutation complementation, mutant construction or introduction of gene clusters bringing new properties to a recipient bacterium. Since the recombinant DNA revolution, several sophisticated in vitro molecular cloning protocols have been devised, based on phage and/or plasmid vectors, mainly for cloning in the model *Escherichia coli* species. A number of cloning vectors derived from broad host range plasmids have been subsequently developed for analysing a large variety of environmental and pathogenic bacteria that do not belong to the enterobacteria [1]. In vitro cloning involves many steps and relies on the efficiency with which the recombinant DNA can be introduced in the recipient strain (usually by transformation or electroporation), a process that might occur at low frequency—especially when cloning large DNA fragments—and with widely different efficiencies in recipient strains of different genera. In this chapter we described a technique of in vivo gene transfer and cloning based on transposable bacteriophage Mu property to rearrange its host

Martha R.J. Clokie et al. (eds.), *Bacteriophages: Methods and Protocols, Volume 3*, Methods in Molecular Biology, vol. 1681,
https://doi.org/10.1007/978-1-4939-7343-9_21, © Springer Science+Business Media LLC 2018

genome (reviewed in Toussaint, Chap. 19). Although developed a long time ago, this technique is rapid and can be easily used in various bacterial genera. It can thus still be useful especially for cloning genes that are not expressed in enteric bacteria.

1.1 Outline of the Method

The method exploits the capacity phage Mu, when it is inserted in a plasmid, to promote the random transfer of the bacterial chromosome and to allow for in vivo cloning by generating plasmids that have captured large chromosomal DNA fragments. To be proficient in gene transfer in as many bacterial genera as possible, Mu derivatives have been constructed that leave the occurrence of replication associated chromosomal rearrangements at a reasonable level while eliminating most of the killing caused by the Mu lytic cycle (http://viralzone.expasy.org/all_by_species/4277.html). Some of these constructs were inserted in a broad host range plasmid that can be transferred to donor bacteria at a reasonable frequency. RP4::mini-Mu, pULB113 is the most widely used derivative among these constructs. It is derived from the broad host range self-transmissible plasmid RP4 [2] and carries a mini-Mu that has an active transposase/DDE recombinase A gene, a truncated transposase activator B gene, and intact Mu termini ([3], see Fig. 1 in Chap. 19).

The mechanism of RP4::mini-MuA$^+$ mediated gene transfer is outlined in Fig. 1. The plasmid first spontaneously and randomly integrates into the host genome by replicon fusion. As a result, since RP4 is self-transmissible, a fraction of the donor population consists of Hfr-like derivatives, each having the plasmid integrated at a different position. The overall probability of any gene being near the RP4::mini-Mu is the same, leading to equal transfer efficiencies for all genes. After transfer in the recipient bacterium, if sufficient homology exists within the transferred region between donor and recipient chromosomes, genes from the donor can be integrated in the recipient chromosome by homologous recombination (Fig. 1a).

Once integrated in the donor chromosome, the RP4::mini-MuA$^+$ can also re-excise by a second mini-Mu-mediated rearrangement (occurring during a subsequent round of replication). This is called "excision-deletion" and leads to the formation of rearranged plasmids that carry a host DNA segment of variable length, originally adjacent to one side or the other of the integrated plasmid and flanked by two mini-Mu prophages in the same orientation. This second event generates R-prime hybrid plasmids that, due to the random location of the integrated plasmids following replicon fusion, have an equal probability of carrying any chromosomal marker (see Fig. 1b).

RP4 DNA is in black, mini-Mus are in red, donor strain chromosome is in blue, recipient strain chromosome is in violin.

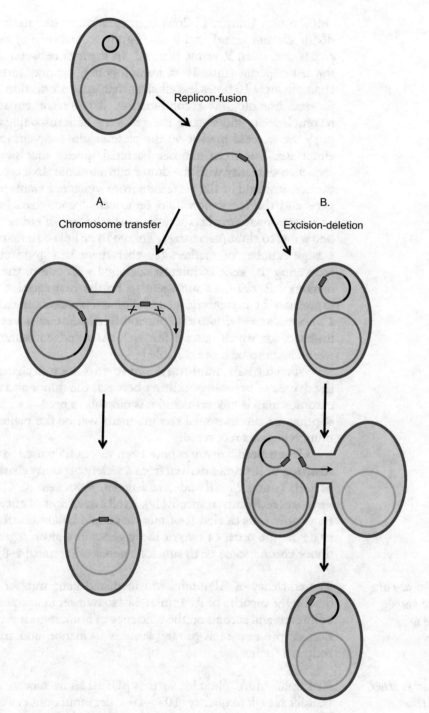

Replicon-fusion

A.

Chromosome transfer

B.

Excision-deletion

Fig. 1 Gene transfer mediated by RP4::mini-Mu

1.1.1 Chromosomal
Transfer Versus In Vivo
Cloning

As discussed above, mating a donor strain with a mini-MuA+
carrying conjugative R plasmid allows for the selection of transconjugants that have acquired a given marker, either as a result of

chromosome transfer (following integration of the plasmid in the donor chromosome) and homologous recombination or by the acquisition of an R-prime plasmid. In matings between strains of the same species (thus 100% homologous), R-prime formation is approximately 10 times less efficient than gene acquisition by transfer and homologous recombination. To prevent chromosomal recombination and ensure the recovery of transconjugants that carry the selected marker on the plasmid, the recipient should be either Rec⁻ or from another bacterial species that shares little sequence similarity with the donor chromosome. In a typical case, the donor would be the bacterium from which one wants to clone a gene and the recipient would be a well-characterized bacterium with appropriate markers to allow for selection of either the gene one wants to clone, or a marker known to be linked to that gene, or a large number of markers (so that there is a good chance of recovering the gene of interest cocloned with one of the selected markers). *E. coli* does not seem to be the best recipient for the expression of nonenteric genes. Some *Pseudomonas* strains and *Cupriavidus metallidurans* (previously *Alcaligenes eutrophus*), for instance, for which quite a few well-characterized mutants exist, were shown to be more suitable [4].

When the RP4::mini-Mu donor strain is not well characterized, the degree of homology existing between the donor and recipient chromosomes is unknown and it is difficult to predict which of the R-prime or chromosomal recombinants will be the major type of transconjugants recovered.

Heterospecific matings have been successfully used in the construction of R-primes derived from a variety of poorly characterized bacteria (using *E. coli* and *Pseudomonas fluorescens* or *C. metallidurans* as recipients, respectively). In all cases, most, if not all, of the transconjugants carried R-primes that could be used further either to define the order of two or more genes in a given region of the donor chromosome or to subclone genes of interest [4–8].

1.2 Efficiency of RP4::Mini-Mu Mediated Gene Transfer

The efficiency of RP4::mini-Mu mediated gene transfer depends first on the capacity of RP4::mini-Mu to transfer in bacterial species of interest and second on the efficiency of homologous recombination or the rate of R-prime plasmids formation and transfer to recipient strains.

1.2.1 Transfer of RP4:: Mini-Mu in Different Bacterial Species

RP4::mini-MuA⁺ plasmids such as pULB113 have been shown to transfer at high frequency (10^{-2}–10^{-3} per input donor) from *E. coli* K12 to many enterobacteria (*Salmonella typhinurium*, provided it is restriction-negative, *Proteus mirabilis*, and *Klebsiella pneumoniae* [3]). In pectinolytic enterobacteria (*Dickeya* and *Pectobacterium* species previously regrouped as pectinolytic *Erwinia*), the frequency of transfer of the RP4::mini-MuA⁺ plasmid from *E. coli* seems to depend both on the subspecies and on the strain used (from 10^{-2} to 10^{-7} per input donor [9]). RP4::mini-Mu also transfers efficiently (frequency from 1 to 10^{-3} per input donor) to other Gram-negative bacteria such as *P. fluorescens, C. metallidurans* [4], *Agrobacterium tumefaciens, Pseudomonas oxalyticus* (F. van Gijsegem and M. Mergeay, unpublished), and *Myxococcus xanthus* [10].

1.2.2   Efficiency of Gene Transfer Mediated by RP4:: Mini-Mu

In different bacterial species (*E. coli* K12, *P. fluorescens*, and *C. metallidurans*), pULB113 mediates the transfer of any chromosomal marker at a frequency of 10^{-4}–10^{-5} and R-primes are formed at frequencies ranging from 10^{-5} to 10^{-8} per input donor in both enteric and nonenteric gram-negative bacteria. These frequencies are usually highest in enterobacteria, such as *Salmonella* Typhimurium, *Proteus mirabilis, Klebsiella pneumoniae* [3], and various *Dickeya* and *Pectobacterium* species (previously *Erwinia carotovora atroseptica, Erwinia carotovora carotovora, Erwinia chrysanthemi*) [5, 6, 8, 9, 11–13], or *Enterobacter cloacae* [7], and lowest in nonenteric strains, such as *P. fluorescens, C. metallidurans*, or *A. tumefaciens* [4, 14].

2   Materials

2.1   Media

1. Rich media for matings: medium composition should be adapted to the bacteria of interest and chosen such that donor and recipient strains grow at about the same rate. The Luria Bertani (LB) medium is suitable for matings that involve most enterobacteria and *Cupriavidus* strains.

   LB medium: Bacto Tryptone 10 g/l, yeast extract 5 g/l, NaCl 10 g/l.
   When required, media are solidified by adding 15 g/l agar.

2. When transconjugants are selected for the complementation of auxotrophic markers, the selective minimal medium should be adapted to the recipient strain used. For enterobacteria, M63 medium ([15]; http://biocyc.org/ECOLI/NEW-IMAGE? type=Growth-Media&object=MIX0-48) is widely used:

   – in 900 ml H_2O, dissolve 13.6 g KH_2PO_4 and 2 g $(NH_4)_2SO_4$.

   – Add 1 ml Fe SO_4 (10 mM).

   – Adjust to pH 7.0 with 10 M KOH.

   – Adjust to 1 l.

   – Autoclave.

   – Add 1 ml 1 M $MgSO_4$ and a carbon source at 0.2%.

For some soil bacteria like *Ralstonia* species for instance, the salt content of the M63 medium is too high but a twofold dilution of the medium works fine.

3. Antibiotics are added at the following final concentrations: Ampicillin (Ap) 100 μg/ml; kanamycin (Kn) 25 μg/ml; tetracycline (Tc) 15 μg/ml.

2.2 Plasmid DNA Extraction

CsCl (irritant).

10 mM ethidium bromide solution (mutagen, irritant).

2.2.1 Material Needed

10% SDS (irritant).

5 M NaCl.

PEG-8000.

A high-speed centrifuge for centrifugation of up to 100 ml volumes at $27,000 \times g$.

An ultracentrifuge.

2.2.2 Buffers

Sucrose solution: dissolve 25 g sucrose in 50 mM Tris–HCl (pH 8) and adjust to 100 ml.

Lysozyme solution: 5 mg/ml lysozyme in 0.25 M Tris–HCl at pH 8.0.

Lysis buffer: 20 ml of 0.25 M Na_2EDTA (pH 8), 5 ml of 1 M Tris–HCl (pH 8), 20 ml of 10% SDS in 10 mM Tris (pH 8), and 55 ml of H_2O.

TE: 50 mM Tris pH 8, 20 mM EDTA.

TES: 50 ml Tris–HCl at pH 8, 5 mM EDTA, 50 M NaCl.

TAE (Tris–acetate–EDTA) electrophoresis buffer (50× stock solution): 242 g Tris base, 57.1 ml glacial acetic acid, 37.2 g $Na_2EDTA \cdot 2H_2O$, H_2O to 1 l.

3 Methods

3.1 Preparation and Testing of the Donor Strain

Plasmid pULB113 is available in the *E. coli* strain MXR (del(*pro*, *lac*), *galE*, *thi*, *recA53*). It's donor and mobilization properties should better be checked before use, for instance by mating according to the protocol below, with an appropriate polyauxotrophic derivative of *E. coli*, followed by selection on minimal media of transconjugants that acquired the wild type alleles complementing one or another of the auxotrophic mutations of the recipient, as well as the pULB113 encoded antibiotic resistances, Kn^R, Tc^R, and Ap^R.

3.2 Mating Protocol (Adapted from Betlach et al. [16])

1. Grow stationary cultures of the donor and the recipient strains in a suitable rich medium and at the optimal growth temperature for the bacteria of interest. For enteric bacteria, *Pseudomonas* or *Cupriavidus* species, this is usually performed overnight at 30 °C (*see* **Note 1**).

2. On a plate containing the same but solid rich medium, spot one drop (20–50 µl) of the donor, one drop of the recipient, and one drop of a mixture of equal volumes of both bacteria.

3. Incubate the plate at 30–42 °C depending of the growth characteristics of the donor and recipient bacteria used. In our experience, 30 °C is a convenient temperature for warm blood organisms-associated commensal and pathogenic bacteria and for many soil, water and plant-associated bacteria (*see* **Note 2**). For enteric bacteria and *Pseudomonas* or *Cupriavidus* species, an incubation of 6–16 h is sufficient but this mating period may be extended depending of the growth rate of the bacteria of interest.

4. Recover separately the bacteria grown in the different spots with a loop or a pipette and resuspend by vortexing in 1 ml of 10 mM $MgSO_4$ buffer (or rich medium).

5. Spread aliquots of the concentrated suspension and a 100-fold dilution of the mating suspension on selective medium. As negative controls, spread aliquots of the concentrated suspension of each partner on this same medium. Spread aliquots of the 10^6 and 10^7 dilutions of the mating suspension on selective media that allow separate titration of the donor and recipient in the mating mixture.

3.3 Characterization of Transconjugants

1. Purify transconjugant colonies grown on the selective medium once on the same selective medium.

2. Verify the antibiotic profile of the transconjugants on rich medium plates supplemented with the corresponding antibiotics. If pULB113 or an R-prime is present, the clones should be Kn^R, Tc^R, and Ap^R.

3. Antibiotic resistant clones may carry either an R-prime or the parental plasmid while having acquired the selected marker in the chromosome by homologous recombination. To discriminate between these two possibilities, using the protocol described in 3.2, transconjugants should be mated with a new recipient that allows for selecting either the marker of interest or plasmid antibiotic resistance. If the selected marker is on an R-prime, the antibiotic resistance and the selected marker are transferred at the same high frequency.

4. To identify the size of the chromosomal fragment carried by the R-prime and for subsequent cloning, the R-prime plasmid has to be purified and analyzed as described below (*see* **Note 3**).

3.4 Preparation of RP4::Mini-Mu–Host-DNA–Mini-Mu R-Prime DNA

Many kits are available today that allow for the purification of low copy number plasmids carrying large cloned DNA fragments (up to 250 kb according to the manufacturers). These kits should be suitable for preparing pure DNA from pULB113 derived R-primes. However, these kits are designed mainly for use with *E. coli*. If you are troubleshooting problems with the use of such kits when using other bacterial genera, the DNA extraction and CsCl gradient purification technique that allowed us to purify R-prime plasmids up to 150–200 kb from either enterobacteria or *C. metallidurans* can still be useful and is presented below.

Please note: CsCl is an irritant; ethidium bromide is a mutagen and an irritant. Safety rules have to be followed for conducting this protocol (*see* **Note 4**).

1. 1–2 g of bacteria is required for this procedure. Grow a suitably sized culture of the R-prime containing strain, in a medium that selects for the maintenance of the chromosomal insert present on the R-prime and eventually supplemented with antibiotics ensuring selection of the plasmid.

2. Centrifuge the bacterial culture for 10 min at $12,000 \times g$.

3. Resuspend the pellet in 50 ml of TE buffer mixed with 100 ml ethanol (*see* **Note 5**) and centrifuge for 10 min at $12,000 \times g$.

4. Resuspend the pellet in 17 ml of 25% sucrose in 50 mM Tris–HCl (pH 8) and keep on ice for 5 min.

5. Add 3.3 ml of a freshly prepared lysozyme solution and keep on ice for 5 min.

6. Add 6.7 ml of 0.25 M Na_2EDTA (pH 8) and keep on ice for 5 min.

7. Add 27 ml of lysis buffer (*see* Subheading 2.2).

8. Incubate at 37 °C until the mixture is clear (usually about 60 min).

9. Add 13 ml of 5 M NaCl and mix by gently turning the bottle upside down several times.

10. Keep on ice for 5–15 h and then centrifuge 30 min at $27,000 \times g$.

11. Recover the supernatant and add PEG-8000 to a final concentration of 10% and keep at 4 °C for one night (or longer).

12. Centrifuge for 30 min at $10,400 \times g$ and resuspend the pellet in 5 ml of cold TE buffer.

13. Centrifuge the large debris for 10 min at $5000 \times g$ and dialyze twice for 2 h against 1 l TES at 4 °C.

14. Recover the content of the dialysis tube.

15. Add 1.1 g CsCl/ml of DNA solution (The volume has to be adjusted depending of the capacity of the ultracentrifuge tubes

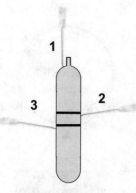

Fig. 2 Collection of plasmid DNA from a CsCl gradient

used). Dissolve. Add 0.4 volume of 10 mg/ml ethidium bromide (*see* **Note 6**).

16. Transfer to sealable centrifuge tubes. If needed, balance tubes with CsCl in TES buffer (1.05 g/ml). Seal tubes. Spin 4 h in a vertical rotor at 500,000 × *g* or 16 h at 350,000 × *g* at 20 °C (*see* **Notes 7** and **8**).

17. Under short-wave UV light, two separate bands are visible in the gradient, the upper band consists of chromosomal DNA and the lower band of plasmid DNA (Fig. 2). To allow air entry in the sealed tube, insert gently a 20G-needle at the top of the tube (Fig. 2, step 1). If the chromosomal DNA band is large, it is preferable to discard it before recovering the plasmid DNA. For this, insert a 20G-needle mounted on a syringe just below the upper band with the bevelled side up and gently suck (Fig. 2, step 2). Let the syringe in place and recuperate the plasmid band the same way (Fig. 2, step 3).

18. Plasmid DNA can then be purified by using desalting columns (*see* **Note 9**).

3.5   Physical Analysis of R-Prime Plasmids

R-primes derived from RP4::mini-Mu carry a continuous random piece of host DNA, flanked by two mini-Mu prophages in the same orientation. The whole structure is located on the plasmid at the site where the mini-Mu was originally inserted [3–6]. In consequence, the number of mini-Mu present on the R-prime and the size of the cloned chromosomal fragment are easily visualized by digesting the plasmid with a restriction enzyme like PstI that cleaves twice into the mini-Mu (*see* Fig. 3a for the restriction map of pULB113). Restriction fragments are visualized on 0.7% TAE agarose gels. An example of such an analysis is shown in Fig. 3b (*see* **Note 10**). The R-prime plasmids should harbor the six bands

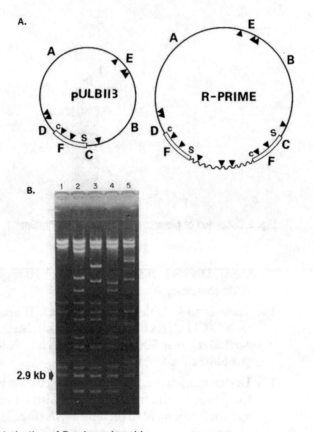

Fig. 3 Physical characterization of R-prime plasmids

present in the parental pULB113 plasmid and a variable number of bands corresponding to the inserted chromosomal fragment. Since the mini-Mu is present in two copies on the R-primes, the 2.9 kb band internal to the mini-Mu is more intense. R-primes carrying the same selected marker share some of the additional PstI fragments which correspond to the chromosomal region encompassing the selected gene.

1. PstI restriction map of pULB113 and a R-prime derivative. Open boxes correspond to mini-Mu DNA, the wavy line corresponds to chromosomal DNA. The PstI sites are marked by arrowheads.

2. PstI restriction profiles of pULB113 (slot 1), R-primes *thy* derived from *E. coli/E. coli* matings (slots 2–4) and a R-prime *leu* derived from a *Salmonella/E. coli* mating (slot 5). The 2.9 kb internal to the mini-Mu is marked by a big arrow. Electrophoresis was performed in Tris–acetate buffer on 0.7% agarose gels at 50 mA for 5 h (adapted from [3]).

4   Applications

Besides isolation of genes by in vivo cloning, gene transfer mediated by the RP4::mini-Mu can be used for the following applications:

1. Mapping by chromosome transfer

   During homospecific transfer, random integration of the RP4::mini-Mu into the donor chromosome results in a nonoriented transfer, which, however, can be used to map genes on the donor strain's chromosome. Indeed, the probability of the plasmid integrating between two given markers decreases as the distance between those markers decreases. Therefore, the closer the markers, the more frequently they are simultaneously transferred to recipients. Once transferred, they will be simultaneously incorporated into the recipient chromosome by recombination with a frequency again inversely proportional to their distance. The mapping thus rests on the overall frequencies of cotransfer of two markers from the donor to the recipient, each individual marker being transferred at the same frequency. This strategy has been successfully used to construct a first map of the chromosome of *Dickeya* and *Cupriavidus* strains [17–19] (*see* **Note 11**).

2. Mapping by Cotransposition

   Because in most cases Mu (or mini-Mu)-mediated transposition of chromosomal DNA results in the transposition of a unique and continuous piece of host DNA of variable size, any individual marker can be transposed onto a plasmid at the same frequency, and any marker linked to it can be cotransposed with a frequency inversely related to the distance that separates the two markers. Consequently, the system can be used to map genes, by looking at the frequencies at which unselected markers are cotransposed on R-primes with a given selected marker. The method has an advantage over cotransduction in that it allows genes to be mapped that are much further apart (genes separated by 3.5 min on the *E. coli* chromosome can still be cotransposed at 16%; [20]). This type of mapping allows the respective order of markers in a rather long region of bacterial chromosomes or plasmids to be easily determined, and the physical analysis of R-prime DNA provides, in addition, a good indication of the physical distance between the markers studied. To perform mapping, once isolated, the R-primes are tested for the presence of unselected markers, either directly if the recipient had the appropriate markers or after retransfer into an appropriate recipient. The cotransposition frequency, i.e., the ratio between the number of clones harboring both the selected marker and a given unselected marker and the number of clones carrying the selected marker, gives a measure of the distance separating the two markers (*see* **Notes 12** and **13**).

3. Identification of transcription units

Because the ends of transposed fragments containing a given marker are not fixed, the method allows for the direct detection of transcriptional units. If, for instance, R-primes are selected that carry a marker located downstream (but not in) a transcription unit, all of the genes belonging to this unit will be expressed only by plasmids that received the unique promoter and the genes it controls. Consequently, some classes of plasmids will be missing [21].

4. Gap closure of genomic sequences

Filling gaps in bacterial genome sequencing projects is sometimes tedious in particular when one has to deal with the presence of large repeated regions at different locations in the chromosome. Although there is no report of such an application in the literature, RP4::mini-Mu mediated in vivo cloning might theoretically be used to fill such gaps simply by selecting R-primes carrying genetic markers located at each side of the gap and sequencing the chromosomal fragment carried by these R-primes.

5   Notes

1. If the RP4::mini-Mu is stable in the donor strain, there is no need to add antibiotics in the donor strain culture. If antibiotics are added, the bacteria have to be washed to remove the antibiotic before mating (2 min full speed centrifugation in tabletop centrifuge and resuspension in rich medium without antibiotics).

   Plasmid stability can easily be checked by plating isolated colonies of the donor strain and testing 50–100 isolated colonies for plasmid presence. Simply test the colonies for antibiotic resistance present on the plasmid (ampicillin, kanamycin or tetracycline for pULB113) by toothpicking on plates with and without the suitable antibiotic.

2. RP4::mini-Mu pULB113 carries the *cts*62 thermosensitive repressor allele. Nevertheless, the mini-Mu promotes gene transfer even at low temperatures. At most, there is a fivefold difference in the frequencies of gene transfer at 42 and 30 °C (our unpublished results).

3. The size of the chromosomal DNA carried by R-primes is variable (from a few to more than 250 kb) and seems to be dependent on the origin of the fragment. For instance, R-primes carrying the *exuT-uxa* region from *Dickeya*

chrysanthemi B374 are on the average 10 times shorter than those carrying the *uxuA-uxuB* region [5].

4. Wear personal protective equipment including lab coat, nitrile gloves for ethidium bromide handling, UV eye protection. Materials contaminated with ethidium bromide must be collected in plastic bags and disposed as hazardous waste.

5. It is quite hard to resuspend the pellet by vortexing. An alternative is resuspension by pipetting up and down 20 ml of buffer and then adjusting the volume to 50 ml.

6. You may see red flocculates in the preparation. This is made by complexion of ethidium bromide with proteins and can be eliminated by centrifugation of the DNA preparation for 5 min at 2000 × g at room temperature. The ethidium bromide/protein complex will form a ring at the top of the tube. The DNA solution below this disc can be recuperated with a Pasteur pipette.

7. The CsCl-DNA solutions may also be centrifuged in swinging buckets rotors for example for 30–48 h at 400,000 × g in Beckman Ti70 rotor or equivalent.

8. The temperature of centrifugation is important because at temperatures lower than 15 °C, CsCl may precipitate at the bottom of the tube causing rotor imbalance.

9. Plasmid DNA can also be purified as follow:

   (a) Eliminate ethidium bromide by extraction with CsCl-saturated isopropanol. Repeat twice after the sample became colorless. Prepare CsCl-saturated isopropanol solution as follows: Take 10–15 g CsCl, dissolve in 5–10 ml TE buffer. You have to see a few crystals of CsCl on the bottom of the tube. Add 20 ml of isopropanol. Keep at room temperature.

   (b) Remove CsCl by using desalting columns or by dialysis overnight against 2 l of TE buffer (10 mM Tris–HCl pH 8.0, 0.1 mM EDTA) or water.

10. In some instances, the R-primes isolated were found to carry a deletion of RP4 DNA at one of the junctions with the mini-Mu [3, 9]. It is not clear whether such deletions occur before or during the formation of the R-prime. There are also a few cases where only one mini-Mu prophage was found on R-primes isolated from *S.* Typhimurium [3] and from *K. pneumoniae* (G.A. Sprenger, pers. comm.), suggesting that factors other than the mini-Mu (IS*1* in the mini-Mu, Tn*1* in RP4) might sometimes be involved in R-prime formation.

11. Retrotransfer/Gene capture

During experiments using RP4::mini-Mu-mediated gene transfer from a prototrophic Sm^R donor to a polyauxotrophic Sm^s recipient, it was first suspected that selection of Sm^R transconjugants that had acquired several wild-type alleles from the donor led to the recovery, not of the expected transconjugants, but of donor derivatives that had received the Sm^R allele from the recipient [17, 22]. In some matings, these "reverse" transconjugants occur at low frequency (about 0.1% of the total number of transconjugants recovered), whereas in others, they can represent the major class of transconjugants (>90%). By introducing a small Tra$^-$ Mob$^-$ plasmid in the RP4::mini-Mu donor strain, it could be clearly demonstrated that a particular class of transconjugants did result from back transfer of one allele from the recipient to the donor [23]. When the donor is an auxotroph, appropriate selection can lead to the recovery, in the donor, of R-primes that have captured the corresponding wild-type allele in the recipient and brought it back to the donor. This may be the major type of event in heterospecific matings where the partner's chromosomes have little homology, so that no chromosomal recombination can occur after back transfer of the recipient chromosome to the donor. In matings between *P. fluorescens* and *C. metallidurans*, where one or the other carries pULB113, depending on the selection used, either one or the other mating partner can be recovered with R-primes that have incorporated genes from the other. This observation opens the way to a new type of mini-Mu-mediated cloning where the donor is also used as the selective recipient strain. This retrotransfer or gene capture process is a property of IncP plasmids that can only be detected if the plasmid has the ability to capture host genes that distinguish it from the resident plasmid, provided for instance by a mini-Mu [23]. As so far other broad-host-range plasmids such as IncN and IncW have not been shown to promote gene capture, shuttle transfer should be avoided by using derivatives of these plasmids with a mini-Mu insertion.

12. Kanamycin-sensitive (Kn^s) RP4::mini-Mu plasmids have been isolated to allow the mapping of Tn*10*-induced mutations (pULB110 [6]). When using these mutations, it should be kept in mind that there are two superimposed transposing systems, the mini-Mu on the mobilizing plasmid and the transposon in the studied mutation. This transposon could also mediate the integration of the plasmid by replicon fusion. The respective frequencies of the two kinds of events should depend on the respective frequencies of transposition of Tn*10* and mini-Mu, which may be quite different in different bacterial backgrounds.

13. *Coincident Transposition*

When selecting for a particular Mu- or mini-Mu-induced rearrangement in the host genome, one should always keep in mind that since rearrangements occur at high frequency during Mu lytic growth, there is a significant probability of recovering clones that have undergone more than one rearrangement. A typical example is what is usually called coincident transposition. A large fraction of the R-primes formed by Mu-mediated transposition carry more than one chromosomal segment, each segment being flanked by Mu prophages, so that the R-prime carries more than two copies of Mu. A given *E. coli* marker can be recovered at a frequency of 1% on any type of selected R-prime, and the frequency at which that marker is coincidentally transposed with any other is not correlated with the distance that separates them. Although such an approach has never been reported, this should allow for the isolation of any marker on an R-prime by simultaneous selection with any other marker. On R-primes derived from RP4::mini-MuA$^+$, the same phenomenon occurs although at a lower frequency. Physical analysis of R-prime plasmids revealed that 1–10% carried more than two mini-Mu prophages and thus, most probably, more than one chromosomal segment [3].

The occurrence of coincident transposition should be kept in mind especially when using Mu/mini-Mu-mediated transposition for mapping distant genes that cotranspose at low frequency, since only plasmids that carry a unique chromosomal segment will provide relevant linkage data.

References

1. Davison J (2002) Genetic tools for Pseudomonads, Rhizobia, and other Gram-negative bacteria. Biotechniques 32:386–401

2. Saunders JR, Grinsted J (1972) Properties of RP4, an R-factor which originated in *Pseudomonas aeruginosa* S8. J Bacteriol 112:690–696

3. Van Gijsegem F, Toussaint A (1982) Chromosome transfer and R-prime formation by an RP4:.mini-Mu derivative in *E.coli, Salmonella typhimurium, Klebsiella pneumoniae* and *Proteus mirabilis*. Plasmid 7:30–44

4. Lejeune P, Mergeay M, Van Gijsegem F, Faelen M, Gerits J, Toussaint A (1983) Chromosome transfer and R-primes formation mediated by plasmid pULB113 (RP4::mini-Mu) in *Alcaligenes eutrophus* CH34 and *Pseudomonas fluorescens* 6.2. J Bacteriol 155:1015–1026

5. Van Gijsegem F, Toussaint A (1983) In vivo cloning of *Erwinia* genes involved in the catabolism of hexuronates. J Bacteriol 154:1227–1235

6. Van Gijsegem F, Toussaint A, Schoonejans E (1985) In vivo cloning of pectate lyase and cellulase genes of *Erwinia chrysanthemi*. EMBO J 4:787–792

7. Seeberg AH, Wiedeman B (1984) Transfer of the chromosomal *bla* gene from *enterobacter cloacae* to *Escherichia coli* by RP4::mini-Mu. J Bacteriol 157:89–94

8. Nasser W, Dorel C, Wawrzyniak J, Van Gijsegem F, Groleau M-C, Deziel E, Reverchon S (2013) Vfm a new quorum sensing system controls the virulence of *Dickeya dadantii*. Environ Microbiol 15:865–880

9. Chatterjee AK, Ross LM, Thurn MEJL, K.K. (1985) PULB113, an RP4::mini-Mu plasmid, mediates chromosomal mobilization and R-prime formation in *Erwinia* amylovora, *Erwinia* chrysanthemi, and subspecies of *Erwinia carotovora*. Appl Environ Microbiol 50:1–9

10. Breton AM, Younes G, Van Gijsegem F, Guespin-Michel J (1986) Expression in

Mixococcus xanthus of foreign genes coding for secreted pectate lyases of *Erwinia chrysanthemi*. J Biotechnol 4:303–311

11. Barras F, Chambost JP, Chippaux M (1984) Cellobiose metabolism in *ERwinia*: genetic study. Mol Gen Genet 197:486–490

12. Barras F, Leopelletier M, Chippaux M (1985) Control by cAMP-CRP complex of the expression of the PTS-dependent *ERwinia chrysanthemi clb* genes in *E. coli*. FEMS Microbiol Lett 30:209–212

13. Hedegaard L, Danchin A (1985) The *cya* gene region of *ERwinia chrysanthemi* B374: Organisation and gene products. Mol Gen Genet 201:38–42

14. Waelkens F, Verdickt K, Van Duffel L, Vanderleyden A, Vangool A, Mergeay M (1987) Complementation by *Agrobacterium tumefaciens* chromosomal genes and its potential use for linkage mapping. FEMS Microbiol Lett 43:329–334

15. Miller JH (1972) Experiments in molecular genetics. Cold Spring Harbor Laboratory Press, Cold Spring Harbor, NY

16. Betlach MC, Hershfield V, Chow L, Brown W, Goodman HM, Boyer HW (1976) A restriction endonuclease analysis of the bacterial plasmid controlling the *Eco*RI restriction and modification of DNA. Fed Proc 35:2037–2043

17. Schoonejans E, Toussaint A (1983) Utilization of plasmid pULB113 (Rp'::mini-Mu) to construct a linkafe map of *Erwinia carotovora* subsp. Chrysanthemi. J Bacteriol 154:1489–1492

18. Hugouvieux-Cotte-Pattat N, Reverchon S, Robert-Baudouy J (1989) Expanded linkage map of *Erwinia chrysanthemi* strain 3937. Mol Microbiol 3:573–581

19. Sadouk A, Mergeay M (1993) Chromosome mapping in *Alcaligenes eutrophus* CH34. Mol Gen Genet 240:181–187

20. Faelen M, Toussaint A (1976) Bacteriophage Mu-1, a tool to transpose and to localize bacterial genes. J Mol Biol 104:525–539

21. Cabezon T, Van Gijsegem F, Toussaint A, Faelen M, Bollen A (1978) Phage Mu-1 mediated transposition:a tool to study the organization of ribosomal protein genes in *E. coli*. Mol Gen Genet 161:291–296

22. Mergeay M, Lejeune P, Sadouk A, Gerits J, Fabry L (1987) Shuttle transfer (or retrotransfer) of chromosomal markers mediated by plasmid pULB113. Mol Gen Genet 209:61–70

23. Szpirer CY, Tp E, Couturier M, Mergeay M (1999) Retrotransfer or gene capture: a feature of conjugative plasmids, with ecological and evolutionary significance. Microbiology 145:3321–3329

Chapter 22

Muprints and Whole Genome Insertion Scans: Methods for Investigating Chromosome Accessibility and DNA Dynamics using Bacteriophage Mu

N. Patrick Higgins

Abstract

Bacteriophage Mu infects a broad range of gram-negative bacteria. After infection, Mu amplifies its DNA through a coupled transposition/replication cycle that inserts copies of Mu throughout all domains of the folded chromosome. Mu has the most relaxed target specificity of the known transposons (Manna et al., J Bacteriol 187: 3586–3588, 2005) and the Mu DNA packaging process, called "headful packaging", incorporates 50–150 bp of host sequences covalently bound to its left end and 2 kb of host DNA linked to its right end into a viral capsid. The combination of broad insertion coverage and easy phage purification makes Mu ideal for analyzing chromosome dynamics and DNA structure inside living cells. "Mu printing" (Wang and Higgins, Mol Microbiol 12: 665–677, 1994; Manna et al., J Bacteriol 183: 3328–3335, 2001) uses the polymerase chain reaction (PCR) to generate a quantitative fine structure map of Mu insertion sites within specific regions of a bacterial chromosome or plasmid. A complementary technique uses microarray platforms to provide quantitative insertion patterns covering a whole bacterial genome (Manna et al., J Bacteriol 187: 3586–3588, 2005; Manna et al., Proc Natl Acad Sci U S A 101: 9780–9785, 2004). These two methods provide a powerful complementary system to investigate chromosome structure inside living cells.

Key words Transposition, DNA replication, Headful DNA packaging, polymerase chain reaction (PCR), Plaque forming units (PFU), DNA melting temperature (Tm), Luria Broth (LB), Chromosome immunoprecipitation (ChIP), Chromosome immunoprecipitation sequencing (ChIP-Seq)

1 Introduction

Bacteriophage Mu is a transposing replicon [1]. As such, it combines the characteristics of a lysogenic bacterial virus with the DNA replication mechanism of an efficient and highly active transposable element. During the lytic cycle, Mu can transpose to hundreds of thousands of different chromosomal positions in a bacterial population. At each insertion site, Mu DNA is flanked by directly repeated 5 bp duplications of the host DNA. After replication, when every cell contains several hundred copies of Mu, the viral

Martha R.J. Clokie et al. (eds.), *Bacteriophages: Methods and Protocols, Volume 3*, Methods in Molecular Biology, vol. 1681, https://doi.org/10.1007/978-1-4939-7343-9_22, © Springer Science+Business Media LLC 2018

DNA is transferred into phage capsids by a double-stranded DNA headful packaging mechanism. The packaging reaction initiates with a double-strand break in host DNA near the Mu left end that enters the capsid and is followed by viral DNA and then 2–3 kb of covalently linked adjoining host DNA beyond the right viral end. Each virus is unique, because the host DNA flanking sequences at the left and right ends of each virus are different in each phage. Thus, the phage retains a record of its insertion during the transposition in the chromosome. In 1 ml of a fresh Mu phage lysate (10^9–10^{10} plaque forming units), there are enough viruses to have integrated into every bp of a 4-mega base pair bacterial chromosome 1000 times over. Using modern tools of molecular biology that include PCR and DNA microarray analyses, it is possible to reconstruct Mu transposition profiles for any region of a chromosome and to compile efficient and highly reproducible patterns of genome-wide transposition under different growth conditions. The results provide important information about host DNA structure and chromosomal dynamics for cells growing at different doubling times and in media composed of different sources of carbon or nitrogen.

2 Materials

2.1 Viruses

Bacteriophage Mu was discovered by Larry Taylor to be the cause of multiple mutations when the virus lysogenizes *Escherichia coli* [2]. Later Mu was shown to cause multiple mutations in a single *E. coli* gene [3], and the recognition that this rare virus is in the transposon family led to many clever genetic modifications designed to facilitate its use in the discovery and mapping of new genes in the bacterial chromosome [4]. Here, we've used exclusively two Mu plaque-forming derivatives of Mu for Muprint and microarray studies. Mu*pAp1* [5] is a plaque-forming derivative of Mu carrying the ampicillin resistance gene from Tn3 and the temperature-sensitive *cts62* repressor [6, 7]. Mu*NXKan* [8] has the *cts62* repressor plus a selectable kanamycin-resistance (Kan) gene in addition to sites for the rare cutting restriction enzymes Not*I* and Xba*I* (Fig. 1). Digestion of purified phage DNA with either enzyme releases the Kan-linked pool of 2 kb host junction sequences, which is convenient for purifying DNA libraries, cloning, or labeling the host sequences for microarray and Muprint experiments.

2.2 Bacterial Strains

Constructing genetic strains for genome analysis is often a time consuming step in a genetic research project. For many strains of *E. coli*, it is possible to carry out a Muprint by infection with a high titer plaque-forming Mu phage stock. However, stocks of phage Mu are often not stable during prolonged storage, so making a Mu

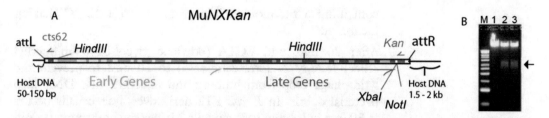

Fig. 1 Map of thermoinducible bacteriophage Mu*NXKan* [8]. (**a**) Mu*NXKan* was designed to make purification of left and right ends of packaged virus convenient for applications involving interrogation of the host sequences covalently attached to the viral left (*attL*) and right (*attR*) ends. It has a thermoinducible Mu*cts*62 repressor to make DNA conveniently from lysogens that grow well at 30 °C switch to the lytic cycle when cultures are shifted to growth at 42 °C. Early genes transcribed left to right (alternating *green* and *yellow* segments) encode proteins required for DNA replication, transposition, and induction of late gene expression. Late genes also transcribed left to right include structural proteins for the viral capsid, for packaging phage DNA, and for cell lysis. A Kan gene introduced at the end of viral sequences allows selection, and the sequence of rare restriction endonucleases XbaI and NotI lie between the genetic right end and adjoining 1–3 kb of host DNA at the right end. (**b**) Cleavage of phage DNA with XbaI (*lane 2*) or NotI (*lane 3*) releases the Kan–*attR*—host DNA fragments that migrate as a somewhat fuzzy band centered at the 3 kb position

lysogen with a *cts62* thermoinducible prophage is a sensible way to make high titer Mu lysates for DNA isolation and genome analysis. One method to make a Mu lysogen is by conjugation, using an F⁺ donor carrying a copy of Mu*pApl* or Mu*NXKan* and an F⁻ test strain. The conjugation method can also be used to introduce Mu into a variety of different gram-negative bacterial species including *Salmonella* Typhimurium, *Salmonella* Typhi, and *Klebsiella*.

Another option for making phage lysates in bacteria that express the Mu tail fiber receptor on the cell surface is to infect and select lysogens. *E. coli* strains can be converted to Mu lysogens by infecting exponential cultures with Mu*cts62pApl* or Mu*NXKan* at a multiplicity of infection of 0.1 and plating aliquots at 30 °C on LB agar plates containing 25 μg/ml ampicillin. This method yields lysogens at a frequency of approximately 1% of infected cells, but such lysogens have a mutation somewhere in the genome. Putative lysogens must be restreaked on LB + ampicillin plates and tested for lysis and Mu phage production after thermoinducing mid-log cultures by shifting to a 42 °C shaking incubator. Starting cultures with a mixture 5–10 independently isolated lysogens can reduce the impact of a specific phage-induced mutation on the experimental outcome.

2.3 Purification of Phage DNA from Cell Lysates

The protocol described below is a simple and efficient method to purify phage DNA from a fresh lysate.

1. Grow a 50 ml culture of a test strain at 30 °C to an optical density at A_{600} of 0.5.

2. Initiate lytic growth by either infection with a high titer phage stock at a MOI of 5 PFU/cell or by shifting a lysogenic culture

containing a thermoinducible prophage to a 42 °C shaking incubator.

3. After 20 min, add EGTA (ethylene glycol-bis(β-aminoethyl ether)-N,N,N',N'-tetraacetic acid) to 10 mM to prevent the lytic phage crop from binding and injecting viral DNA into bacterial debris. In *E. coli* K12 derivatives, lysis usually occurs in 50 min in LB, but lysis can take 2 h in some host mutants or in minimal medium containing a range of different the carbon sources.

4. After lysis, add NaCl to 0.5 M and remove cell debris by a 10 min centrifugation at 10,000 × g in a Beckman J21 centrifuge rotor or the equivalent.

5. Add polyethylene glycol 8000 (Sigma) to 6% w/v. Mix at room temperature until the solution clears. Then place at 4 °C overnight.

6. Collect the flocculent phage precipitate by centrifugation at 4000 × g at 4 °C. Suspend the pellet in 1 ml of SM buffer (0.1 M NaCl, 0.05 M Tris–HCl, pH 7.5, 0.01 M MgSO$_4$, and 0.01% gelatin).

Phage stocks made with this method generally yield 10^{10}–10^{11} PFU/ml. Mix the phage solution with 1 ml of chloroform to kill any remaining viable bacteria. Store the upper phage layer at 4 °C.

7. Disrupt phage particles in 0.1% SDS and 5 mM EDTA. Mix this solution with phenol saturated with 0.1 N Tris base. Then extract again with a 1:1 mixture of phenol and chloroform.

8. Precipitate phage DNA with 2 volumes of 95% ethanol. Suspend the DNA in TE buffer (50 mM Tris–HCl, pH 7.5. 0.1 mM EDTA). Store at −20 °C.

2.4 Purification of Chromosomal DNA for Muprinting (See Note 1)

Muprints can also be synthesized using the DNA template isolated from cells prior to phage lysis. The advantage of Muprinting chromosomal DNA is that one can analyze long PCR products from the *attL* end of phage insertions. When using phage, only the *attR* end gives long (1–2 kb) PCR products. With *attR*-only reads, one misses the Mu insertions with the opposite orientation (about half of the insertions). At many positions in the chromosome a strong bias exists for insertions in one orientation [9].

The purification described below produces cellular DNA from bacteria that is suitable for PCR reactions.

1. Initiate bacterial cultures (30–50 ml) with a fresh overnight culture by a 1:500 dilution in fresh LB.

2. Start the Mu lytic cycle at a cell density of 0.5 A_{600} by either phage infection or temperature shift of a Mu *cts62* lysogen from 30 to 42 °C.

3. When the lytic cycle is well underway (30–40 min in LB), chill the culture to 4 °C. Pellet the cells by centrifugation at 10,000 × g for 10 min at 4 °C.

4. Suspend the cells in 1 ml of a solution of 15 mM EDTA, 150 mM NaCl, 1 mg/ml egg white lysozyme. Incubate at 4 °C for 20 min.

5. Lyse the bacteria by adding 0.15 ml solution of 0.1 M Tris, pH 8, 0.1 M NaCl, 0.1 M SDS. Add 10 μl of a 10 mg/ml Proteinase K solution. Incubate the reaction for 1 h at 60 °C.

6. Shear the DNA by five passages back and forth through a 22 gauge needle. Extract the solution twice with an equal volume of a 1:1 mixture of phenol and chloroform.

7. Precipitate DNA by adding 1/10 volume of 3 M sodium acetate, pH 7.0, and 2.5 volumes of 95% ethanol.

8. Collect the DNA by centrifugation or by spooling DNA onto a glass capillary pipette. Rinse by dipping into 70% ethanol and dry briefly by blotting on a Kimwipe. Then suspended the DNA in TE buffer at a concentration of 500 μg/ml.

2.5 Oligonucleotide Primers

Oligonucleotides for Muprint experiments should be carefully designed and pretested in PCR reactions with host DNA to ensure that they do not create artificial bands (i.e., no bands with only one primer) and that they yield pure PCR reactions of the desired length in control reactions using a purified host DNA template. For this purpose the program Primer3Plus (http://www.bioinformatics.nl/cgi-bin/primer3plus/primer3plus.cgi) is valuable because it gives choices of multiple locations in which a 20-mer has optimal predicted primer properties with a Tm near 70 °C. The left and right ends of Mu are AT-rich, so we recommend two primers that work well for PCR reactions from left and right viral ends. For Mu right end PCR reactions, synthesize the MuR 25-mer 5′-TTCGCATTTATCGTGAAACGCTTTC and use an annealing temperature of 67 °C. The MuL left end primer is a 22-mer with the sequence of 5′-TTTTTCGTACTTCAAGTGAATC, and PCR reactions can be performed using an annealing temperature of 59 °C.

3   Methods

3.1 Muprinting PCR and Gel Electrophoresis

Muprints are controlled amplifications of segments of bacterial DNA fused to the right or left end of Mu. The Muprint outline for analyzing chromosomal DNA is shown schematically in Fig. 2. Each Muprint reaction includes one labeled oligonucleotide primer that matches a sequence in bacterial DNA and an unlabeled MuR primer. When this technique was developed, Taq polymerase was

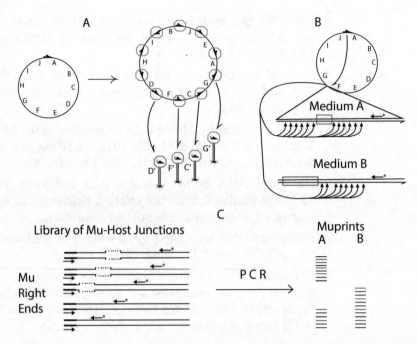

Fig. 2 Analysis of Mu transposition patterns at bp resolution in a desired chromosomal field [10]. The *circle* in (**a**) represents the chromosome of a lysogen containing a Mu prophage (indicated as a *thick arrow*). Following thermoinduction, which leads to replicative transposition and headful packaging, the phage progeny contain a library of host junction fragments. (**b**) Distribution of phage insertion sites and how they differ between cells cultured in medium A or B. The transposition patterns can be derived from phage DNA using a ^{32}P-labelled primer matching a specific host sequence (indicated as an *arrow* with *asterisk*) and PCR reaction conditions. (**c**) The oligonucleotide on the *left* primes synthesis out of the Mu right end, and the ^{32}P-labelled oligonucleotide on the *right* (with *asterisk*) primes synthesis from the host junction fragment. *Dotted boxes* indicate a region of restricted transposition in the chromosome. Following gel electrophoresis and autoradiography, the distribution of ^{32}P-labeled PCR products reveals the transposition patterns for different physiological states

widely used for most PCR reactions. This enzyme lacks an editing function and can reliably make products only 1–2 kb in length. Today, high fidelity enzyme like the Phusion polymerase (e.g., New England BioLabs; https://www.neb.com/products/m0530-phusion-high-fidelity-dna-polymerase can efficiently generate PCR products of 10 kb and longer. It is feasible to scan large segments of bacterial chromosomes on agarose or composite agar-acrylamide gels to identify regions with significantly different accessibility to Mu transposition in vivo. The protocol described below is designed specifically for Taq reactions and resolution of products on denaturing sequencing gels.

Each Muprint PCR reaction includes a radiolabeled chromosomal primer, a nonradioactive ("cold") MuL or MuR primer, and bacterial genomic DNA. Synthetic nonphosphorylated oligonucleotide primers can be radiolabeled using phage T4 poly-nucleotide kinase in 10 μl reactions as follows:

1. Assemble a mixture containing: 2 μl of oligonucleotide (10 μM stock), 1 μl of T4 kinase buffer (600 mM Tris–HCl [pH 7.8], 100 mM MgCl$_2$, 150 mM KCl); 1 μl of fresh 150 mM dithio-threitol; 0.5 μl of T4 polynucleotide kinase (1-U/ml); and 5 μl of [γ-^{32}P]ATP (NEN; 3000 Ci/mmol).

2. Incubate at 37 °C for 30 min followed by heating to 80 °C for 5 min. The radiolabeled primer from this reaction can be used in four Muprint PCRs.

3. Assemble PCR reactions containing 2.5 μl of radiolabeled oligo-nucleotide (described above); 2.5 μl of MuL or MuR primer (10 mM); 5 μl of Taq PCR buffer (100 mM Tris–HCl [pH 8.3]; 500 mM KCl); 6 μl of 25 mM MgCl$_2$; 1 μl of Taq polymerase (2.5 U); 2 μg of bacterial genomic DNA; and water to make the final volume 50 μl.

4. Carry out 20–30 thermocycling rounds with steps of 1 min at 94 °C; 1 min at 55 °C; and 2 min at 72 °C.

5. Precipitate PCR products by adding EDTA to 10 mM and 2.5 volumes of ethanol. Air dry and suspend in 10 μl of sequencing stop solution (95% deionized formamide, 10 mM EDTA [pH 8.0]; 0.1% bromophenol blue; and 0.1% xylene cyanol.

6. Load 2 μl of each Muprint reaction into one lane of a 6% polyacrylamide denaturing sequencing gel, and electrophorese at a constant 1600 V. Save the rest for reanalysis on a second gel.

7. After electrophoresis, dry the gel and make an autoradiograph for quantitative analysis.

3.2 Genome Wide Microarray Scans of In Vivo Mu Transposition Insertions (See Note 2)

Transposition efficiency at each gene can be measured from a ratio of fluorescent signals from two input DNAs. Random priming generates the first label using genomic DNA purified from the test strain. The second DNA is made with a Mu specific primer that labels host genes linked to the Mu right end. The method uncovers transposition hot spots and cold spots that have particular interest for chromosome function (Fig. 3). A range of microarray platforms can be used to detect and quantify transposition sites and frequencies, including oligonucleotide arrays and PCR chips that include all genomic open reading.

3.2.1 Host Chromosome Labeling

1. Shear genomic bacterial DNA to 2 kb fragments. One method is to use the Branson Sonifier 450 (Emerson Industrial Application; http://www.emersonindustrial.com/en-US/branson/Products/Sonifiers/Pages/default.aspx) with output control

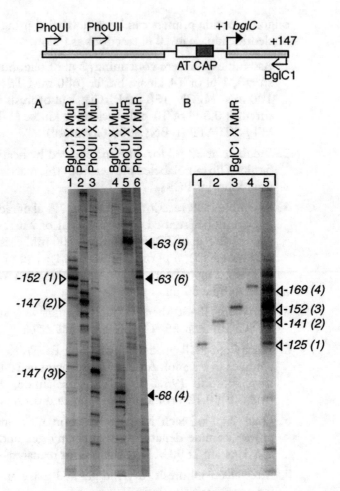

Fig. 3 Mu hot spots in the transcriptional control region of the *bglC* gene of *E. coli*. At *top* is a map of the *bglC* control region along with DNA primers used for Muprint PCRs from genomic DNA. The cyclic AMP protein DNA-binding site (CAP), and the *bglC* transcription start site are indicated. (**a**) Muprints were developed using BglC1 and MuR (*lane 1*), PhoUI and MuL (*lane 2*), PhoUII and MuL (*lane 3*), BglC1 and MuL (*lane 4*), PhoUI and MuR (*lane 5*), and PhoUII and MuR (*lane 6*). The strong hot spots are indicated to the *left* and *right* of the autoradiographic image. (**b**) Examples of cloned Muprint bands run alongside the Muprint PCR. *Lanes 1* through *4* show PCRs (using primers BglC1 and MuR) of four cloned Muprint bands, with insertions determined by sequencing (*See* **Note 3**)

setting at 5, the duty cycle setting of constant, and sonication bursts of 10 s.

2. Mix 2 µg of sheared DNA with 20 µl of 2.5× random priming buffer mix from RadPrime DNA labeling kit (Invitrogen). Assemble a 50 µl reaction for each chip containing 5 µl of 10× dNTPs (1.2 mM each of dATP, dTTP, and dGTP, and 0.6 mM dCTP); 3 µl of 1 mM Cy3-dCTP (Amersham Pharmacia Biosciences); and 2 µl (50 U) of the Klenow fragment of DNA polymerase I (Invitrogen).

3. Incubate at 37 °C for 2 h. Stop the reaction by addition of 5 μl of 0.25 M EDTA.

3.2.2 Host DNA Labeling from Mu Phage DNA

This protocol generates a red probe complementary to the sequences presence in Mu phage DNA.

1. Assemble a 50 μl reaction for each chip containing the components: Cy-5 dCTP (3 μl of 1 mM stock); 5 μl of 10× PCR buffer (Sigma); 2.5 mM MgCl$_2$; 2 μg Mu phage DNA; MuR primer (5 μl of 10 mM stock); 0.2 mM each of dATP, dTTP, and dGTP; 0.1 mM dCTP; 5 U of Taq polymerase (Sigma); and H$_2$O to make a 50 μl reaction volume.

2. Carry out 30 cycles of linear DNA synthesis in a thermocycler with steps of 1 min at 55 °C; 15 s at 94 °C; and 2 min at 72 °C.

3. Remove unincorporated nucleotides from labeled DNA probes using a Microcon 30 filter from EMD Millipore (Billerica, MA, USA; http://www.emdmillipore.com/US/en/product/Amicon-Ultra-15-Centrifugal-Filter-Units,MM_NF-C7715).

4. Mix the red and green probes and apply to a microarray chip, following procedure for conditioning the chips, denaturing the mixed probes, loading chip, hybridization, washing, drying and reading the chip according to procedures that recommended by the manufacturer.

The reproducibility of microarray scans of two independent phage experiments is illustrated in Fig. 4.

4 Notes

1. In many cases, the Muprint technique offers an inexpensive alternative technology to ChIP, ChIP-SEQ, or in vivo DNA footprinting for investigating the structure of chromosomal regions inside living cells. Transposition patterns are influenced by transcription and can indicate remodeling of sites up- and downstream of a transcribed operon that accompanies gene expression. Muprints of the *lac* and *bgl* operons in *E. coli* under induced and repressed conditions demonstrate the local impact of transcription on structures at the promoter and downstream in the transcribed tract [9, 10]. The importance of a transcription effect was confirmed with genome wide scans of *E. coli* and *Salmonella* that demonstrated the presence or movement of RNA polymerase on highly transcribed genes insulates these regions from transposon insertions [13, 14]. And Muprints illustrated the dynamic local impact of transposition immunity, which prevents transposons like Mu and Tn3 from inserting into their own sequences [15]. Surprisingly, the major hotspots were not shared between *E. coli* and *Salmonella*, which

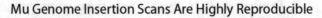

Mu Genome Insertion Scans Are Highly Reproducible

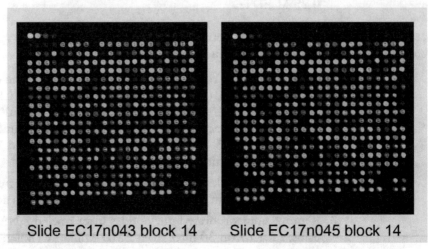

Slide EC17n043 block 14   Slide EC17n045 block 14

Fig. 4 Two blocks (400 pin spots) of a whole genome chip hybridized to mixed *E. coli* probe demonstrate reproducibility of the technique. Mu insertion scans for *E. coli* were carried out using cells grown in LB at mid log phase. Phage Mu DNA was prepared from two independent cultures and was thermoinduced on different days. The host probes [11] and phage Mu R single-stranded probes [12] were synthesized, mixed, and applied to two microarray slides that were hybridized and developed on different days. Aside from imperfections in some of the spots, the two patterns are identical to the naked eye

confirms the finding that species differences are important factors in the phenotype of identical mutations in homologous genes [16, 17]. One general conclusion is that transcription, translation, and DNA replication machineries are highly integrated systems that must work together seamlessly. The common core of housekeeping genes are critical, but the distribution of the workload among these proteins can differ significantly between species [16, 18]. Techniques that can be carried out to compare different species are essential for understanding how evolutionary forces customize the gene expression systems.

2. The two techniques described here are complementary and can be applied to many current problems in molecular genetics. The structure/function mechanism of the ubiquitous protein H-NS is one example. This protein regulates gene expression in many different organisms. However, the phenotype of a null mutant is quite different between *E. coli* and *Salmonella*. One property of H-NS is its ability to bind to a high affinity site and then spread out upstream and downstream of a control point to silence gene expression of 2–300 genes in *E. coli* and *Salmonella*. Muprints in *E. coli* provided a model for how H-NS binding/spreading works in the *bgl* operon, and cooperative protein spreading along DNA has become a major theme for proteins that control DNA replication and movement during the cell cycle. Examples include the mechanism of DnaA acting to control initiation of

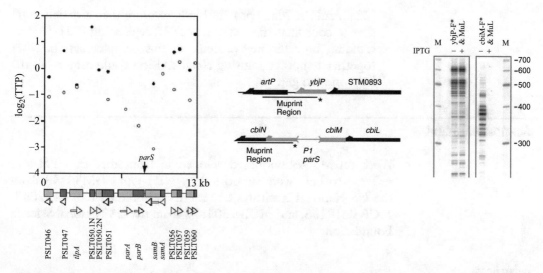

Fig. 5 Discovery of a Mu cold spot by microarray scanning (*open circles on left*) and creation of a cold spot on the *Salmonella* chromosome by inserting a parS site into the *cbiM* gene that becomes a cold spot detected by Muprinting when ParB is expressed (*right*). *Lanes 1* and *8* on the *right* contain a 5′-^{32}P-labelled 100 bp ladder

replication at *oriC* [19], the binding of the MuB protein during Mu transposition [11], and behavior of the P1 ParB plasmid system that controls segregation of newly replicated DNA to sister cells [20]. A microarray experiment in *Salmonella* identified a cold spot on the pSLT plasmid located near the *parS* site of this plasmid partitioning gene *parB* (Fig. 5 open symbols on left). When the *parB* gene was disrupted, the cold spot disappeared (Fig. 5 closed symbols on left). And when an ectopic P1 *parS* site was introduced into the *cob* operon of *Salmonella*, Muprints showed this to be a hotspot when ParB protein was repressed, and it changed to a >500 bp cold spot after ParB expression (Fig. 5 on right). Muprinting can clearly be useful for finding targets of spreading proteins and defining their limits in vivo.

3. There are critical regions of the bacterial chromosome that seem suitable for the Muprint analysis. For example, the distribution of proteins and the cold spot created by transcription across the highly expressed ribosomal RNA operons inform understanding of long-range supercoiling effects. Also, the *E. coli* chromosome replication termination Ter domain, which has multiple complex systems that coordinate sister chromosome segregation is a good target for Muprinting. Locations of interest include the region near the *dif site*, which has a high affinity site for Topo IV and XerCD resolvase. These enzymes work together to decatenate sister chromosomes and resolve chromosome dimers before cell division [21]. The FtsK motor protein displaces proteins as it moves across the Ter domain in the direction toward *dif*

[22]. And the MatP protein binds to more than a dozen *matS* sites to coordinate final stages of DNA replication [23]. Understanding how the key proteins in this complex system work together temporally during a cell division could require all the help one can get.

Acknowledgment

Work on Mu techniques developed in laboratory of NPH and referred to here were supported by NIH grant GM33143 from the US National Institutes of Health and grants MCB 9122048, MCB 9218153, and MCB 9604875 from the US National Science Foundation.

References

1. Higgins NP, Manlapaz-Ramos P, Gandhi RT, Olivera BM (1983) Cell 33:623–628

2. Taylor AL (1963) Proc Natl Acad Sci U S A 50:1043–1051

3. Bukhari AI, Zipser D (1972) Nat New Biol 236:240–243

4. Faelen M (1987) Useful Mu and mini-Mu derivatives. In: Symonds N, Toussaint A, van de Putte P, Howe MM (eds) Phage Mu. Cold Spring Harbor Laboratory Press, Cold Spring Harbor, NY, pp 309–316

5. Leach D, Symonds N (1979) Mol Gen Genet 172:179–184

6. Vogel JL, Li ZJ, Howe MM, Toussaint A, Higgins NP (1991) J Bacteriol 173:6568–6577

7. Vogel JL, Geuskens V, Desmet L, Higgins NP, Toussaint A (1996) Genetics 142:661–672

8. Manna D, Deng S, Breier AM, Higgins NP (2005) J Bacteriol 187:3586–3588

9. Manna D, Wang X, Higgins NP (2001) J Bacteriol 183:3328–3335

10. Wang X, Higgins NP (1994) Mol Microbiol 12:665–677

11. Greene EC, Mizuuchi K (2002) Mol Cell 9:1079–1089

12. Coplin DL, Frederick RD, Majerczak DR, Haas ES (1986) J Bacteriol 168:619–623

13. Manna D, Breier AM, Higgins NP (2004) Proc Natl Acad Sci U S A 101:9780–9785

14. Manna D, Porwollik S, McClelland M, Tan R, Higgins NP (2007) Mol Microbiol 66:315–328

15. Manna D, Higgins NP (1999) Mol Microbiol 32:595–606

16. Champion K, Higgins NP (2007) J Bacteriol 189:5839–5849

17. Cameron AD, Stoebel DM, Dorman CJ (2011) Mol Microbiol 80:85–101

18. Higgins NP (2014) Curr Opin Microbiol 22C:138–143

19. Erzberger JP, Mott ML, Berger JM (2006) Nat Struct Mol Biol 13:676–683

20. Schumacher MA, Funnell BE (2005) Nature 438:516–519

21. Graham JE, Sivanathan V, Sherratt DJ, Arciszewska LK (2010) Nucleic Acids Res 38:72–81

22. Lowe J, Ellonen A, Allen MD, Atkinson C, Sherratt DJ, Grainge I (2008) Mol Cell 31:498–509

23. Mercier R, Petit MA, Schbath S, Robin S, El Karoui M, Boccard F, Espeli O (2008) Cell 135:475–485

INDEX

A

Adsorption 4–10, 12–20, 22,
    33, 34, 43, 44, 46, 197
Adsorption rate 4, 9, 13–15, 19, 20, 23
Annotation 121, 142, 147,
    186, 187, 192, 197, 201, 202, 204–207, 209,
    217–228, 231, 232, 244, 246, 247, 253

B

Bacteriophage 31, 41, 49,
    59–68, 71–86, 89, 90, 97, 98, 109–124, 139,
    141–145, 147–162, 165–167, 176, 179,
    197–211, 217, 224, 226, 249
Bacteriophage Mu 279–285, 287, 303
Burst size 4, 6, 17–19, 38, 41, 44

C

Cesium chloride (CsCl) 53, 59,
    61, 76, 179, 292, 294, 295, 299
Chromosome immune-precipitation (ChIP) 112, 310
Chromosome immune-precipitation
    sequencing (ChIP-Seq) 310
Chromosome mobilization 287
Coding DNA sequences (CDS) 147,
    192, 202–205, 207, 225, 243, 246, 247, 253
Comparative genomics 110

D

DNA Master 124, 204, 217, 224, 226
DNA melting temperature (Tm) 307
DNA replication 140, 147, 149,
    234, 303, 305, 312, 314

E

Efficiency of plating (EOP) 4, 6, 7, 23
Electron microscopy 7, 199, 200

F

Functional annotation 234, 237

G

Gene expression 186, 264, 266, 305, 310, 312
Gene mapping 264, 287

G (continued)

Gene prediction 186, 187,
    218, 222, 233
Genomes 109, 115, 117,
    120, 127–132, 134, 135, 139–162, 165–177,
    179–183, 185–188, 191–193, 197, 201,
    203–205, 208, 210, 217, 224, 226, 231, 232,
    239–253, 255, 258, 263, 264, 266–269, 272,
    279, 287, 288, 298, 301, 303–314
    annotation 147, 209,
        217, 218, 231, 232
    termini 139, 141–145, 147–162

H

Headful DNA packaging 272, 304, 308
High frequency read sequence (HFS) 148,
    149, 152, 153, 162
High-occurrence reads as termini
    (HORT) theory 140, 144, 147–162
High-throughput sequencing (HTS) 140,
    144, 146–149, 152–155, 160–162

I

In vivo (gene) cloning 264, 268,
    287–290, 297, 298
Ion-exchange chromatography 60
Iron chloride flocculation 49–56

K

Killing titers 5, 7–9, 12, 13, 23

L

Latent period 4, 16–20,
    38, 41, 42, 44, 46
Library preparation 56, 111,
    123, 146, 180, 191
Listeria monocytogenes 72, 91, 93,
    97, 99, 102–104
LISTEX™ P100 .... 97, 99, 103
Locus tag 202–204, 225

M

Methylase 127
Methylome 133

Martha R.J. Clokie et al. (eds.), *Bacteriophages: Methods and Protocols, Volume 3*, Methods in Molecular Biology, vol. 1681,
https://doi.org/10.1007/978-1-4939-7343-9, © Springer Science+Business Media LLC 2018

Mini-Mu 265, 266, 268, 269, 272, 274, 275, 283, 284, 287–301
Mini-Muduction 269
Multiplicity of infection (MOI) 8–10, 12, 13, 15, 305

O

Ocean viruses 49, 50
One-step growth 41–46
Open reading frame (ORF) 147, 186, 192, 202, 217, 224, 232–234

P

PacBio 128–130, 133, 135
Phage 3–28, 31–35, 38, 41, 42, 44, 46, 59–64, 66–68, 72, 74–86, 89–94, 97–104, 109–117, 119, 121–124, 127, 128, 132–136, 140, 150, 166, 170, 172, 179–183, 185, 188, 191, 197–211, 217, 218, 220, 231–237, 239, 241–243, 247, 248, 250, 251, 253, 263, 283, 287, 304–306, 308–310
    ecology 27
    Mu 266, 288, 304, 312
    population growth 4, 17–23
    therapy modeling 4, 9, 16, 22, 23, 27, 72
Plaque forming units (PFU) 5, 6, 24, 77, 80, 81, 85, 304
Poisson distribution 4, 5, 8–12, 15, 16, 129, 192
Polymerase chain reaction (PCR) 146, 152
Promoter 32, 35, 71, 192, 209, 218, 265, 266, 269, 272, 298, 310
Purification 59, 64, 66, 76, 144, 146, 168, 170, 175, 176, 199, 272, 294, 305–307

R

RAST 142, 147, 205, 231, 232
Ready-to-eat meats 97
RNA 31, 34, 35, 166–169, 173, 174, 176, 185–194, 202, 206, 209, 231, 234–236, 310, 313
RNA sequencing (RNA-Seq) 185, 188, 191

S

Sequence analysis 128, 140, 154, 162
Single-molecule real-time (SMRT)
    sequencing 128–135
Software 37, 61, 63, 68, 110, 112, 114, 129, 130, 142, 147, 187, 192, 202, 204, 211, 218, 219, 226, 256, 257

T

Taxonomy 197, 201, 203–205, 208, 210
Tea extract 98
Terminator 121, 191, 192, 194, 209, 266
Titer determination 23, 26, 27
Transcription unit identification 298
Transcriptome 185, 188, 191
Transposable phages 263, 264
Transposition 263–266, 268, 269, 272–275, 279, 297, 298, 300, 301, 304, 305, 308–310, 313
Transposition technology 279
Transposons 280–285

V

Viral ecology 50
Virucide 98, 102, 103
Visualization 5, 202, 239–258

Printed in the United States
By Bookmasters